GLOBAL GEOGRAPHIES OF POST-SOCIALIST TRANSITION

The collapse of the Eastern bloc states in the late 1980s triggered considerable interest in, and debates on, the ways in which subsequent developments took place in the region. Research has tended to concentrate on Eastern Europe and the former Soviet Union, as they had been stylised as 'the enemy' during the Cold War. In addition, the Iron Curtain running through Europe had, in all its symbolism, further contributed to the distinct Euro-centric perspective on 'post-socialism'.

Global Geographies of Post-Socialist Transition provides a unique synopsis of the diverse meanings of communism around the world, interacting with European, Asian, African and Latin American traditions, and the particular impact of colonialism outside Europe. It discusses in detail:

- the nature and legacies of communist states and the subsequent post-communist developments;
- the meanings of 'socialism' and 'communism', terms that are frequently being used as mere labels without much consideration of the particular legacies they represent;
- the main identified global regions of post-communist development, Central and Eastern Europe, the former Soviet Union, China, Africa and its particular interpretation and application of 'communism', and Cuba as an almost iconic bulwark of post-colonial socialist ideology in Latin America directed against US hegemony;
- the role of identity under the impact of post-communist changes and diversities, and an overall résumé of observations.

Global Geographies of Post-Socialist Transition acknowledges the many different 'versions' of socialism and, subsequently, post-socialism that have developed in other parts of the world. By taking such a comprehensive perspective, the many different facets of socialist and communist rationales become apparent, reflecting a wide range of histories and legacies, in all their particularities, across the post-communist 'Second' and 'Third' Worlds.

Tassilo Herrschel is Senior Lecturer in Economic Geography and Director of the Centre for Urban and Regional Governance at the University of Westminster, London. He has also been a founder member, and Chair, of the Post-Socialist Geographies Research Group of the Royal Geographical Society (with the Institute of British Geographers).

ROUTLEDGE STUDIES IN HUMAN GEOGRAPHY

This series provides a forum for innovative, vibrant, and critical debate within Human Geography. Titles will reflect the wealth of research which is taking place in this diverse and ever-expanding field.

Contributions will be drawn from the main sub-disciplines and from innovative areas of work which have no particular sub-disciplinary allegiances.

GLOBAL GEOGRAPHIES OF POST-SOCIALIST TRANSITION

Geographies, societies, policies

Tassilo Herrschel

Routledge
Taylor & Francis Group

LONDON AND NEW YORK

First published 2007
by Routledge
2 Park Square, Milton Park, Abingdon, Oxon OX14 4RN

Simultaneously published in the USA and Canada
by Routledge
270 Madison Ave, New York, NY 10016

Routledge is an imprint of the Taylor & Francis Group, an informa business

© 2007 Tassilo Herrschel

Typeset in Baskerville by Book Now Ltd
Printed and bound in Great Britain by
TJI Digital, Padstow, Cornwall

British Library Cataloguing in Publication Data
A catalogue record for this book is available from the British Library

Library of Congress Cataloging in Publication Data
Herrschel, Tassilo, 1958–
Global geographies of post socialist transition: geographies, societies,
policies / Tassilo Herrschel.
p. cm. – (Routledge studies in human geography)
Includes bibliographical references and index.
1. Communism and geography. 2. Post-communism. I. Title. II. Series.
HX550.G45H47 2006
909'.097170829–dc22   2006008456

ISBN10: 0–415–32149–2 (hbk)
ISBN10: 0–203–10279–7 (ebk)

ISBN13: 978–0–415–32149–5 (hbk)
ISBN13: 978–0–203–10279–4 (ebk)

CONTENTS

CONTENTS

CONTENTS

FIGURES AND TABLES

Figures

Tables

PREFACE

The inspiration for this book came from teaching a post-graduate course on globalisation and post-socialist transition, which attempts to offer a broad sweep of the various experiences with post-socialist transformation processes and outcomes across the worlds – 'Second' and 'Third'. After most of the regimes labelled as 'socialist' or 'communist' came to an end at the beginning of the 1990s, some exceptions, notably Cuba, China, North Korea, Vietnam and Laos, have been carrying the can for socialism, albeit in a rather less ideologically focused way than during the height of the Cold War. The nature of the taught course meant that many of the students who had opted for it came in fact from formerly socialist countries, whether Eastern Europe, Central Asia, Africa or Cuba. It was during the many discussions we had in class, when students compared notes about their respective experiences back home, that interesting parallels, but also distinct differences, emerged. The result was a truly global perspective. Concentrating on individual global regions alone, such as Eastern Europe, Russia or Africa, for instance, would not have revealed these interesting insights.

It was then that the idea of a much broader approach to post-socialism emerged, to include the many different 'varieties' of socialism that had existed in the first place. These ideas then led on to a number of papers presented at conferences. With the idea of a book on this topic held in the back of the mind, views of other participants and members of the Post-Socialist Geographies Research Group of the Royal Geographical Society, including Craig Young, were sounded out. After its inception, the way the book actually 'came about' changed, leading to a more detailed discussion of the meaning of socialism and communism in order to offer a background against which post-communism can be discussed. Frequently, it seems to be assumed that there is a general understanding of its principles, and characteristics, but this is not always the case.

The publishers continued to support this project, despite the many extended and missed deadlines, and I would like to take this opportunity to thank them wholeheartedly for their patience and continued support. Many thanks also to Judy Budnick, Ben Gore and Sean Cleary for their competent proofreading at very short notice and with little time to spare. My thanks go also to friends and family for their continued support and putting up with my prolonged 'invisibility'.

1

INTRODUCTION

Variability, legacies, outcomes of post-communist transitions

'Transition' and 'communism' have been closely associated terms over the last decade and a half, following the spectacular, dramatic collapse of the Eastern bloc at the end of the 1980s, including the 'Evil Empire', as former US president Ronald Reagan defined the Soviet Union at the beginning of the 1980s when justifying his 'Star Wars' programme. Much has been speculated, written and argued about the nature and possible outcome of these changes and their likely end. Indeed, some commentators have observed an obsession with 'end-isms' as the *fin de siècle* was approaching, whether announcing the 'end of history' (Fukuyama, 1989, 1992), 'end of revolution' (Touraine, 1990) or, in particular, the widely proclaimed end of socialism (Pearce, 1993). And, just into the new millennium, Carothers (2002) proclaims 'the end of the transition paradigm'. This corresponds with the view expressed by the Polish ambassador to London during a brief personal exchange with the author two or three years ago. He emphatically pointed out that post-communist *transition* had certainly ended, as the goal of an equal footing between all European states had been reached in terms of societal and state structures.

While perhaps on purely formal criteria this may be argued, supported by the eastward expansion of the European Union in 2004, there are many legacies, both physical and mental, that continue to reflect the particular nature of the communist period. Pridham and Ágh (2001), referring to the Hungarian experience, point here to the necessity, as they see it, of distinguishing between the actual transition period, with its mixed economy and transformation recession, and the subsequent period of consolidation which is less 'messy'. In addition, most of the debate and observations have been focused on Europe and Russia (as the only really internationally visible successor to the demised Soviet Union). But there have been other experiences with 'communism' too, and thus other changes since. One obvious such 'other' example is China. While officially still adhering to the communist teachings of Mao Zedong, there are fundamental changes taking place under the mantle of communist doctrine: the conquering of the global market with products 'made in China'. Hong Kong, Shenzhen and Shanghai have become glittering symbols of the 'new' international and capitalist-minded China, although this picture may be rather unrepresentative of the rest of the country. Cuba is another cause célèbre of the Cold War, having developed its own version of communism in the backyard of the United States. This includes distinct post-colonial overtones, and these have been combined with a clearly nationalist agenda over the last decade or so. Cuba sits at the intersection between 'north–south' and 'east–west' global relationships.

With the East–West dichotomy of the Cold War era now gone, the contrast between the developed and developing parts of the world has become the dominant feature of global, western-driven politics, which seek to spread 'democratisation' and 'liberalisation' as expression of the 'universalization of Western liberal demo-cracy as the final form of human government' (Fukuyama, 1989, p. 4). The ongoing struggles in Afghanistan and Iraq continue to attract attention to the feasibility of externally driven 'democratisation' and 'liberalisation'. And these agendas, symbolised by the Washington Consensus (Chapter 3) were also prevalent at the time of the collapse of the dictatorial communist regimes of 'Eastern' Europe. Since then, the difficulty and unpredictability of regime changes and policy transfers have become increasingly evident from the very different routes taken by the post-communist governments. And taking other cultures and economic contexts into account, such as those of post-colonialism, the picture gets even more complex and unpredictable in its likely outcome.

The process of 'transition', central to this book, is a multi-dimensional, complex phenomenon, shaped by a set of overlapping and intersecting variables. These include country-specific 'internal' variables, such as historic legacies and geopolitical histories and circumstances, as well as wider external factors, especially international political and economic parameters. And it is the particular combination of, and interface between, those two sets of factors that shape both the course and nature of a country's post-communist transition.

In their 1998 analysis of the Vietnamese model of post-authoritarian transition, and with reference to Taiwan's development model, Wu and Sun identify three key factors shaping smaller countries' institutional choices: hegemonic dependency, economic imperative, and elite idealism (Wu and Sun ,1998, p. 397). In particular, they focus on the interaction between external and internal factors, and their relative importance. External dependencies under hegemonic political and/or economic conditions are seen as the main factor circumscribing the ceiling – actual or perceived – for individual states to devise 'their' political-economic regimes. The greater the hegemonic dependency, the less likely will be independent forms of governance. Instead, the hegemon's institutional principles of governance will be incorporated. Within that framework, following Wu and Sun's argument, it is the political skill of the domestic political elite that can carve out some leeway for independent policy-making.

Obviously, personalities and personal political ambitions and assessments of risks and opportunities will be important determinants of policy-making, and these will, inevitably, be influenced by particular national (and sub-national) legacies in state-building, societal arrangements and aspirations. The perceived value of a government's activities for the national course, including economic prosperity, geopolitical security and domestic societal coherence, circumscribe its legitimacy and thus moral authority and acceptance among the people. A low rate of acceptance will encourage the search for alternative arrangements, even under the heavy hand of an authoritarian regime. Impulses for change may come from (underground) grassroots pressure, or may be initiated by the ruling elite itself, if existing conditions appear 'hopeless' and to the detriment of their interests. Both routes to change have been adopted in the shift away from communist regimes, as will be discussed in this book. While Central and Eastern Europe (CEE) illustrate the former route to bringing down communist regimes, the former Soviet Union represents the latter. Process, and outcome of any such process, is thus far from uniform, with each country following its particular route of change, with its own pace, depth and result of post-communist transition (see also Dryzek and Holmes, 2002). It is at this point that an international comparison becomes so useful, as it not only allows study of the varying importance of factors, but also their very composition, especially in different global regions. The legacy of colonialism, combined with the impact of globalisation, for instance, has had a major impact on the coming and going of communist ('socialist') regimes in Africa, and in Latin America, with new alliances emerging between the respective developing countries.

Against this background, this book adopts and explores an understanding of 'transition' that distinguishes between four main sets of factors influencing the process and outcome of transition from one political-economic regime to another. This is shown in Figure 1.1. These four main determinants consist of external and internal factors: hegemonies, societal-economic conditions, leadership qualities within the regime, and legacies. While the first three find their references in Wu and Sun's 'hierarchy of hegemonic dependence, economic imperative and elite idealism' (Wu and Sun ,1998, p. 397), 'legacies' have been included as a separate category. This reflects both the aim of this book to take a global approach to analysing post-communist regime changes as a particular 'version' of post-authoritarian transition, and the outcome, which demonstrates the importance of place- and country-specific conditions. Political-historic legacies, such as colonialism, or major social-cultural differences, such as between Africa, Asia, Latin America and Europe, may be expected to shape the way communist and post-communist regimes developed across the world and their characteristics, and this includes all countries, whether global powers or not. Legacies also involve collective social memories, be they at the national or sub-national (ethnic) level, including possible antagonisms and anxieties about past and potential future hegemonic ambitions by neighbouring countries. Such are clearly prevalent among the CEE countries bordering Russia (see Chapter 4).

Legacies also include past experiences with democracy, which could offer a point of reference for a move from a communist authoritarian to a post-communist democratic system. Indeed, as will be discussed later, pre-communist democratic experiences have been used as a bridgehead to connect to the current democratic ambitions, and thus bypass or 'airbrush' out of history' the communist period altogether. In the case of the CEE countries, past links to 'Europe' have also been re-emphasised and re-constructed to highlight their 'Europeanness' as a contrast to Sovietism and, indeed, 'Russianness'. Then, there are also legacies of mismatches between national identities and their territorialities, especially in post-colonial

3

Africa. There, colonial-era European notions of territorially defined 'statehood' have been superimposed on existing ethnically based senses of belonging. The resulting states have been anything but nation states as the European tradition generally views statehood.

To some extent, the concept of 'African socialism' tried to reconcile the two traditions – African communitarianism on the one hand, and European class-based societalism on the other (Babu, 1981). Perhaps not entirely surprisingly, these attempts had a rather mixed result, mainly because of pressures from economic problems, which, in themselves, were largely the result of colonial dependencies (see Chapter 7).

Global factors, especially hegemonic economic dependencies, have increasingly come to shape national development prospects. Success or failure at the economic front has become the critical test for the viability of a country's political regime, with the 'big fish' obviously in a better, more influential position, than the 'small fry', although all governments face the challenge of economic globalisation.

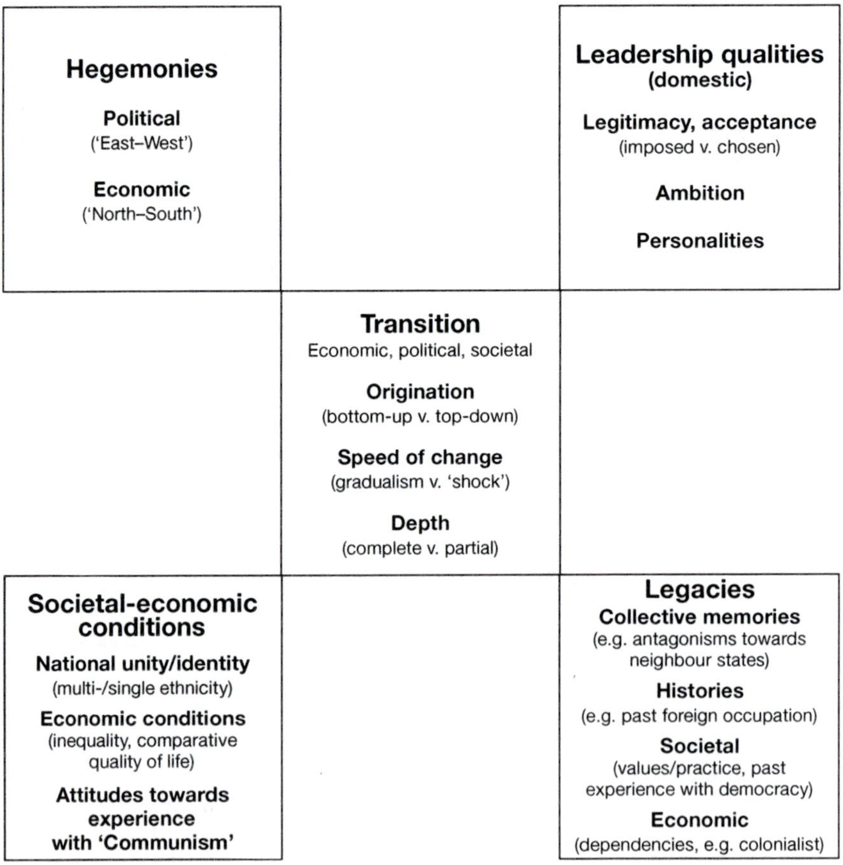

Hegemonies **Political** ('East–West') **Economic** ('North–South')		**Leadership qualities** **(domestic)** **Legitimacy, acceptance** (imposed v. chosen) **Ambition** **Personalities**
	Transition Economic, political, societal **Origination** (bottom-up v. top-down) **Speed of change** (gradualism v. 'shock') **Depth** (complete v. partial)	
Societal-economic **conditions** **National unity/identity** (multi-/single ethnicity) **Economic conditions** (inequality, comparative quality of life) **Attitudes towards** **experience** **with 'Communism'**		**Legacies** **Collective memories** (e.g. antagonisms towards neighbour states) **Histories** (e.g. past foreign occupation) **Societal** (values/practice, past experience with democracy) **Economic** (dependencies, e.g. colonialist)

Figure 1.1 Key determining factors of 'transition'

Economic factors were instrumental in developing the rationale for socialism and communism, but they were also the main factor behind the ultimate failure of the communist regimes, whether in Africa or Europe. Indeed, the handful of countries still claiming adherence to a socialist/communist ideology – China, Laos, Vietnam, North Korea, Moldova and Cuba – all had to make some concessions to their economic regimes by introducing, albeit rather tentatively in some cases, elements of a market economy, while maintaining the official doctrine of a communist state. North Korea has been coyest about any deviation from the path of Marxist–Leninist doctrine, including the economy, although it depends on support from China to stay 'afloat'.

The economic dimension of the 'hegemony' variable revolves around the continued 'North' versus 'South' debate, substantially a colonial legacy, replacing the previous more politically focused 'East' versus 'West' arguments of the days of the Cold War. Both have had major implications for the ways in which communist regimes were established, maintained, and then brought to an end. The fourth main set of factors impacting upon the course of 'transition' is largely the 'personality factor' among the leadership. Rather than institutional structures *per se*, it is the use of personal initiatives, managed through the political leaders' personal characteristics and ambitions, that seem to have been particularly effective in shaping the nature and viability of the various communist regimes. For instance, the end of the Soviet Union and the beginning of Russia's existence as an autonomous state, were fundamentally shaped by the initiatives of individual personalities and their competing ambitions.

Transition itself is discussed here according to its three main dimensions – economic, political and societal. This threefold division will be the backbone of the analysis of the five main global regions in each of the relevant chapters in this book: the political, economic and societal factors in transition, both as outcomes, and as stimulants of the observed changes. The three strands of transition are recurring themes of the analysis of post-communist transition, thus providing the common reference points for the comparative discussion. Economic change, especially privatisation and liberalisation, political change, (i.e. mainly democratisation) and societal change (i.e. largely rising income inequalities) emerged as the main drivers of 'transition'. Their dynamism varies, reflecting national circumstances and policies. Thus, while in most CEE states all three drivers experienced fundamental changes, the situation was less comprehensive elsewhere. In China, Vietnam and Cuba, for instance, government policy tries to keep the economic and political-societal arenas strictly separate, in an attempt to improve economic performance without threatening the continued existence of the one-party regimes. Economic performance has become the main criterion by which these governments' legitimacy of being continuously in office is measured. Good economic performance is more likely to bolster the governments' standing.

The *process* of change and the outcome of 'transition' are in themselves the results of three intersecting dimensions: the origin of the changes (bottom-up or top-down) and thus their degree of legitimacy, then the speed of change, summarised under 'shock' versus 'gradualism', and, third, the depth, or comprehensiveness, of change.

The various combinations of all these indicators of transition constitute the particular national (and sub-national) versions of transition for each participating country. The different elements together constitute the most comprehensive regime change among all transitions. They involve not merely amendments to an existing state structure but, uniquely to post-communism, establishing a completely new state structure, including government, economy and society (see also Carothers, 2002). It is this comprehensive new start that marks the main difference between general post-authoritarian shifts and the post-communist 'variety'.

The notion of 'transition', which had gained broad prominence in the late 1980s against the backdrop of the spectacular collapse of the Eastern bloc regimes, draws on five core assumptions (Carothers, 2002): (1) it is dynamic and depicts a change in conditions; (2) it will lead to democratisation as a 'natural' outcome of these changes; (3) it gains legitimacy through elections, whereby elections are often seen as equal to democratisation *per se*; (4) 'transition' is universal, unaffected by place-specific conditions and legacies; and (5) states are perceived as the main actors, thus presuming their continued existence. These core assumptions were based on the experience with post-authoritarian transition in Latin America and southern Europe in the 1970s (see Chapter 3), and were simply written into the future. The last 15 years of post-communist development have shown that these assumptions do not hold true. The broad comparative approach in this book clearly demonstrates the importance of legacies and particularities for the ways in which post-communist transition evolved in the different countries, including the very notion of what 'communism' means and how it should be implemented 'on the ground'. Against this background, claiming the 'end of the transition paradigm' (ibid.) would perhaps appear like 'throwing the baby out with the bath water'. Certainly, the simplistic, linear understanding of the transition paradigm *per se* has been questioned by the events on the ground. Changes have been much 'messier' than predicted and outlined by the model. In other words, democratisation and transition have occurred in many different shades of grey. 'Of the nearly 100 countries considered as "transitional" in recent years, only a relatively small number – probably fewer than 20 – are clearly en route to becoming successful, well-functioning democracies, or at least have made some progress towards a democratic process and will enjoy a positive dynamic of democratization' (ibid., p. 9). But the outcome may not be the democratic ideal projected and propagated by the western advisers and institutions, who jumped onto the post-communist transition bandwagon as it unfolded in Central and Eastern Europe in 1989, and in the former Soviet Union shortly thereafter. In effect, 'transition' became firmly associated with the European scenario of the shift from communism to a democratic regime. Developments elsewhere, including in China, were barely visible on the radar screen of 'transition' discussions. Only recently, China and its economic transition has entered the headlines, mainly because it is discussed as a 'threat' to western interests by western observers. Outcomes of these transition processes mostly include imperfect democracies, either in the form of 'feckless pluralism' (Carothers, 2002), which makes consensual politics all but impossible, or as 'dominant-power politics' (ibid.). The latter frequently involves 're-launching' the previous dominant, authoritarian elite under a different umbrella, such as nationalism, while continuing restricted party political competition.

The following chapters explore the conceptual backgrounds to 'transition' and 'post-communism', the two main pillars of this book. Chapter 2 examines the notion of 'transition' and its applicability to the developments in those countries that had officially subscribed to 'Marxism–Leninism' or 'communism', but faced the collapse of their regimes at the end of the 1980s/early 1990s. Much of the discussion will focus on the economic aspect of transition, reflecting the importance of that theme for policy makers and the academic debate. A successful economic transition was widely regarded as the first step towards successful western-style democratisation, but both method and outcome varied considerably: adopting either an 'all-out', 'revolutionary' form of post-communist transition, abandoning the old regime and adopting a new one wholesale – an approach dubbed 'shock therapy' – or following a more gradual, evolutionary approach, shifting to a new regime through modifying the existing system. In both instances, the resulting challenges have been huge in terms of disruption to social, economic and societal conditions, and ways of doing things. Post-communist transition turned out to be much less straightforward and predictable than initially believed, reflecting, after all, different circumstances, past experiences, social and political structures, and so on. It is in this respect that the global perspective of post-communist change, through the inclusion of so many different local/national paradigms, legacies, and experiences, helps to shed light on the multi-faceted nature of 'transition'. Looking at individual countries or individual global regions alone cannot provide the necessary breadth of analysis to catch the many different natures of 'transition'.

Similarly, it is only through taking a global perspective, and thus including both 'Second' and 'Third World' communist countries, that the diverse notions of first 'communism' and then '*post*-communism' become apparent. Indeed, as the cases of Africa, Asia and Latin America will show, the usage of the terms 'socialism' and 'communism' had quite different connotations there than were held in Europe, and 'resulted in a conceptualization of human nature divergent from both the Western liberal view and from the more traditional Marxist view, at least as articulated by Second World socialist states' (Pollis, 1981). Indeed, 'African Socialism' (Friedland and Rosberg, 1964a) and, later, 'Afro-Communism' (Ottaway and Ottaway, 1986), sought to detach the principles of Marxism–Leninism from their European (and thus colonial) background, and fuse them with pre-colonial African values and social practices. In fact, Marxist socialist principles were 'perceived as consistent with the preservation of [traditional African] communal values' (Pollis, 1981, p. 10). Thus, the negative colonial associations would be removed from the teachings of Marxism–Leninism, allowing it to be portrayed as a truly anti-colonial battle cry. Consequently, socialism and communism were associated with quite different values and experiences from those of eastern Europe: liberation from external domination, the regaining of political autonomy and, at least seemingly, control of national resources. Doctrine was much less important than the new idealised experience of regaining *de jure* independence. The label 'socialism' or 'communism' also served as a convenient label for an authoritarian single-party regime, usually including the revolutionary leadership. Similar underlying values of post-colonial independence underpinned the socialist/communist regimes in Latin America, with their distinct Latin features, such as the political 'strongman' and a visible military presence in politics.

In China, as in the other developing countries, Marxist doctrine had to be adapted to the pre-industrial, rural rather than high-industrial economy from which it had been derived. Initially following Moscow's advice and example, Mao Zedong increasingly shaped his Chinese way of communist development, with its much greater emphasis on personal improvement, and an individual's contribution to society, rather than reducing the individual to a minute cog in the big machinery of state politics, as under Sovietism. Inevitably, these systems left different legacies from the ones in Eastern Europe, where Soviet oppression and occupation were the paramount experience and association with 'communism'. This translated into quite different values for '*post*-communism'. While in Eastern Europe, and also many parts of the Soviet Union, it was seen as an act of liberation and regaining of national autonomy, a sensation not dissimilar to that of post-colonialism, no such clear positive associations developed in the other global regions experiencing the change, or the end, of their communist regimes. There, communist regimes ended either because of economic mismanagement or political indifference, allowing strong leaders and elites to retain power, albeit under a different, often nationalist label. In most cases, 'communism' had been more a label of convenience than a dogmatic mission. Abandoning it when the *raison d'être* had disappeared was, therefore, not particularly difficult for most of them. Alternatively, as in the case of China or Cuba, a 'halfway house' approach has been adopted, seeking to combine one-party authoritarianism with a market economy to counteract the politically corrosive effects of continued economic struggles.

Chapters 4 to 8 explore the five global regions of post-/communist development distinguished in this book: Central and Eastern Europe, the former Soviet Union, China and Southeast Asia, Africa, and Latin America, especially Cuba. The distribution of 'communist' countries across the global regions varies, as illustrated in Figure 1.2. The largest concentration, not surprisingly, is in Eastern Europe and the former Soviet Union, with the latter having been the 'epicentre' of communist ideology and its implementation. China, by its sheer size, together with a handful of smaller communist countries following the Chinese model, is also well represented in the geography of communism. In Africa, there are several smaller countries claiming to be 'communist', but there is no 'big fish' in the pond, reflecting the relatively small-scale geographic pattern of African states. Lastly, in Latin America, 'communist' states were largely concentrated in Central America, although there were some socialist governments outside, such as that under Allende in Chile in the early 1970s (Moss, 1973).

The distinction being made between 'Second' and 'Third World', or between European and non-European, or 'colonisers' and 'colonised', suggests that there are deemed to be some similarities between the five regions' socialisms/communisms, and thus post-communisms. And the plural is used here deliberately to reflect the many different 'versions' that developed within and between those

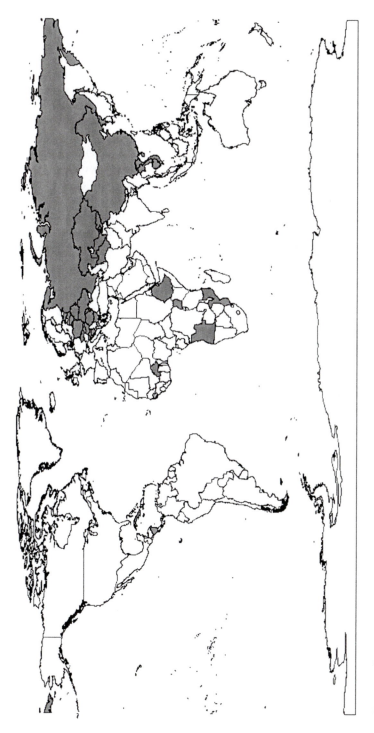

Figure 1.2 Communist countries

regions. Africa, Asia and Latin America share their colonial legacies, the memory of external domination, suppression, imported 'alien' societal and political-administrative cultures, and externally defined territoriality and statehood. In this context, communism was associated with freedom fights, liberation, greater egalitarianism and control of national assets, and thus a sense of empowerment and independence. The end of formal communist regimes was therefore associated less with ideological considerations than with fundamental economic concerns of survival in an increasingly more difficult and 'anti-communist' political-economic environment. There was no longer a communist hegemon to offer economic assistance, and western institutions were the only remaining source. Claiming communist credentials was less helpful here in obtaining favourable consideration. Meanwhile, in Central and Eastern Europe, the end of Soviet-imposed communism was something akin to decolonisation, that is, the end of external control and imposed alien societal and institutional-administrative practices and values. As under decolonisation in the developing world, adopting the political-ideological antithesis was considered a sign of liberation; a liberal, free market economy, democratisation and internationalisation were seen as just the opposite of what communism had meant, and that made all-out liberalisation and marketisation so attractive. It was an act of emancipation and throwing out the old memories and practices. Table 1.1 summarises some of the key characteristics of the five regions' communisms and, accordingly, *post*-communisms, especially their origins – 'imposed from outside' against 'chosen from within'.

The interaction with outside forces and influences, whether perceived or real, has thus played a fundamentally important role in the attitudes towards communism – seen either as part of a chosen, empowering path to self determination and independence, or as the opposite, an instrument of oppression and subjugation by an external hegemon. The CEE countries and many of the former Soviet Republics fall into the latter category, while the developing countries fit into the former. Their experiences are thus all but diametrically opposed, and this has shaped their attitudes towards, and expectations of, post-communist changes.

Chapters 4 and 5 focus on the European dimension, although this extends into central Asia through Russia/the former Soviet Union. During the Cold War, the more general, popular use of the term 'Eastern Europe', or 'Eastern bloc' included Central and Eastern Europe (CEE) and the Soviet Union, whereby this was largely seen as identical with 'Russia'. But there are important distinctions between the two systems and the ways in which post-communism developed. Thus, for instance, the former Soviet Union had been 'installed' some 30 years before the rest of the Communist bloc, thus adding another generation brought up under the communist regime. This allowed for greater deepening of communist doctrine in people's (including bureaucrats') perception, thinking and action. In the CEE countries, there was also a continued tradition of tentative democratisation, and a European focus from before the Seond World War. This, together with the moral authority of the Polish-born Pope, facilitated the grassroots movement that, in the end, triggered the collapse of the communist states. These pressures for change were decidedly bottom-up and started through informal lobbying.

Table 1.1 Some key features of the five global regions' post-/communisms

Features	Central/Eastern Europe	Former Soviet Union	Asia (China)	Africa	Latin America (Cuba)
Legitimacy	imposed through hegemonic USSR	developed from within through elite	developed from within (mobilised peasants)	developed from within through independence leaders	developed from within, revolution mobilised by elite
Communism associated with	occupation, suppression, shortages, enforced Sovietisation, stagnation	imperial status, global power, Stalinist suppression, shortages, stagnation	modernisation, suppression during Cultural Revolution, 'new market economy'	independence, anti-colonialism, pan-Africanism, new African identity, economic struggles	independence, anti-colonialism, national identity and autonomy (anti-US domination), economic struggles
End of Communist regime triggered through	'bottom-up', grassroots movements (environmental groups, church)	top-down, elite-instigated reforms led to 'inadvertent' end of Soviet Union	'top-down', cautious reforms of economic sphere only, continued use of Marxist doctrine	little concern with Marxist doctrine, loss of external support and 'usefulness' of 'socialism' label, economic crises	'top-down', cautious reforms of economic sphere only, continued use of Marxist doctrine, egalitarian society
'Post-communism/socialism' viewed as	liberation, regained autonomy, end of Sovietism, high economic costs and inequality, democracy, return to Europe	mixed feelings, end of empire, unfair privatisation, inequality, regained autonomy from Russian domination	partial post-communism only, restricted to marketised economy, new inequalities between regions and within society	indifferent, mixed results with democratisation, new tribal warfare undermining states, power held by (military) strongmen	partial post-communism, restricted to limited market economy, ('dollar apartheid'), return to Latin traditions, national independence

This contrasts with the Russian model of an elite-driven, top-down facilitation of change, even if soon developing its own dynamism and far exceeding the original agenda. Another important distinction is the multi-ethnic nature of the Soviet Union's territoriality, with many borders running through historically ethnicity-defined areas. The end of the Soviet Union sharpened these dividing lines, as they became international borders. In Central and Eastern Europe, by contrast, statehood, nationhood and territoriality largely coincided, giving these states a clearer and stronger starting position than those of the former Soviet Union (FSU). The former Yugoslavia was the only state where internal divisions between nationalities burst into the open, destroying the state as a consequence. These variations in legacies between the CEE countries and those of the FSU resulted in different transitional outcomes. While most central and many eastern European countries have made clear strides towards a successful establishment of democratic principles with popular involvement – that is, with clear signs of a developing civil society – the situation is much less clear in Russia and, especially, the other former republics of the Soviet Union. The difference is highlighted by the strong emphasis on the 'returning to Europe' paradigm pursued by most of the CEE states and formalised through their membership of the European Union. This also provides further support for their economic development, and thus the acceptability of the new conditions to their population. While the then Soviet General Secretary Mikhail Gorbachev propagated the 'Common House of Europe', within which he counted the Moscow-centred Soviet Union, such an affinity is less clear now. In fact, the former Soviet Union finds itself beyond the reinforced, higher European Union external border, with authoritarian streaks increasingly more prevalent, thus bringing into question the degree to which its claimed democratisation is more than a formal arrangement 'to be seen to be doing it'. In the FSU, the top-down-instigated changes towards democratisation have produced much less of a sense of ownership by the people than has arisen in the CEE countries, where pressure for change had been 'from below'. Only now, as evident from events in Ukraine at the end of 2004 and, albeit rather tentatively, in Azerbaijan almost a year later, are popular claims emerging for more democratic involvement in politics and national affairs. In a way, this suggests a move towards the CEE version of post-communist development, with evidence of an emergent, nascent civil society. The way in which democracy 'comes about', especially whether it is installed as a *fait accompli* by the ruling elites or chosen and struggled for by the people who are meant to be at the centre of a democratic state, seems crucial for the quality of the resulting democratic regime. Is it a merely technical or, indeed, practised form of democracy?

China illustrates a quite particular form of regime transformation – dispensed 'from above' by the incumbent political elite at varying doses. Having witnessed the rapid disintegration of the Soviet Union, the up to then perceived bulwark and spiritual and political bedrock for communist regimes around the world, the Chinese government has sought to retain power by vigorously guarding the established political status quo of one-party rule. Citing the doctrine of Mao Zedong Thought serves as public legitimation for keeping the status quo in politics. At the same time, market elements were carefully inserted into the economic sphere of the regime. The political elite is well aware of the vital importance of the economic factor for the viability of the communist principle of government, drawing on the European experience. China is thus effectively trying to ride two horses simultaneously, one representing the economic and the other the political-societal sphere. Both are kept rigorously separate. While increased consumption seems to serve as a substitute for a greater say in political matters, there are signs of devolving economic responsibility to a more local level of decision-making to increase productivity and responsiveness to the market. Pragmatic *economic* reasons are thus the main driver of modifications to the existing system, not a wider, principally idealistic interest in regime change towards a fully fledged democracy. In other words, economic interests outweigh concern about full democratisation.

So far, the system has been very successful in terms of economic output, to the extent that China is increasingly being portrayed as a potential major challenge to established western economic interests and presumed certainties. But there are signs of difficulties ahead, especially the widening and deepening inequalities at the national, regional and local levels between those benefiting from the changes and those who are not. So far, the 'lid' has been kept on any signs of emerging popular political ambitions, including through limiting access to non-domestic sources of information and, at times, the use of force. The big question is, therefore, whether this course of pursuing two essentially contradictory systems – a liberal market economy and an authoritarian one-party state whose political ideology rejects private property and the principle of 'the market' – can continue to be maintained. Certainly, the notions of 'communism', 'market' and thus 'capitalism' have lost their seemingly clear, mutually exclusive, vehemently opposed nature as reinforced during the Cold War. In China, they are forced to cohabit, uneasily perhaps, but so far, economically, quite successfully. Chinese transition has thus followed a two-track approach, surging ahead in one while steadfastly trying to stand still in the other. At the same time, this has challenged the models of post-communist transition propagated since the late 1980s in which marketisation and democratisation are the two main, inseparable ingredients *sine qua non*. The Chinese model has so far suggested otherwise.

Chapter 7 focuses on a much less widely reported global region, in terms of regime change from communism to post-communism, or 'socialism'. Compared with the rapidly growing amount of work on China, reports on transition in Africa are much fewer. The main focus of these studies is the emergence of 'African socialism' (Friedland and Rosberg, 1964a) immediately after the end of colonial rule in the late 1950s–early1960s, followed by suggestions of 'Afro-Communism' in the 1980s (Ottaway and Ottaway, 1986). The end of communist rule in Africa, and subsequent developments, have been largely subsumed by more general discussions on democratisation and regime politics on the African continent. But there are a handful of countries that have attracted more attention, because of their more visible stance on following a Marxist–Leninist, Moscow-friendly path of development, in particular Mozambique, Angola and Ethiopia (Babu, 1981). Others have proclaimed adherence to 'socialism', such as Tanzania (Saul, 1985a) or Kenya or, in Northern Africa, Libya and Algeria. But the majority of those countries claiming to follow socialist principles used 'socialism' as a label of convenience for authoritarian regimes. The emphasis was on the formal arrangements of one-party rule, promising a ruling elite able to claim power in perpetuity, rather than on the underlying ideology. In effect, it was Leninism without the Marxist ingredients.

There were attempts at Africanising the essentially industrial and European nature of Marxist doctrine, adjusting it to pre-industrial, agrarian societies, as well as connecting it to African pre-colonial societal values. Immediately after decolonisation, in the 1960s and early 1970s, 'African socialism' was propagated as part of attempts at pan-Africanism, seeking to develop an African identity *vis-à-vis* the former colonial masters, but also the world in general. Driven by idealist new African leaders, more pragmatic considerations of maintaining control of the inherited state structures, with their difficult ethnic legacies and needed economic development, increasingly moved to the fore. This included using 'socialism' as a means of playing one superpower off against the other to obtain development aid, both economic and military. The end of the Cold War removed that opportunity and thus the incentive to use the label 'socialist'. Even the few more convinced followers of Marxism-Leninism were eventually resigned to the reality of a one-superpower world, and the paramount paradigm of democratisation and liberalisation as the precondition for obtaining development aid.

Claims to 'socialism' have come to be seen as anachronistic and ultimately doomed. Proclaiming democratic principles has thus seemed more opportune for those seeking to stay in power and gain economic and financial support from international agencies. After an initial 'wave of democratisation' in the 1990s, used by many African leaders to claim or retain power, there now seems a growing realisation among them that such systems may 'turn real', and people may vote them out of office. There is a growing tendency among them, therefore, to renege on the democratic principle to avoid such a fate (see *The Independent*, 16 Nov 2005). Transition to a post-'socialist' or post-communist regime have thus been much less profound than in the European context, being more part of a general 'muddling through' way of governance than a pursuit of grand idealist strategies. Staying in power is a much more immediate concern, a not untypical feature of politicians, but to keep changing the system to suit that ambition seems more of an African characteristic.

Similar ambitions, albeit in a somewhat different historical-cultural setting, can also be found in Latin American countries, although perhaps less so in those proclaiming adherence to socialism. Chapter 8 explores the nature of changes to regimes that pursued a socialist agenda, most of them in Central America. The most idealistically committed regimes have been those in Nicaragua and Honduras and, still maintaining its idealism, Cuba. Others, like Chile under Allende, proclaimed socialism, but more as a political programme than as an ideology on the basis of Marxism–Leninism. Cuba is clearly the most prominent representative of a communist regime in Latin America, with its position just off the North American coast adding to the poignance of its situation. It is not surprising, therefore, that Cuba was seen by the 'West' as a Soviet outpost. Nevertheless, the Cuban government has sought to maintain a visible distance, emphasising national autonomy and independence. As in Africa, socialism, or communism, has been closely associated with 'the revolution' – that is, the break with external control and domination. An important particularity is the personal authority of Fidel Castro who, as original revolutionary leader, has gained legendary status. While seeking to maintain the status quo of post-revolutionary authoritarianism under the umbrella of Marxism-Leninism – suitably interpreted for Cuban conditions, the Cuban government had to adjust its policies, if not so much the ideological rhetoric, to the new, post-Soviet conditions. The end of especially economic but also political support has brought about a shift in Cuban policy towards an emphasis on its Latin culture and traditions, rather than Marxist–Leninist values. Cuba keenly observes the Chinese model of separating economic development from political ideology and government, and seeks to stimulate its own version of a dual economy – one for foreign investment, one for domestic activities. All these changes have somewhat undermined the rationale for maintaining the political status quo, especially the harsh economic conditions in the early 1990s, but shifting the arguments towards nationalistic themes, especially the notion of defending Cuban independence against an imperial, hostile USA, has so far managed to keep the population largely behind the government. Fidel Castro's personality, of course, has helped to maintain the regime's authority. But it remains to be seen how a generational change, both within government and 'on the ground', will affect attitudes and allegiances.

But what has become clear is the importance of the positive association of 'liberation' and 'national independence' with the communist regime and its claim to power. This provided a sense of natural legitimacy for the government, albeit with continued, considerable *de facto* external economic dependency. It is the history and memory of the revolution that has allowed the government, so far, to ride out economic dissatisfaction and disillusionment. But whether a new generation, more detached from past events, will continue to give the government that reverence and 'benefit of the doubt' remains to be seen. So far, 'transition' has been a rather stop–go affair, reluctantly permitted by the regime rather than actively pursued by it. New personalities, affinities and loyalties may emerge in the future, in the aftermath of a generational shift

Chapters 9 and 10 take a genuinely global perspective, exploring the different global regions' post-communist transitions from a thematic perspective – that of cultural and national identity, and economic development and inequality, respectively. The end of strict one-party control has meant the re-emergence of suppressed ethnic and social identities, be they cultural, religious, or class-based. Often, these identities are conflicting and contradictive, seeking to straddle 'old' and 'new' realities. Cuba's dual economy, with 'dollar apartheid', exemplifies this division. These economic differences translate into new cultural and social divisions, both spatially – most visibly within the cities – and socially, such as between generations. It is these new differences and inequalities that are likely to pose major challenges to the existing, often still emerging, new forms of post-communist government.

As the communist experiences fade into more distant (collective) memory, reappraisals of those conditions may occur, leading to different reactions and attitudes to those dominant in the immediate past. Already, there are signs of resurgent, more affirmative, less apologetic noises about the former communist state, such as in the former East Germany, as the initially expected rosy realities failed to materialise. The increasingly more obvious, growing inequalities between the winners and the losers of the transformation process may well harbour considerable challenges for policy makers in maintaining national and societal coherence. In a way, this may be viewed as the second, more critical and informed phase of post-communist transition. Changing the system and structures can be done quickly and extensively. Changing people's mindsets, expectations and memories is much more difficult and takes much longer. In that respect, it seems, 'post-communist transition' is far from 'done and dusted', but is still an ongoing process, if at a different level and format than the rather simplistic state-centred perspective would have suggested.

2

SOCIALISM, COMMUNISM AND AFTER

With a contribution by Tomasz Zarycki

After the end of the Second World War, the new geopolitical division of the world into a Soviet dominated 'East' and an American-centred 'West', and the associated Cold War rhetoric, very much shaped the public discourse on the meaning of 'socialism' and 'communism'. This discourse varied depending on whether seen from 'within' or 'without' the countries experiencing socialist/communist regimes in practice. From an outside, western perspective, little difference was made between the two concepts. Both were associated with 'Moscow' and the 'Eastern bloc', and were thus by definition negative, inferior and hostile, or, in other words, the antidote to a democratic market-based society. And it is the Soviet context that socialism was closely associated with, 'both by anti-socialist ideologists in the West and by many in the East too' (Nove, 1991). State domination, authoritarian excesses, Siberian Gulags and continuous shortages (Carson, 1990) in almost every aspect of consumption were the most frequent images projected about the perceived reality of a communist regime. Of course, there were also socialist parties in the West, campaigning on the platform of fighting against ruthless capitalism on behalf of the exploited workers, and even communist parties, proclaiming allegiance with 'Moscow', not least in the form of the weekly *Marxism Today*, but they were largely kept outside mainstream politics by the political elites. The key justification for an envisaged 'regime change' from capitalism to socialism by this group was a 'morally informed vision of a better life' (Luntley, 1989, p. 3). In this 'armchair' socialism, greater morality, rather than a free market system, counted as the main driver of improving people's quality of life and 'happiness', however defined (see, for example, Miliband, 1977; a good collection of Marxist and recent 'post-Marxist' papers is provided by Sim, 1998; and a much earlier collection by Miliband and Saville, 1974). But 'socialism is not a moral theory that offers a particular vision of the good life, instead it is a theory about how the good life is possible' (Luntley, 1998, p. 15). In other words, this understanding views socialism as an instrument, as a means of getting to a 'better' form of society shaped by 'good' moral values, such as equality. Consequently, Luntley contrasts two complementary theses about the nature of 'socialism', one emphasising its pro-active nature shaping society through a set of moral values, and one taking a more passive role, where it is threatened as a morally superior form of society by the emergence by capitalism. It is this latter understanding – the contrast in social values and the nature of society – that has underpinned the discussions on post-socialist transition and the contrast between the 'communist' starting point and 'neo-liberal' end point of this shift.

Despite, or perhaps because of, the vivid debates during the Cold War and immediately thereafter, the very meaning of 'socialism' as a state regime has remained far from clear (Sik, 1991a), and been made even more complex and uncertain by its often interchangeable use with 'communism', especially when referring to the developments since the collapse of the Iron Curtain. This is despite the differences in the words' underlying meanings: '"Socialism" has come to stand for so many things to so many people that the radical cause against capitalism has been thoroughly blunted' (Luntley, 1989, p. 2). Three main strands of argumentation may be identified at a more general level: for one, there is the difference between 'socialism' and 'communism' as attributes of the 'Eastern bloc' states. There were those with a 'Communist Party', like the Soviet Union, or a 'Socialist Party', such as East Germany. There are different suggestions as to how to differentiate the varying use of terminology in a more systematic way. Sik (1976) focuses on the term 'socialism' and its varied usage: on the one hand, it 'applies to the really existing' version of socialism, as evident from the various countries claiming to follow its ideology, while on the other, there is the idealistic, philosophical version propagated in the 'West' as idealising 'Travel Brochure socialism' (Luntley 1989). The latter, also dubbed as 'armchair socialism', may be viewed as part of a backlash against the New Right politics of Thatcherism and Reagonomics with their emphasis on economic advantage at the expense of more direct societal concerns. But much of this debate was in intellectual, academic circles of the affluent, capitalist 'West', not in politics as part of a genuine debate on the respective virtues of the two contrasting social-economic models – socialism versus capitalism. Perhaps tellingly, the self-proclaimed 'Western' socialists were jokingly caricatured as keeping Mao's Little Red Book in the glove compartment of their Jaguar cars. The venerable *Marxism Today* weekly newspaper may be seen as a beacon of this western 'socialist establishment'. A number of left-wing academics followed a serious Marxist analysis of economic and historic trends, inspired by the evident geographic, social and political inequalities produced by 'the market', especially through globalisation, and looked for alternative models of societal development. 'Restructuring for capital' and the 'spatial division of labour' became trademark terms in the spatially focused analysis of that time (Massey, 1984, 1988; Massey and Allen, 1988). But the reality of socialism in practice, usually represented by highly authoritarian regimes, looked quite different from the ideal portrayed by the western 'socialists' who always could resort to their established western lifestyles.

Differentiating between an 'inside' and 'outside' perspective raises connotations of imprisonment – and, although not initially intended when choosing them, these contrasting terms reflect a basic and most important feature of most communist states, especially those following the model of the Soviet Union: the much-restricted scope for residents to leave their country at will. Likewise, access from the outside was strictly controlled. In this way, it functioned like a prison, and the features of the Iron Curtain and the Berlin Wall very much highlighted, symbolically and effectively, this sense of imprisonment. Most of the affected residents felt it that way, and the scenes witnessed when the Berlin Wall and the Iron Curtain came down memorably illustrated this. From an 'inside' perspective, 'socialism' possessed different connotations from those held 'outside', in the West. For once under communist regimes, 'deviations from socialism' were seen as all things negative, and thus 'socialism' was *the* criterion of 'good' public life – living and working – that means first and foremost the individual's contribution to building up a socialist society, rather than the striving for personal financial gain (see also below the discussion by Zarycki).

Then there is the confusion about the usage of the terms 'communism' and 'socialism'. The terms, as well as their respective 'post' versions, are used in most instances almost interchangeably (e.g. Kornai, 1992). There are some attempts at distinguishing between the two in reference to their different historic meanings and origins – Mandelbaum (1996), for instance, prefers the term 'post-communism' to 'post-socialism' – although both emphasise the dual nature of the term. Looking both ways, back and forward, they link the legacies of socialism/communism to the path of future development after the demise of the socialist/communist state systems. Mandelbaum's definition of 'post-*communism*' implies a Marxist–Leninist implementation of the theory of 'socialism' through a strong role for an authoritarian Communist Party, usually through top-down pressure, resulting in 'a political system that was "totalitarian" in aspiration, aiming to control every aspect of social life'. Sik (1976) sees this as part of the 'perversion' of the idea of socialism through excessive bureaucratisation. 'The instrument of control was the all-powerful, hierarchically organized, self-selected, and self-perpetuating Communist Party' (Mandelbaum, 1996, p. 1). Claiming a unique predisposition to implementing the ideas of Marx (the idealist) and Lenin (the pragmatic politician), the Party claimed a monopoly of power as a logical conclusion. As such 'it was responsible for, among other things, managing the country's economy, which was "planned" in the sense that important decisions were made by administrative fiat. Party officials, and not the ebb and flow of supply and demand, determined what would be produced and grown, in what quantities, and the prices at which what was produced would be sold' (Mandelbaum, 1996, p. 2).

This central control of the economy, through exclusion of the market, is unique to the communist system among all other totalitarian regimes, and was a *sine qua non*. Any private activity, whether political or economic, or even cultural, was strictly controlled and censored, and, if deemed potentially to undermine the Party's and the apparatchiks' claim to power, suppressed. It is in this sense that the terms 'socialism' and 'communism' will be used here: 'socialism' and 'post-socialism' refer to the theoretical ideas as developed by Karl Marx, arguing for an ultimate redundancy of the state *vis-à-vis* a homogenous, essentially egalitarian society. The terms 'communism' and 'post-communism' are used for the practical application of these ideas through an authoritarian regime to the practical organisation of a state, following Lenin's model of applying Marx's ideas.

Despite the general common 'standard' characteristics of 'communism', there were considerable differences in its national implementations, not only between the countries of Europe and those in other parts of the world – largely pitching 'industrialised' against 'developing' countries – but also between the 27 countries of 'Eastern bloc' Europe and the former Soviet Union, especially Central Asia (see e.g. Lock, 1994). These differences reflect national particularities in culture, tradition or development status, but also positions in the global framework as between Berlin and Vladivostock, for instance. Thus, in Poland, the Catholic church gained considerable influence, especially after the papal election of a Polish bishop, and farmers maintained their non-collectivised peasant structure, despite governmental pressures, reflecting a stronger commitment to Polish nationalism than to Moscow's orders. Hungary introduced market reforms for the agricultural sector, and Tito balanced Yugoslavia between Moscow and western Europe, allowing Yugoslav workers to go to Germany for temporary employment as 'guest workers'. Even greater differences existed in comparison with the Central Asian Republics, for instance, where powerful, autocratically ruling local clans were often more important than the far-away Supreme Soviet in Moscow, and the black economy was widely tolerated and accepted throughout the Caucasus region. Thus, established local/regional practices and ways of doing things mixed with particular national/regional circumstances, which then produced various versions of communism in practice. The already considerable differences at the Eurasian level became even more diverse when other global regions and their particular legacies and practices were included, especially the African countries running under the banner of socialism, or China and Cuba.

With such diversity in the use of the term of 'socialism', it is not surprising that Balcerowicz (1991, 1995) asks what 'socialism' actually means. Is it social democracy, as in Sweden, based on a genuine idealistic concern with social welfare, egality, and communitarianism, or is it merely a label of political convenience for an authoritarian one-party state system? He believes that behind all these various concepts of 'socialism' stands the idea of 'universal happiness' (Balcerowicz, 1995, p. 20), and any deviation from that, whether in a democratic or an authoritarian socialist system, is thus seen as by definition aberrant. Nevertheless, the use of the term 'socialism' *per se* is diverse, suggesting a fuzzy underlying rationale. Balcerowicz (1995) distinguishes four broad categories: (1) the idealistic quality, where 'socialism' refers to socialist ideals about the desired egalitarian state of society; (2) the economic dimension of socialism, with the main focus on economic management as centrally regulated system; (3) the doctrinal usage of 'socialism', usually derived from Marxist–Leninist teaching and seeking to legitimise the aims and rationales for the development (and, if needed, imposition) of socialist systems as superior to other (capitalist) systems; and (4) the programmatic use of the term, where 'socialist' has been employed as a label for political parties and their programmes.

This vagueness of the understanding of 'socialism' has been exacerbated by the increasingly broadened definition and use between and within the communist countries. For instance, African communism was quite distinctly different in its ideology and rationale from Cuban, Chinese or Soviet communism, although, from western perspective, they were all seen as part of the amorphous 'Eastern bloc'. Within these countries, the use of the terms 'socialism' and 'communism' also varied over time, owing to 'amendments' and modifications to justify particular policy measures. This 'almost unlimited expansion of the concept of socialism' (ibid., p. 20) has two main causes. First, it evolves from a necessary response to a perceived crisis in the countries that adopted socialism as the dominant ideology, and the term was stretched to justify these policies and their outcomes. At times, this went as far as arguing that liberal economy and socialist principles are essentially in harmony, despite the fundamental contradiction of liberalism to Marxism in terms of the efficient workings of 'the market'. Whether issued by the current Chinese government, or in the 1987 resolution of the Hungarian Workers' Party, these are attempts by reformist, non-orthodox communist governments to explain the inevitable contradictions between 'plan' and 'market' (Sik, 1967) to the people in an attempt to justify the continued upkeep of the communist system altogether. The second reason was to pave the way for the envisaged changes, and to blunt likely attacks by traditionalists within the Party on these changes as a betrayal of socialist/communist ideals.

But delivery of these achievements was through centrally controlled development only, excluding private initiative. The inevitable result was a massive increase in bureaucratisation, with all its costs and inefficiencies, undermining its very functioning. Indeed, 'in Eastern Europe official propaganda has for decades equated socialism with the system imposed by the communist party, and so with poor quality, neglect of consumer interests, inefficiency, empty shelves and poor service' (Nove, 1991, p. 87). It is not surprising, therefore, that the word 'communism' is being used from 'within' the CEE states to sum up the rejection of the socialist experience *per se*. Referring to the example of Poland, this is discussed in more detail in the section below by Zarycki.

The lack of democracy made a crucial difference to the form of socialism envisaged by Marxist ideology, with the Communist Party claiming an irrevocable absolute grip on power. In reality, therefore, instead of democratic control through the masses, as envisaged by Marxism (Berki, 1988), a new minority dictatorship was created, which was a far cry from the officially propagated 'workers' democracy'. Lenin's supporters admitted as much, when referring to the 'dictatorship by the proletariat' as the foundation of the state. Any attempts to claim more influence by the population were immediately and, if necessary, violently suppressed. As a result, public life, and with it an active civil society, withered away, replaced by a new political and bureaucratic elite, recruited selectively from the 'proletariat' as the stalwarts of the new envisaged society. Many of the created 'democratic' institutions were little more than mere loincloths for an authoritarian regime. Infamously, the saying goes, the leader of East Germany, Erich Honecker, once questioned about the democratic credentials of his regime by his own apparatchiks, said that 'it needs to look like democracy', and nothing more.

It is this domination of all aspects of life, from general politics, via the economy, right down to individual organisations and clubs, even if for nothing more sinister than pigeon fanciers, that crowded out any signs of independent activity. Apart from the direct costs of maintaining a huge bureaucracy to monitor all those activities, there was the indirect cost of lost economic opportunities through state-imposed inertia and bureaucratisation. Given the importance of 'functioning' within the state/party to set targets, bureaucracy mattered over competence when seeking promotion. As a result, the economic interests of the bureaucrats were focused primarily on their immediate position in the apparatus and how decisions could promote them. Actual economic improvement through genuine productivity gains was not the main interest. Instead, achieving centrally set targets, usually of a quantitative rather than qualitative nature, were seen as a ticket to higher office (see also Sik, 1976). Quantity looked good in statistics, whereas there was no officially relevant feed-back on quality. In the absence of choice, quality was almost irrelevant. This focus on delivering on quantitative targets applied to all levels of the administration right down to the business manager. Manipulating figures and reports so as to suggest steady improvements to output and achievement, at least, if not over-achievement of production targets, was in the interest of everyone within the state system. It all hung together. Good results helped legitimise the system's way of operating and with it the underlying Marxist–Leninist ideology in general, and individual local apparatchiks, more specifically.

Back in the 1970s, when the communist world still looked firmly entrenched and secure, Sik (1976) examined the conceptual-theoretical meaning of 'socialism' against the background of socialism in practice, as implemented by the authoritarian 'communist' regimes established across the 'Eastern bloc'. He emphasised that it is the 'communist parties' that, with repeated, dogmatic reference to Marxist–Leninist theory, propagate their hold on absolute power in 'their' states. They use this reference as their legitimation to maintain power, and suppress any attempt at challenging this primacy by denouncing such challenges as 'anti-socialist', 'reactionary' and 'counter-revolutionary'. This aggressive defence is regarded by Sik (1976) as an attempt to disguise the obvious discrepancies that had emerged between Marxist–Leninist theory and its implementation as applied in state socialism 'on the ground' (Lane, 1996; see also Snooks, 1999). In this respect, Sik (1976) identifies three key points:

1   The basic principles of the regime, such as centralised economic planning and the unchallengeable leading authority of the communist party, are propagated as 'essential' socialist ingredients. The main concern is effectively instrumentalist, focusing on technocratic concerns with production and modernisation through a strictly centralised regime (Kautsky, 1998); people are valued primarily as functioning parts of the productive economic system, rather than as individuals with a concern for improving their quality of life. In particular, Leninism's emphatic modernisation agenda for a largely agriculturally dominated economy appealed to many post-colonial developing countries, as it combined a strong modernisation agenda with a state-centred, *dirigiste* tradition (ibid.).

2   The second principle is the selective, instrumental use of Marxist–Leninist theory to justify party political policies and ambitions, preaching the liberation of people from state control and suppression, as long as the state is 'capitalist', but propagating no such thing if the system is 'socialist'.

3   The third key feature is a refusal to critically review the validity of Marxist–
    Leninist economic analysis in the light of the actually experienced difficulties
    and, instead, clinging dogmatically to the nineteenth-century-inspired
    ideology. As a result of the supreme importance of dogma, evident weaknesses
    and shortcomings in the chosen development path were ignored and denied,
    including costs in peoples' welfare. With no official voicing of criticism, let
    alone opposition, permitted, there was no mechanism for challenging estab-
    lished practices in policy-making, once established. Consequently, Marxist–
    Leninist theories evolved into a dogmatic, static credo propagated by the
    communist states, whose main ambition was to maintain the communist party
    state.

Overcoming these considerable systemic deficiencies was, according to Sik
(1976), the precondition for transferring the actually existing form of socialism into
a genuinely socialist system. That, he continues, would include the application of
democratic principles; indeed, 'socialism cannot be realised without democratic
conditions' (ibid., p. 15). Accordingly, the so-called socialist system of the Eastern
bloc cannot really be called 'socialist' in the conceptually intended sense. Rather, it
should be labelled as 'Communist'. He thus chose the term 'communist' to label the
ideology-driven, dogmatic state-socialist system practised in Central and Eastern
Europe. 'Communist' is also used to indicate the leading roles of the various
countries' Communist Parties in establishing and maintaining those systems and
their own hold on power. In practice, these are provided by the power of the
bureaucratic apparatus and the state-monopolistic economic structure (Sik, 1976).

Looking at the socialist economic system, its main distinctive feature is its base in
social ownership of the means of production. By definition, it excludes private
ownership, and even contradicts it. Social ownership is the opposite to private
ownership. The main rationale for socialisation was, first, income from private
ownership, especially rents, was considered 'unearned' and inherently exploitative
and against the working classes. This sort of income was seen as the epitome of
'greedy capitalism', and translated into the abandonment of much of the old
housing stock as the owners were vilified and then 'socialised'. Second, private
ownership of the means of production (the second stand of capitalist domination)
was condemned as inherently wasteful because of its competitiveness and inherent
lack of coordination. Instead, the rational, detailed and overarching planning
would avoid such inefficiencies and maximise productive potential, in effect saving
the economy from the capitalist in-built road to self-destruction. Private property
makes such central planning impossible and thus jeopardises efficiency gains in the
economy. The logical consequence is to abolish private ownership and replace it
with public ownership and thus total control of all aspects of the economy in the
interest of its best efficiency.

This essentially follows Marx's doctrine/reasoning. 'Therefore, regardless of
what Marx's dreams were, the only real form of social ownership compatible with
his vision of socialism is a market-less economic system, is centralized state
ownership. Real socialism . . . is thus in this respect in basic agreement with what

Marx's vision of socialism really contains' (ibid., p. 25). But increasingly, 'new' models of socialism emerged in response to the evident sub-optimal performance of socialist economy, as in Hungary's 'Goulash communism' (Kornai, 1990, p. 255), moving rather towards the principle of a more subtly regulated 'socialist market economy' (Kovacs, 1991; Pierson, 1995) – although that would appear a contradiction in terms. This reflects a reduced emphasis on ideology among the political elite, realising the need to face, and respond to, the reality of economic deficiencies and a relative weakening *vis-à-vis* the western market-based system. But 'socialist' is understood primarily as 'social ownership', and thus the avoidance of the inefficiencies inherent in free markets is seen as a good enough reason to continue the socialist approach. Indeed, socialism may be seen as following communism (Pierson, 1995), effectively implying a return to the idealistic virtues of socialism after their seeming betrayal through the perceived aberrant misappropriation of the idealist principles by communist rule.

Brus (1991) echoes the three key elements identified by Sik (1978) for the Marxist system presented above: (1) socialism as the legitimate outcome of historic development and thus successor to capitalism; (2) socialisation of the means of production as guarantor of wider participation in its outcomes, essentially a wider 'shareholding'; and (3) the perception of socialism as an 'end' of historic processes (societal evolution), its undertone revolutionary and disruptive to existing structures. '*Socialism was thus the first pre-designed sociol-economic order, a mega-experiment in social engineering*' (Brus, 1991). The underlying notion is thus democratisation of the economy and means of production. In reality, however, '"real socialism" is evidently lacking – by almost universal admission nowadays [that's 1991!] – the democratic political component. For many this is a flaw which makes any generalisations about socialism based on this experience illegitimate' (Brus, 1991, p. 51), but this, Brus argues, is more or less throwing the baby out with the bath water. With state ownership as central plank of the ideology, 'real socialism' was applied socialism. Indeed, he continues, 'if one looks for socialism as a bounded system, a distinctive socio-economic formation, it would be hard to find anything else of such a close fit' (Brus, 1991, p. 51).

Not surprisingly, therefore, China seeks to hold on to some form of state control of the economy to maintain the mantle of 'socialism' for its regime. But it also serves to protect the political elite's position in power. Essential, however, for the functioning of the system, was keeping control of state bureaucracy and the central planning apparatus. Consequently, in those cases where reforms of socialism were pushed, including China, 'The arbitrariness of central planning was to be checked ... by the postulated pluralism in the political sphere' (Brus, 1991, p. 52). It was thus about changing the ways of defining targets, rather than the nature of the control system *per se*. The principle of top-down managed bureaucracy was sacrosanct as a distinctive feature of socialism. Inevitably, this 'fiddling on the edges' could not really address the problems emanating from the system. All it could do was change from direct to indirect bureaucratic control (see also Kornai, here referred to in Sik, 1991a), whereby the allocation of investment was the main and most effective instrument of control and encouraged an extension of control into ever smaller issues. But 'letting the capital market in has ... profound consequences for the very concept of a socialist economic system' (Brus, 1991, p. 53). This is because the capitalist market challenges the very existence of the main pillars of socialism: planning as *ex ante* design of economy and society, the mechanism of distributing resources, and the dominance of political considerations (see Brus, 1991, p. 54). It is the planning aspect that is particularly important, and crucial for shaping and maintaining the essence of socialist principles. A mixed economy, combining a sizeable non-state sector of cooperative and private/quasi-private enterprises, seems to be the main element in establishing market and state control (ibid.). And it is political pluralism that is indispensable for managing the transition 'from the old to the new economic system' (ibid., p. 55). This shift acknowledges the no longer valid justification of a command economy as being most effective in overcoming (initial) underdevelopment, and favours instead the adoption of a social (market) democracy (ibid.).

But was there scope to modify the system and improve its economic perform-ance? By the same token, Sik maintains that this does not necessarily have to mean that socialism cannot work at all, but merely that adjustment to the system and less dogmatism are necessary, leading to his 'third way' model, and coining that notion long before New Labour in Britain adopted it in 1996/7. This 'Third Way' he primarily understands as a shift away from the now discredited term 'socialism' as implemented by the Eastern bloc states ('real socialism'). Effectively, he advocates a model that includes a good dose of market principles, while allowing continued 'adjustment' and 'tweaking' of the system to deliver the goods (literally). The outcome is a social market economy, not unlike the post-war German model or Sweden's, hailed by many as the most successful example of 'true' socialism, or 'Humane Economic Democracy' (Sik, 1991a, p. 17). 'In fact, Sweden may be the nearest form of quasi-socialism which might be acceptable to a West European electorate' (Nove, 1991, p. 83). But it is only 'quasi' in Sweden, because there is no *planned* economy, and it is thus missing an essential element of a state-socialist economy.

Summing up, state-socialist systems – and it is the organisation of the economy that fundamentally shapes the concept of 'socialism' – are characterised by the all-dominant role of the state and, by extension, of the communist party as the only party seen as possessing the historically derived legitimacy to rule for the benefit of the 'proletariat'. Drawing on the nineteenth-century exploitative nature of the newly established industrial capitalism, the relationship between capitalists, seen as owning all means of production, and the workers, with no access to these means, is portrayed as inherently antagonistic. The representation of those disempowered 'masses' to help them to their 'fair share' of the gains of industrial capitalism, is portrayed by communist ideology as a historic responsibility. While Karl Marx envisaged the state as ultimately becoming superfluous after overcoming the domi-nant role of capitalists, the actual, implemented communist systems worked through an all-powerful state under the communist party's hegemony. The resulting one-party state controls all aspects of public, and many of private, life. Rejecting the principles of capitalism, and thus the workings of 'the market', its main distinctive feature, is its base in social ownership of the means of production, thus eliminating the key driver of markets as mechanisms of distribution and allocation of resources. Social ownership contrasts with private ownership. The main rationale for socialisation was the cutting out of rent-seeking behaviour: income from private ownership, especially rents, was considered 'unearned', inherently exploitative and essentially unethical, epitomising 'greedy capitalism'. Adopting a strict planning regime instead of the 'unfair' market was thus presented as a natural response to the inequalities generated by the market. In practice, this 'plan' exceeded the economic sphere of resource allocation and became a controlling instrument of society and politics in general. Because of its roots in nineteenth-century industrial society, its rationale and objective reflected the conditions then, and it was the resulting ideological conclusions and policy responses that were continuously reproduced over the subsequent centuries, irrespective of the changing techno-logical, economic and social conditions in 'industrial society' elsewhere. The result was the maintaining of a time-warped world and world view, which, eventually, was overtaken by the forces of discrepancy between conditions and rhetoric 'inside', and social-economic conditions 'outside' the communist world. In some countries, the realisation of this discrepancy had led to tentative 'tinkering' with the – economic – aspects of the system by their respective governments. Yugoslavia, for instance, sought a less dogmatic, more pragmatic and technocratic approach in a 'socialist market economy', although that is in essence a contradiction in terms, of course. But it was not until the Chinese government responded to the discrepancy visibly and increasingly determinedly at the end of the 1970s, that evident shortcomings in the implementation of communist ideology were admitted and discussed, leading to progressively bolder changes towards marketisation. But these changes – and permitted debates – were almost exclusively restricted to the economic side of the communist state (see Chapter 6). The outcome has been various forms of 'post-communism', as discussed in the next sections.

The 'socialist market economy' has been promoted as an alternative 'third way' between a centrally planned socialist economy and a free market economy. It is thus seen as different from capitalism, although the nature and extent of that difference are not so clear. In the author's view, 'it is an economy where the market mechanism is the dominant mode of coordination in the sphere of private goods, that is, goods which can be distributed among individual users' (Sik, 1991a, p. 28). The very nature of a market mechanism as a mode of coordinating supply and demand would make any interference thus seem counter-effective, especially centralised decisions affecting local transactions. The rationale (and justification) of the socialist market economy is that there should be no centralised control, but decentralised facilitation of a demand-to-supply mechanism. This, at least, is the rationale of the 'in between approach', but 'muddling the concept of socialism is probably thought to be a price worth paying for the – hopefully – increased chance of introducing new arrangements' (ibid., p. 29).

Types of communism/socialism

Discussions on possible different meanings of, and approaches to, 'communism' in different countries go back to the creation of the Soviet-centred Eastern bloc after the Second World War, because it encompassed so many different national cultures and circumstances within Europe and, especially, beyond. Thus, for instance, China set out to develop its own 'variation' of socialism, with distinct differences from the original Soviet version, which it sought to copy at first. China saw itself increasingly as the pioneer of an approach to building communism that was much more geared to the particularities of a developing, and post-colonial, pre-industrial country, than was Russia. The result has been a competition between the two largest states following communist doctrine for influence in Asia, and the emergence of the Maoist version of communism with its roots in the rural parts of the country. But following the death of Stalin, and particularly since the Polish and Hungarian rebellions of 1956 against the imposed Soviet communist system, scope for diversity and individual national paths of communist state building were continually rolled back by Moscow for fear of losing control.

While after Stalin's death the Soviet Union was ready to take a more pragmatic approach to diversity in implementing communism – or socialism, as some of the countries, and their parties, preferred to call it – they were also concerned about disunity and 'Titoisation' potentially challenging their hegemony within the Eastern bloc. Instructively, the Moscow Conference of 81 Communist Parties in 1957 very much revolved around the debate on the essence of a socialist state, and how much leeway there should be to find and implement national interpretations of it. The outcome was a package of minimum *sine qua non* requirements for constructing a socialist state (see Zagoria, 1963, p. 12). These included guidance by the working class as expressed by a Marxist–Leninist Party (effectively proletarian dictatorship through the Party), replacement of capitalism through public ownership of the basic means of production, gradual socialist reconstruction of agriculture (collectivisation), planned development of the national economy to raise people's living standards, effecting a socialist revolution in culture and ideology, and solidarity between socialist countries as 'proletarian internationalism' (ibid., p. 13). Noteworthy is the rather general, broad phrasing of these minimum requirements, refraining from seeking to impose the particular *Soviet* interpretation, and giving scope for variation and diversity within the bloc. But this applied primarily to the international or, rather, inter-continental dimension of implementing socialism, and here the difference was between largely agricultural, post-colonial countries, and the industrialised countries of Central and Eastern Europe. The varying degrees of interest taken by Moscow in different parts of the world were reflected in their concern with ensuring control, and the CEE countries were economically, and politically and strategically, the main asset. They were also, with distinct national histories and identities, likely to be the most 'unruly', and thus running a tight ship with them seemed opportune in order to maintain control. Correspondingly, China, the other 'big fish in the pond', saw itself as better equipped to respond to the developmental issues of countries outside Europe, because of its own colonial history and pre-industrial developmental state.

'So long as the process of socialist construction is led by a Communist party and includes nationalization of industry, collectivization of agriculture, and loyalty to the Soviet Union, it would seem that the attitude toward diversity reflected in this document is quite permissive' (ibid.). However, it is this very diversity, so obviously seen as necessary in the late 1950s/early 1960s, that later became ignored and ruled out by doctrine and the strong hegemonic, superpower ambitions of the Soviet Union, especially under Breshnev. Diversity was seen as weakness and the beginning of a possible disintegration of the empire. But these concerns became more immediate amid growing signs of economic weaknesses, structural problems and public disillusionment during the 1960s and later. It is this more global outlook of Moscow's, especially in looking at western Europe and the USA as competitors and challengers, that distinguishes its views and policies from the ambitions and perspectives of China's communists. Their main focus has, until recently, been much more concerned with the Asian continent first and foremost, rather than on ambitions as a global political player. Only now have its interests, not least driven by its economic ambitions, moved to a global scale.

The two countries reflect their different historic traditions and ambitions. 'The Maoist attempt to distinguish the Chinese revolution as a model for all backward areas of the East, from the Russian Revolution, which was the model only for the more advanced capitalist countries, can be traced back at least as far as 1940, when Mao's *On New Democracy* was published' (Zagoria, 1963, p. 16). He contrasted his version of socialism with the Soviet model of dictatorship by the (largely urban) proletariat. China's much greater dependence on peasant agriculture with low degrees of urbanisation and industrialisation caused it to project itself as the champion of socialism for all other ex-colonial, underdeveloped (agrarian) countries. China was much less developed economically than Russia, with a lower degree of mechanisation and urbanisation even in 1917, while possessing a higher population growth rate than the Soviet Union until the 1960s. Because of this low level of development in China, any disruption to the social and economic order in agriculture during a revolution would be much less costly for the peasants affected. There was not much more to lose, whereas Russia's rural economy collapsed and impoverished the rural population, especially under Stalin (e.g. Carson, 1990). In addition, China adopted a much slower pace of collectivisation than Russia under Stalin, allowing the rural population to get used to the new situation. The Chinese government's drive to collectivise also needed to take the greater fragility of the new collectivised system into account, unlike Russia/Soviet Union. The lower degree of technological input in China also meant that individual peasant farmers quite easily could, at least technically, leave the collective without losing their means of economic survival. They were not dependent on centrally provided equipment. Large parts of the population in China thus have a quite different memory of the emergence and implementation of communist rule than in Russia with its much more traumatic events. Whereas Russia's approach was truly revolutionary, China's way into communism was more a gradual evolution, although with varying speeds and degrees of intensity. Worst of all, probably, is the memory of the period of the Cultural Revolution of the 1960s, particularly for the urban popu-lation, showing distinct parallels to Stalin's methods in the Soviet Union. But generally, China could, and did, draw on the Soviet experience and sought to avoid the mistakes, including those leading to the eventual break-up of the Soviet Union in 1991. In any case, China's experience with, and implementation of, socialism was distinctly different from that of Russia, and this seems to have had a major impact on the two countries' approaches to post-communist transformation and reform.

Twenty-five years after Zagoria's assessment of communism and its varied implementation, and just before the collapse of the communist state systems, Smith (1987) offers an interesting analysis of the emergence of 'modifications' from Marxism via Leninism, to Maoism and to 'Fidelism'. He points to the common origins of the emergence of communism, because 'in Russia and China alike a large peasant population subjected to severe economic hardship was ruled over by a bureaucratically cumbersome government badly shaken by foreign defeats' (ibid., p. 19). But this apparent similarity is largely superficial only. Underneath, there are distinct differences, translating into today's differences in the nature of develop-ments since 1990. Smith characterises the nature of communism as shaped by the underlying idealism of a more harmonious, cooperative, rather than competitive society. This links to the notion of class struggle as the second key feature of socialism. Common idealistic roots with the humanist movement of the nineteenth century are obvious, and this includes a distinct dose of inherent paternalism. Following on the heel of the changing nature of society towards a more communally focused arrangement, Marx concluded that subsequent political change would make the state redundant as a mechanism of control and coercion. As soon became evident, however, this ideal was quite different from the implemented reality, where state and control were writ large. This 'adjustment' to reality was essentially the input of Lenin, who revised the idea into a practical programme of implemen-tation, thus extending the concept of socialism into 'communism' (Smith, 1987, pp. 24–25). In other words, state socialism is essentially a contradiction in terms, and would be better termed 'communism' in the Leninist sense. This, again, links the first realisation of Marx's ideal to the particular process of implementation through Sovietisation. It is, therefore, not so surprising that in Central and Eastern Europe 'socialism' is widely referred to as 'communism', emphasising the reference to the Leninist-Soviet version, rather than the idealistic idea from which it emerged (see Zarycki's excurse below). It is the party leadership and its totalist claim to represent all proletarian interests for the better that has come to define socialism in its really existing version.

> Marx envisioned liberation as a decidedly popular affair, but he never set forth the exact function of communism as a vanguard . . . Thanks to Lenin, however, the party's functions have been enormously expanded and spelled out in some detail. Its leadership of the masses has now become total, accountable and permanent, while internally it is structured along

the lines of 'democratic centralism', a form of military regimentation in which the party rank and file is obedient to the will of those at the top of the party hierarchy.

(ibid., p. 24)

The communist party has been instrumental in promoting/facilitating communist states, yet at the same time, 'the party's subsequent monopoly of power has stifled cultural life, handicapped economic development after a certain level of growth, and created a series of rigid and often cynical political systems that are beset by succession crises, personality cults, and the threat of terror' (ibid., p. 25).

Taking a retrospective look, Kornai (1992) identifies five main competing paradigms of communist behaviour: (1) communism as a totalitarian system; (2) the utopian nature of communism, promising rewards after the struggle before; (3) communism as a radical modernisation agenda through forced industrialisation (Kautsky, 1998); (4) the communist party as agent of 'engineering' a new society and political consciousness; and (5) the planning-focused nature of communism, excluding the market as allocative mechanism. In addition to those 'standard' five leading paradigms, Kornai (1992) suggests a sixth, the 'reconstructionist' or externally oriented, competitive paradigm, that is an inherently expansionist, militarily focused agenda, contrasting with the internally directed 'modernising' 'developmental' agenda.

Its inherent rigidity, and unwillingness to learn institutionally, ultimately caused the communist system's downfall, as it failed to admit the evidence of, and permit modifications and adjustments to, the rigid, technocracy- and bureaucracy-dominated political-economic structure. The Solidarity movement in Poland in the 1980s, and the 1970s move of establishing Eurocommunism as a more CEE-based version of Marxism–Leninism, attempted this, but they were rebutted by the defenders of the status quo. Leninist parties vehemently resisted adjustment and change, concerned about losing the ideological clarity of Marxism–Leninism as their main source of legitimation, and seeing the argument for justifying their continued claim to absolute power weaken. Not surprisingly, therefore, the perceived challenge and threat by the western system to their own created a siege mentality among communist regimes, which served to justify 'crackdowns' on dissidents and any sign of rebellion or even criticism. This included declaring a state of emergency and martial law, thus abandoning the last bit of pretence of a 'democratic' state.

The nature of 'post-communism' and 'post-socialism'

Speaking of post-socialism or post-communism as a singular case obscures the fact that there are many different versions of post-communist development. These originate from the many variations in the conditions surrounding the formal end of the communist regimes, and the legacies left behind by them. Referring to the situation in eastern Europe, Fowkes (1999) distinguishes between four main features of post-communist transition:

1   The first characteristic involves distortions inherited from a malfunctioning socialist system, including such typical features as a 'shadow economy', as reflection of attempts to bypass the state. This derives from a deeply rooted sense of irresponsibility to the state and society. The latter is a result of the division between public and private spheres, where only the latter permitted expressions of views about the system and state without danger of persecution. The resultant destruction of any sense of community has become a major negative legacy for the development of civil society to make democracy genuinely work as a community-based project.

2   The second feature includes the role of legacies, that is re-emergence of pre-communist inheritances, divisions and inequalities, such as religious, ethnic or, indeed, economic differences in a societal and geographic sense.

3   The third factor refers to the ways in which the post-communist systems were inaugurated, especially the re-creation of a market economy. In most cases, encouraged by a belief in 'shock therapy' (see Chapter 3), the principles of the newly established market economy itself contributed to the destruction of large parts of the national economies, with their inherited inefficient means of production, and ideology rather than economy-driven geographic distribution of places of production. They were dependent on state support, and the return of market principles meant that their lack of competitiveness was quickly exposed and they either collapsed immediately (as in eastern Germany) or had to be artificially kept alive through state intervention, at least for a transitionary period (e.g. Czech Republic).

4   Fourthly, a new middle class, small entrepreneurs, has been emerging, albeit to different degrees between countries, consisting partly of *nomenklatura* (civil servants) who used their insider knowledge to obtain control of ex-state-owned, usually larger size, businesses, illegal operations (Mafia controlled), and smaller service sector businesses. Apart from ethical considerations, the control of considerable parts of the restructured economies by illegal operations, particularly in the former Soviet states, weakens the economies' growth prospects, because much of the generated profit is invested abroad. Given the close involvement of communist-era elites, Fowkes (1999) speaks of 'late socialist' business ownership. This prominent role of elites in the transformation process suggests a lesser, or even lost, role for popular movements as part of a democratisation movement (see Adler and Webster, 1995).

Admittedly from the perspective of hindsight, Sik (1991a) argues that fundamental errors in the concept of socialism led to its inevitable demise. Essentially, 'Realist-socialist practice as implemented under state socialism, was founded on a false theoretical premise' (ibid., p. 9), largely as a result of oversimplification and ignoring three key development factors:

1   the importance of the market as a platform on which consumer interests can be expressed and thus forces of production be stimulated;

2   an underestimation of the role of entrepreneurialism in shaping production and productivity, a stimulant that bureaucracy cannot mimic; and

3   a misjudgement of the capitalist system's continued inherent growth impulse without evidence of 'running out of steam' as a system *per se*.

Inevitably, given the many different stories of post-communist developments, interpretations of post-communist events vary, especially with regard to the so much trumpeted outcome in terms of democratisation and marketisation, although experience suggests that there is no automatic connection between marketisation and democratisation (Kurtz and Barnes, 2002). While some, taking a state-centred, rather technocratic perspective, believe in automatic success if only the right form of institutions can be established and operationalised (Holmes, 1996), effective state-building is seen as crucial to generate a strong government able to establish certainty and predictability in state development. In most cases, this happened through elite competition over policy-making authority (Sutter, 1995; Grzymala-Busse and Luong, 2002), rather than input from the various popular movements (Appel, 2001; Yoder, 2001). Depending on the relative impact of the competing elites, one of four main ideal types of state-building will emerge: a democratic or autocratic, or fractious or personalistic state. Still, there seems to be a general lack of public participation, mainly owing to relative apathy, rather than exclusionary pressure by the elites (Poznanski, 1999). For Skidelsky (1996), following a neo-liberal understanding, the opposite is the case, too much state and bureaucracy being seen as the main obstacle for a freely functioning market economy and thus leading to sub-optimal outcomes of transition. Markets alone, therefore, are seen as the guarantors of successful transition. In fact, Skidelsky sees the failure of communism more as a further, if somewhat extreme, case of the failure of state-led/influenced economic management. This, he concludes, reinforces the argument in favour of liberalism with minimal state intervention in market processes. And this, he argues, was made very clear by the total failure of the 'all state' model applied under communism. Essentially, both authors represent the instrumental, technocratic view of communism and the time after, with suggested predictable outcomes, 'if the terms are right'.

More recently, this technocratic, instrumentalist view, driven by economic modelling and evaluations of the costs and benefits of transition (e.g. Feng and Zak, 1999), has increasingly been challenged by those emphasising legacies and the role of culture in shaping the process (and progress) of transition. These authors focus on attitude, established practices and skills, and historically rooted cultures that determine social and political life (see Szarvas, 1993; Seleny, 1994; Gati, 1996; Mueller, 1996). Inevitably, the 'success' of post-communist tradition will be seen and portrayed as variable, reflecting the relatively easily verifiable institutional 'post-communist' arrangements. This is easier to define in its 'achievement stage' than it is with cultural processes with less clearly defined development directions.

Referring to the crucial role of the economy in distinguishing communist states from those with other authoritarian regimes as heralding the end of communism as an applied form of state-societal organisation, Mandelbaum (1996, p. 11) observes that 'communist economies were not, strictly speaking, *under*developed, they were *mis*developed'. Consequently, he argues, transition will need to include not only building new structures, but also destroying old ones. This, of course, raises the question of how to account for the Chinese model, that is, the constructive transition. This 'transition recession' (ibid., p. 12) poses a problem, as it may well erode the basis of support for democracy and market reforms because of the destruction of existing familiar structures, however inadequate they may have been. As a result of these differences, post-communist states are dominated either by a desire to establish a legitimate government and state in the first place, or finding ways of making the established state system work properly. In the latter case, the central issue is the balance between public and private domains in economic policy and development. For the former, the main focus is on establishing legitimate borders for the newly established nation states. But effective marketisation does not simply equal rolling back the state. Well-functioning institutions are required for an effective operation of liberal markets, because they, too, require reliable and predictable operating conditions, especially guarantee of private property (Comisso, 1991). Instead, institutional knowledge was lost through a zealous purge of 'communists' within the civil service, a fast-track privatisation process at knock-down prices (Róna-Tas 1998; Poznanski, 2001), and the creation of new small national economies, mainly through the independence of the former Soviet Republics (Bicanic, 1995).

Geography of post-socialist transition: where is it happening?

The geography of post-communist transition has been dominated by a Eurocentric, Russia-focused perspective. This is a direct result of the post-war political-ideological dichotomy between 'East' and 'West', with the dividing line, the Iron Curtain, as epitome of the divided world, running through Europe. The division of Germany and, especially, of the city of Berlin became potent symbols of the opposing two halves of the world. The reference in the early 1980s, by the US president Ronald Reagan, to the Soviet Union as the 'Evil Empire' highlighted the Soviet–US antagonism and political-ideological juxtaposition, and Europe was the most important and symbolic battleground for domination. There were other arenas of that competition for global ideological and political influence, some fought in a form of proxy wars, such as in Korea in the 1950s and Vietnam in the late 1960s. As a result, socialism or communism are immediately associated with eastern Europe, however vaguely defined in its territorial extent. Only then, the focus shifts to East Asia, especially China, and the Central Asian Republics. The latter are not always seen in the context of the Soviet Union, which tends to be more closely associated with Russia to the extent that both terms are used interchangeably. Countries in other global regions with a period under a socialist/communist regime enter public awareness – for example, Cambodia, Vietnam and, albeit less well known, Laos. The wars fought in the former two countries raised their presence in public awareness; this also applies to North Korea which has become a symbol of a now anachronistic Stalinist system. Countries that experimented with socialism of one kind or another in Africa or Latin America are, by comparison, visible only on the margins of public awareness: Mozambique, Kenya or Nicaragua, for instance, do not immediately feature on the list of 'post-communist' countries. Cuba, by contrast, has become something of a cause célèbre.

But the relative fuzziness of the geography of 'communism' and 'post-communism' does not apply on the scale of global regions alone. It has also become an important issue within these regions, especially Europe, in response to the variations in political ambitions and historic legacies *vis-à-vis* western Europe. Fowkes (1999) discusses the complexities of defining 'eastern Europe', in relation to 'the West', referring to particular historic, political or geographic rationales for the various groupings suggested in the literature. 'Eastern Europe' is often used with a negative undertone, especially in public discourse, referring to/implying underachievement, backwardness and a need for 'catching up' with the 'advanced' West as the 'normal case'. Since joining the EU, 'eastern Europe' has also become a byword for low-cost production and cheap labour (Chirot, 1989). Historic legacies play an important part in the way in which the end of communism has generated differences and divisions within eastern Europe: generally, a lesser degree of urbanisation and industrialisation, manifesting a relative backwardness and lesser economic capacity compared with western Europe; the differences between the Ottoman-influenced and the Roman Catholic- and Christian Orthodox-influenced parts of southeastern Europe (Balkans). This effected not only different economic conditions, but also different forms of civil society and, in particular, experiences with democracy. Fowkes (1999, p. 10) attributes the eastern European relative economic backwardness to a politically lesser experience with democracy and a tendency towards more authoritarian regimes, a social characterisation by landowning (feudal) and bureaucratic elites and, culturally, a thin veneer of western-influenced intellectual life. Tanase (1999) describes the period of the Soviet occupation of eastern Europe as 'a return to a *sui generis* feudalism that denied liberal democracy and modernity'. But, of course, there are variations within eastern Europe of this development, which sharply contrasted with that in western Europe. Such variations suggest a subdivision made earlier into central and southeastern 'eastern Europe'. Many of these differences, at least in economic terms, were inherited and then more or less preserved under the communist system, with the more backward countries remaining in that position of low economic performance. In the 1980s, taking Austria's economic performance (GDP per capita) as 100, the GDR achieved 59 per cent, Czechoslovakia 55 per cent, Hungary 41 per cent, Bulgaria 37 per cent, Poland 27 per cent and Romania 20 per cent (Fowkes, 1999, p. 11).

In reality, there are many overlapping connotations of post-socialist European regionalisation: 'eastern', 'central', 'southeastern', and so on. None of these are clearly defined and are used by different authors with reference to varying geographic entities. As Fowkes (1999) points out, the Czechs, the Poles and the Hungarians, for instance, dislike being subsumed under 'Eastern Europe', and prefer the term Central Europe instead (ibid., pp. 1–2). One of the reasons is, of course, the signal sent out about their degree of belonging to 'western' and 'eastern' Europe and the associated connotations of an 'advanced West', and 'backward East'. In addition, there is the notion of the 'East' being closely associated with Russia/the Soviet Union, while the 'West' is associated with western Europe and, by extension, the United States. Belonging to the latter group has been the main political and philosophical-ideological ambition of these countries since the end of communism. Their keenness to join the European Union and NATO underlines this attempt at geopolitical repositioning. Against this background, Central Europe sounds closer to the 'West' than 'Eastern Europe'. Fowkes makes the suggestion to 'retain "Eastern Europe" as an overarching geographical definition, and to use the term "East-Central-Europe" to embrace Poland, Hungary, the Czech Republic and Slovakia, with the possible addition of the three ex-Soviet Baltic countries of Estonia, Latvia and Lithuania and the two northernmost ex-Yugoslav states of Slovenia and Croatia' (ibid., p. 2).

Such geographic groupings are inevitably crude and do not allow for divisions that may exist within the countries. This applies to Poland and Slovakia, for instance, where the eastern parts have more 'eastern European characteristics' than the western parts. Such divisions also apply, in a rather interesting way, to the unified Germany, where former East Germany (the GDR) was 'attached' to western Germany. The relocation of the capital to Berlin, some 80 kilometres west of the Polish border, may suggest a shift to the east for Germany, from 'western' to 'central', but such somewhat simple interpretations ignore the continued divisions that exist between the two halves – the western part continuing to feel more 'western' than the eastern part with its central European traditions. These are not just a result of the post-1945 divisions, but go back to earlier spheres of influence between the French-influenced Rhineland, and southwest Germany and Bavaria with their western European traditions, and Prussia/Saxony with their central European roots.

Not least because of these embedded differences does the unification process prove much more difficult and incomplete than had initially been assumed and expected. It is the failed presumption that East Germany could simply be incorporated into western Germany, and would thus simply 'disappear', that makes discussions on post-socialist legacies so interesting and important (e.g. Anheier et al., 2000). Frequently, the East German case has been excluded from debates on 'eastern Europe' as too 'special'. In reality, however, many of the difficulties and processes observed in eastern Germany with integration into an established institutional, cultural and political-economic framework apply to the other central European countries, too, especially when considering their preparations for joining the Euro-pean Union. Here, too, an established regulative framework, with set institutional and policy-making provisions, was presented as the norm to which the new aspirant states had to adjust.

The political implications of this externally driven change are not yet clear. But experiences with eastern Germany suggest possible new divisions emerging in the wake of a reaffirming 'eastern' European identity and confidence. Politically, judging by the latest *general* elections in Germany in September 2005, those political parties that emphasise a separate eastern (post-socialist) identity and agenda gained considerably in approval throughout eastern Germany (ibid.). This reflects a growing divide in the mutual understanding between East and West, with the main 'western' parties viewed as pursuing first and foremost western German interests, losing their at first dominant position to the played-out advantages of a 'home team' of reconstituted neo-communists (PDS) and hard-left former Social Democrats (including former western German politicians), who gained about a third of the eastern German electorate's votes.

This response is very much an expression of a dissatisfaction and disappointment with the political and, especially, economic outcome of post-communist developments when compared with the form of society dreamt of by the dissidents in the mid and late 1980s. This did not envisage an all-out, wholesale westernisation to the extent that any 'eastern' legacies be brushed aside, to be replaced by a cloned version of western political, economic and societal conditions. Many had envisaged some form of a 'third way', effectively combining the 'best of both worlds', with a newly reconstituted, strengthened civil society as the centrepiece of a new state-societal organisation. But western pressures, derived from a lack of a genuine understanding and knowledge of their eastern cousins' legacies and, indeed, a sense of superiority towards them – developed as part of Cold War propaganda – soon made this ambition appear illusionary. 'What in fact happened was the transfer of political issues from the realm of civil society to that of political parties, more easily controllable by the elite' (Lagerspetz, 2002, p. 9). In eastern Germany, for instance, the Bündnis 90 (Association 1990) of various dissident groups formed in the run-up to the first general elections there in 1990 was a direct outcome of these grassroots movements and ideas (Appel, 2001). But their role and vision in the post-socialist transition process were short-lived both economically and politically, soon out-manoeuvred by the politically more experienced western-based parties and, during the 1990s, the reconstituted former communists (see e.g. ibid., p. 153; Yoder, 2001). In addition, the less than egalitarian and democratic privatisation process, with often predatory acquisitions of economic capacity through *nomenklatura* and former managers, contributed to disillusionment and even resentment towards the 'West' and the new 'Western-backed' elite, viewed as a 'sell-out' of national assets both in an economic and cultural-historic sense.

Global regions of post-communism: beyond Soviet-centric and China-centric regions of post-communist transition

Asked to draw up a geography of post-communist states, most people will immediately refer to Russia and 'eastern Europe', followed by China. This is clearly a legacy of the Cold War when 'the democratic West' was – literally and politically-strategically – represented by the United States (Washington), and contrasted with the communist East represented by Russia (rather than the Soviet Union) and thus 'Moscow'. China, by contrast, took something of a back seat in public geopolitical discourse, being associated with developing countries and thus seen as less threatening to western interests and security concerns. It is only since China's reform policies have yielded its attention-grabbing economic success that its challenge to western political-economic hegemonic ambitions has entered public consciousness. China has become a serious competitor for the 'West', but it is a reforming China that is seen, with its proclaimed communist political-societal ideology viewed as an internal affair with little effect on external relations.

Pickel (2002) places China next to the former Soviet Union as foci of the two largest global regions of post-communist transition, but he also acknowledges distinct differences between – and within – these two macro regions:

1   The Soviet-centric region includes the former SU and its then satellite states in eastern Europe. But there are clear variations between the different countries' subsequent paths of development after the formal end of the *communist* regime (Offe, 1996). Here, it is the distinction made between those states believed to follow the transition to market democracies as was projected (and expected) by the 'West'. Pace, extent and final (?) destination of change varied, though. These differences refer back to 'historically rooted civilizational, religious, and cultural differences ... which make the Western model less universally applicable than initially assumed' (Pickel, 2002, p. 108). The differences are shaped by real agents and sources of change, which had been ignored by the generalising projections propagated by the advocates of neo-liberalism. In some countries, for instance, like Russia, Mongolia, Kyrgystan and Moldova, neo-liberal reformers gained power and influence under the approving eye of the international community and organisations. But they failed to convince 'their' people at home who had to deal with the immediate political-economic impact of the transformation process. Nationalist groups emerged in response to the newly gained independence from 'Moscow', but also as a backlash against a perceived 'internationalisation' of the post-communist development process and thus loss of national autonomy. In some cases, this resulted in antagonisms, fragmentation and the emergence of new divisions and borders, such as in former Yugoslavia (Brown, 1994) or Czechoslovakia.

2   The China-centric region of state socialism, embracing the Asian communist countries, especially China, Vietnam, Cambodia (Pei, 1996). China, and the former communist countries of East and Southeast Asia, have adopted less instant forms of transforming the communist system – maintaining communist rhetoric and political control in the public political field, while slowly introducing liberal market conditions in the economic strand of the state. This model, developed in China, has also found its way to Vietnam, Cambodia and Laos, also reflecting China's influence in those countries.

   While they, too, are going through transitional processes (or transformations), it is not about attempting a carbon copy of '*western* capitalism', but rather the Asian variant of growth-based liberal economy, if under the auspices of a prevalent, autocratic state. Existing state-socialist political-societal arrangements are maintained under the claimed continued auspices of 'communism'. Politically, because of its particular combination of 'old' and 'new' instead of 'new' replacing 'old', China's approach could be seen as a case *sui generis*, seeking to run a dual track approach of economic liberalism and political state control. It thus does not follow a sequential approach to regime change, where one model replaces the previous one altogether. Liberalisation and democratisation are thus not seen as two sides of the same coin. China does not follow the 'standard' model of economic liberalisation combined with political democratisation, as advocated by western New Right advocates and pursued under the Washington Consensus (see Chapter 3), and thus does not pursue a textbook neo-liberalism as a package. Overall, therefore, Pickel (2002, p. 111) concludes in his account of the changes, 'post-communist transformation is a – variously definable – set of practical problems', and explanative theory needs to encompass this diversity.

Africa and Latin America, including Cuba, are not represented in Pickel's distinction between global paths of post-communist transition (Pickel, 2002). They have no dominant state of visible global presence as champion of an African or Latin American 'model', bringing it to the attention of a wider audience. In the absence of a large, globally active player, their presence and particular nature is much less reported and present in public, and academic, consciousness. These regions tend to appear as part of more general discussions on post-colonialism (see, e.g. Post and Wright, 1989) and development or Third World development, where their link with socialism is seen more as a peculiarity than as the basis of categorisation and comparison with other formerly socialist countries. Socialist legacy is rarely used as a conceptual backbone in its own right, with the overview by Holmes (1997) one of the few exceptions. But there have been some attempts at linking the issue of post-communist democratisation with those of a more general democratic shift away from authoritarianism (Linz and Stepan, 1996; Pickles and Smith 1998; Lavigne, 2000).

The paradoxes of Central and Eastern Europe: post-communism or post-socialism? Post-colonialism or post-imperialism?

Tomasz Zarycki

Were the countries of the Soviet bloc communist, or rather socialist? Consequently, should we call them today post-communist, or post-socialist? Some of them were officially called 'socialist' republics, others 'people's' republics, others, like East Germany, 'democratic' republics. Thus, none was formally labelled 'communist'. The situation was different, however, in the case of the names of ruling parties. Many of them were indeed called 'communist', like the Communist Party of the Soviet Union or the Communist Party of Czechoslovakia. Nevertheless, others bore the name 'socialist': for example, the Hungarian Socialist Workers Party or the German Socialist Unity Party. In other cases, such as in Poland, the ruling party was labelled simply as 'workers party'; its full name was the Polish United Workers Party (*Polska Zjednoczona Partia Robotnicza* – PZPR).

At the same time, the relations between the notions of 'socialism' and 'communism' were explained in different ways in different countries and in different periods. The classic approach in the Soviet Union was, for example, to present communism as the future, final and ideal stage of the development of the Soviet society. The actual system was usually referred to as 'socialism', which was essentially viewed as a transitory, imperfect phase in the journey from capitalism to communism. But there is also an inherent difference in the respective roles of state and society. In Kautsky's words (1998), 'Marxists . . . hoped to advance and empower the working class'. They thus suggested a subordinate role of the state, while 'Leninists were modernizing revolutionaries in underdeveloped countries seeking their rapid industrialization. They relied on centralized revolutionary movements, which, in power, formed centralized bureaucracies that advanced industrialization through mass persuasion, regimentation and terror, as well as central planning' (ibid., p. 379). The state machinery and, ultimately, the Communist Party, are clearly superior to the people. This ambition shaped the communist regime's policies, seeking to limit evident social differences and creating a more homogeneous 'classless society'. Post-communist societies inherited the effects of these policies and were subjected to renewed change (Słomczyński and Shabad, 1997), effectively bringing back and reinforcing social difference.

Thus, it seems that in the Soviet context one of the main reasons for using the label of 'socialist' rather then 'communist' was an attempt to present the various difficulties of state and society as temporary. Allegedly, they would disappear after reaching the phase of 'communism'. In this case, 'communism' was supposed to be something better than 'socialism'. In many other contexts the situation was the opposite: it was, and still is, often assumed that 'socialism' is something better (more civilised, European, democratic, etc.) than 'communism', with which all the more totalitarian and 'barbaric' forms of the Soviet-type societies were associated. The fact that the notion of 'communism' was, for many reasons, deeply unpopular in many parts of Soviet-controlled Europe was forcing the ruling parties to avoid the term. While communism and Marxism–Leninism were the official doctrines of all the ruling parties of the Soviet bloc, in many countries the notion of 'communism' was so deeply resented that it was often simply avoided in official discourse. In Poland, which is a good example of such a situation, words like 'communism', 'Bolshevik' or 'Soviet' had such deeply negative connotations that since the first days of the Soviet-imposed government in Poland these usually liberally used terms have been largely replaced by less ideologically loaded synonyms. The word 'Communist' was, at least in the last decades of the 'People's Republic of Poland', used only in internal party discourse and in contacts with the Soviets. In the speeches of the party's first secretary, Wojciech Jaruzelski, the words 'communism' and 'communist' appear very rarely. The supposed positive essence of the system has been identified by the official propaganda with the notion of 'socialism'. In the second half of the 1980s, an emblematic propaganda slogan read 'Socialism – Yes, Distortions –No'.

At the same time, the legacy of the use of the label 'communism' in Poland includes the name of the Communist Party of Poland (KPP), which existed in the inter-war period as an anti-system organisation directed from Moscow and aimed at unifying Poland with the Soviet Union. Since this time, 'communist' has become in Poland a synonym for traitor and renegade. 'Bolshevik' and 'Soviet' had a similarly bad connotation. This is why in 1944 the Soviet-installed regime invented a new Polish synonym for the adjective 'Soviet', based on the literal translation of the word which means 'council' in Russian (the new Polish version was '*radziecki*' in the adjective form, instead of the previous '*sowiecki*').

Although the old system collapsed some 15 years ago in Poland, the ambivalence regarding the use of notions of 'post-/communism' and 'post-/socialism' persists. In the particular Polish context this semantic uncertainty is very clearly linked to the main cleavage of the political scene of the country. This fundamental cleavage, often called the post- versus anti-communist conflict, was previously based on the different views of the Soviet domination of Poland and is transformed today into the conflict over the political interpretation of the Soviet-dominated period. On one side of this cleavage we find the pragmatically oriented collaborators with the regime and their supporters who today defend their choices as optimal, given the geopolitical realities of the epoch. On the other side are the hardliners, called at that time 'the opposition', rejecting cooperation with the Soviet-dependent government and the ruling party. Today they are insisting on recognition of the period of the existence of the People's Republic of Poland as a form of Soviet occupation and deny the former leaders of the party moral rights to occupy any public posts. The two parts of this key political cleavage are conventionally labelled respectively as the Left and the Right. It is unsurprising that it is the Left that insists on calling the pre-1989 Poland, as well as other counties of the Soviet bloc, 'socialists', which implies a much more positive view of the period. The Right on the other hand, insists on the use of the negative label 'communism'. Consequently, the main parties of the Left, having their roots in the structures of the Soviet-times Polish United Workers Party, are called 'post-communist' in the discourse of the political right. But this is resented by the Left, as they consider this an insult to their political agenda and tradition. As the Left's representatives argue, there were practically no 'real communists' in the Polish ruling party, at least since 1968. Indeed, in 1968, during an internal conflict in the Workers Party, the more ideal-istic 'traditionally socialist' faction was smeared with an anti-Semitic campaign instigated by the less idealistic, Soviet-inspired sections in the Party. In effect, the 'last real communists' either emigrated or started to drift towards unofficial opposition, losing their communist identity in the process. At the same time, the Party became dominated by technocratic, Moscow-obedient careerists. As the present day representatives of the Left would argue, the party of the 1970s and 1980s was a party of modernisation, irrespective of what current Western-oriented technocrats and pragmatists are claiming. For the Right, the Soviet-appointed administrators of Poland were simply opportunists or even nihilists. And precisely these qualities are implied by their deliberate use of the term 'communist'.

In other words, the term 'communism' in this context does not suggest that Polish 'communists' ever had ideological communist sympathies, or any plans for building 'communism' in Poland. The use of the word 'communism' has here first of all the dimension of moral judgement. 'Communist' here means 'cynical' and 'opportunist', and often 'totalitarian', 'immoral', 'undemocratic' or even 'barbarian'. When the Left protests against being labelled 'post-communist', the Right counter-argues that, whatever their qualities and views are, and whatever 'communism' really does or should mean, members of the Left parties did in fact call themselves 'communists'. They should therefore not try to rewrite history and pretend loss of memory of the relevant events.

At the same time, the debates regarding the right to use the word 'socialism' continue. Most of the transformed previous ruling parties, with the notable exception of the Communist Party of the Russian Federation, prefer to use the labels 'socialist' or 'social democratic'. These allow them to retain some of the old 'Left' identity, while at the same time presenting themselves as 'modern', democratic, Western-style parties of the Left. This is of course criticised by their opponents. They argue that neither the social model advanced by them during the Soviet era, nor their policies after the fall of the Soviet system, could justify the use of the 'socialist' label. This applies in particular to countries where the former ruling parties retained their unity and strong influences, such as in Poland, Hungary or Lithuania (interestingly, most of them are predominantly Catholic). When obtaining power in the post-1989 period, they appeared to be much more pragmatic, rather than ideological, and often implemented essentially liberal policies that could hardly be called socialist or even 'left'. Critics use this seeming abandonment of ideological principle as evidence of the former ruling parties being first and foremost concerned with securing the existing perks for their members. In any case, for a number of reasons, the 'left' governments often introduced liberal free-market measures, despite their contradiction of left-wing principles. Their continued socialist rhetoric thus appeared to be largely merely electoral campaign tactics.

In analysing the meanings of the notion of 'socialism' in Central and Eastern Europe one could also take a look at the history of the anti-Soviet movements, especially in such countries as Poland, where the tradition of the indigenous socialist movement is quite old. A very good example would be the Polish Socialist Party (PPS), which was established in the late nineteenth century and became one of the most important parties of the inter-war Polish Second Republic. The Polish hero of that era, Józef Piłsudski, was originally one of the leaders of the party, which was a mass left-wing movement that also developed a strong pro-independence (that is, nationalist), anti-Bolshevik force. After 1944, the Soviets annihilated the party by forcing it to merge formally with the (Soviet-backed) Polish Workers Party (PPR). The result was the Soviet-controlled Polish United Workers Party (PZPR), which retained a hegemonic political position throughout the communist period. To avoid possible internal opposition, many of the old PPS activists were arrested and also deported. It is not surprising, therefore, to find that for many Poles identifying themselves with the left, the heirs of the PZPR, is ethically and morally not acceptable. In effect, the ex-communists' use of the word 'socialist' or 'social democratic' destroyed the credibility and acceptability of the political left, and of 'socialist labelling' with it, including the original, 'genuinely socialist' movement in Poland. After several decades of living with 'socialist democracy', 'socialist rule of law' and 'socialist elections', many Poles use the adjective 'socialist' mainly with ironical undertones, where 'socialist' often simply means 'false'. Not surprisingly, this is quite frustrating for many people on the political left, and it is also a political problem. As they point out, this legacy is an important aspect of the current political climate, which is so unfavourable to the organization of social and political action around truly left ideals. Another intriguing aspect of the problem is the loss of legitimacy of the political left and, in particular, (post-) Marxist language in the mainstream discourse of the social sciences in the countries of Central Europe. This makes it much more difficult, for example, to present the situation of the region in the terms of world-system theory or other theoretical explanations of the mechanisms of dependence in the contemporary world. And this challenges the modernisation paradigm. As it is argued by some, this may make the affected countries more vulnerable to exploitation by stronger economies as they lack the appropriate intellectual debate and critical questioning of their position.

Another interesting question and semantic paradox is the application of the post-colonial and post-imperial labels to the Soviet and post-Soviet universe. As it appears, the question is no less ambiguous than the socialism versus communism debate, and has an important political dimension, too. Thus, on the one hand, as is well known, the Soviet Union, together with its allies, presented itself as an active supporter of the anti-colonial movement around the world. It proved its commitment through numerous economic and military operations in support of anti-colonial rebellions in several countries, especially in Africa, as well as 'socialist revolutions', and thus effectively accelerated the emergence of many new post-colonial states. The Soviet anti-colonialism was thus not merely a useful rhetoric and a strategy to legitimise its involvement in political conflicts abroad *per se*, but was also a pragmatic programme of challenging Western interests around the world as part of competitive Cold War power politics.

At the same time, however, as is increasingly argued, the Soviet Union itself, despite its use of anti-colonial slogans, could be seen as a colonial state, with its own huge empire consisting of zones of different degrees of control from the centre.[1] In particular, the countries of Central Europe under Soviet control could be perceived as actual Soviet colonies. Viewing them in this way is not merely an expression of an ideologically rooted (anti-Soviet) critique, but seems to be a useful pragmatic tool for analysing the many different aspects of the ways in which the societies of these countries functioned before and after 1989. One could give the example of political principles in the Soviet-dependent countries. They were usually organised in such a way as to minimise the potential for conflicts in the relationships with central – that is here, Moscow's – rule. In a truly colonial fashion, the emphasis was on a smooth and thus effective running of the 'empire', with the colonies acting as willing executors of centrally dispensed orders. The apparatchiks, as links between the centre and the dependent countries, faced similar dilemmas of loyalty and identity as are found among 'traditional' local colonial elites. The economic sphere also provides many interesting parallels and illustrates the core-dependent relationship. The same is true of the cultural dimension, where Soviet efforts in Russifying their zone of influence are well reported.[2] Literature is a fascinating sphere, and not only in peripheral regions, but also in the Russian centre. As argued above, the deconstruction of the colonial way of perceiving peripheral regions could be effectively adapted to the analysis of the Soviet and earlier Russian literature dealing with the peripheries of the Russian and Soviet Empires.

By the same token, one could point out the inherent paradox of applying Western perceptions of 'Eastern Europe' to the post-Soviet world. It lies in the 'ambiguous' geographic location and political-economic and historic relationships of several of the contemporary Central European states. Thus, historically, most of them functioned as eastern peripheries of western Europe (in some periods some of them could be even seen today as forms of colonies of the western states, as for example Bohemia under Austrian rule, Estonia under Swedish rule or, indeed, Poland under Prussian rule). In any case, the dependence on the western European economies and cultures was strong in many of them, and in non-political forms persisted even in the Soviet period. This affected the seemingly paradoxical sense of 'superiority' among some of the countries of Central Europe over the 'colonial centre' in Moscow. Even the Baltic republics of the Soviet Union were well known for their much higher standards of living than were found in Central Russia during Soviet times. The cases of Czechoslovakia and East Germany were even more striking. Poland, on the other hand had been known for the influences of its culture on Russian elites. Even if Polish was first of all attractive as a tool providing access to western European culture through translations, the popularity of this peripheral language among the intellectuals in the Soviet metropolis was truly paradoxical, taking into account that Poles, despite their formal obligation to learn Russian, tried to avoid doing so as far as possible. Of course, most of these effects can be attributed to the attractiveness of Western culture to the Soviet (Russian) elite, access to which was more readily available in the western peripheries of the Soviet bloc. It was here, too, that Soviet attempts at discrediting Western culture and way of life were resisted most determinedly. Against this background, it may be argued that Soviet 'hard' military and political colonialism could be seen as the loser in the battle against the Western modern 'soft' cultural-economic colonialism, effectively undermining the rationale and legitimacy of Soviet rule. The argument of the relative effectiveness of 'hard' versus 'soft' forms of colonialism and dependency has become an important part of the analysis of the current policies Moscow is seeking to manifest, and thus extend control in its zone of geopolitical influences.[3]

When discussing colonialism in the post-Soviet space one could also point out the third aspect of the problem, that is, the historical heritage of some of the Central European countries like Hungary or Poland. Both have important historical experiences of playing the role of regional powers, and could be seen in the past (in particular during the fifteenth and sixteenth centuries) as fully fledged colonial states in their own right. Although they have largely overcome the post-imperial neurosis, so characteristic now of a Russia seeking to come to terms with its recent loss of influences in the region, and abandoned any territorial ambitions, a heritage of geopolitical thinking and way of perceiving their neighbours is still in place. A very good example is the way of seeing Russia, Ukraine or Belarus in Poland.[4] Convinced of its own civilisational and moral superiority, there is a continued belief in a special role to be played by Poland in these countries. A Polish-centred sense of bequeathed cultural heritage to the region still features clearly in aspects of Polish attitudes towards its eastern neighbours, which may be seen as intriguing vestiges of a Polish post-imperial syndrome. And this happens exactly at the time when Poland is suffering so many forms of the typical post-colonial syndrome that emerge from the various disappointments of the so-called transition, or period following the perceived liberation, from Soviet colonialism (or maybe communism, or maybe . . . socialism?).

Notes

1　See for example Carey, Henry F. and Raciborski, Rafal (2004) Postcolonialism: A valid paradigm for the Former Sovietized States and Yugoslavia? *East European Politics and Societies*, 18(2): 191–235.

2　Inside the borders of the Soviet Union the Russification efforts were best manifested in the programme of the so-called Soviet National Policy. Although for some periods it permitted the relatively free development of national cultures, it assumed their eventual merger into the 'Soviet people'. On this journey toward their Soviet identity, three main stages have been proclaimed (there are also three stages in the shift towards the final state of socialism): (1) rastvet (the flowering of nations), (2) sblizhenie (the growing closer together) and (3) sliyanie (the merger of nations); see for example. Conquest, Robert (ed.) (1967) *Soviet Nationality Policy in Practice*, London. For an analysis of the geographical dimension of the Russification process, see for example Wixman, Ronald (1981) Territorial Russification and linguistic Russianization in some Soviet Republics. *Soviet Geography: Review and Translation*, XXII (10): 667–75.

　　Outside the border of the USSR, the development of independent national cultures was usually permitted and even supported (to ease control). They have, however, been significantly redesigned to be compatiblE with Soviet culture. At the same time, Russian became the first foreign language taught compulsorily in all Soviet-dependent countries, from about the age of 12 until at least to the beginning of university-level education.

3　A good example is the recent debate around the programme of building of a 'Liberal Empire', proposed by Igor Chubais of the SPS party.

4　See for example Zarycki, Tomasz (2004) Uses of Russia: the role of Russia in the modern Polish national identity. *East European Politics and Societies*, 18(4): 595–627.

3

THE MEANING OF POST-COMMUNIST 'TRANSITION'

Post-authoritarianism, democratisation and liberalisation

Introduction: the meaning of transition and 'transitology', and the relationship between state, society and economy

'Transition' and 'transformation' have been at the centre of many discussions in politics and academia for the last 15 or so years. The unexpected, spectacular collapse of the communist 'half' of the world raised the public profile of discussions on democratisation and 'westernisation'. Since the end of the 1980s, transition, post-socialism/communism and democratisation have become closely associated and much discussed phenomena. Geographically, post-communist transition in both literature and public discourse has largely been associated with Central and Eastern Europe and the former Soviet Union. This very much reflects the post-war division of the world into 'East' and 'West', and Eastern Europe/the Soviet Union very much epitomised the 'evil forces' of communism. Communist countries elsewhere played a much less visible role in the public debate, with China and Cuba among the few exceptions; others included Cambodia, North Korea and Vietnam. The latter's visibility was largely a result of the high-profile American intervention there after the Second World War. Other countries with a regime change from socialism have gained much less visibility, although the changes there have been of fundamental importance. But the concepts of transition and transformation, especially since the latest Iraq War/regime change, have a wider connotation than that.

Prior to the end of state socialism, transition referred primarily to developing countries and the notion of their progress towards a western-style market democracy. This was projected as the default final stage of state-societal organisation. Such a rather one-dimensional understanding of the outcome of societal development was strongly advocated during the 1980s, with the struggle between communism and liberal market democracies now seen as won by the latter. It is against this notion of westernisation that transition and democratisation have been conceptually associated with each other so closely. The term 'developing', *per se*, highlights a transitional, transformational process on the way towards an envisaged desired 'final' stage of development (as embodied by the 'First World' industrialised

45

countries). For this to be achieved, the widespread notion had been the need for satisfying certain formal, structural pre-conditions to allow democratisation to take a foothold (see Rustow, 1970, and others below). The implicit understanding is that the relevant countries need to be prepared, that is, shaped by an external force/ influence to get them on track for democracy. This understanding, projecting the notion of a superior West against inferior non-western-style state-societal regimes, was still distinctly evident in the West's response to the events in Eastern Europe, for naturally having won the battle between the systems, the western paradigm of state-societal organisation, summarised in democratic market economy, was to be adopted in the East wholesale. This paradigmatic doctrine, reinforced by the New Right ideologies of Reaganomics and Thatcherism, had already been questioned in the 1970s, and influenced mid to late 1990s policies. However, as the most recent events and debates around the Iraq War of 2003/4 have demonstrated, a new triumphalist notion of western-style market democracies as the only path to happiness seems to continue to shape international political paradigms.

The phenomenon of 'transition' has attracted a considerable amount of interest, in both political and academic debates (Murrell, 1992; Altvater, 1993; Gowan, 1995). Arguments revolved around the nature of change: whether the process would be ad hoc and complete, as suggested by Lipton and Sachs (1990), or more gradual like 'transformational recession' (Kornai, 1992, 1994) or 'rebuilding the boat in the open sea' (Elster et al., 1998).The collapse of communism, or state-socialism, depending on one's ideological position, generated a plethora of publications on forms, processes and mechanisms of 'transitions' and, especially, their direction, including whether there was an inherent tendency to 'catch up' with the West (Kolodko, 2001a). While the end of the Iron Curtain has brought about a massive increase in the popularity of the issues of democratisation and marketisation, post-socialist developments, it is argued here, need to be seen in the wider context of regime change, from authoritarianism towards more democratic principles of state management. These were much less 'spectacular' in their changes, and thus attracted much less public attention than 'post-socialism', going back to democratisation movements in Latin America and southern Europe in the late 1970s (see also Linz and Stepan, 1996).

Despite this multitude of publications, two main strands of argumentation may be identified: those focusing on the market as main arbiter of social and economic development (with the emphasis on 'economic'), and those with a stronger 'statist' perspective, seeing the state as a major actor, or manager, of transition towards democratic principles and, especially, the development of civil society. Depending on one's position, the market approach attracts undertones of globalisation, capitalism and corporate domination, while the state-centred arguments are often derided by the advocates of markets as idealists seeking to hang on to communist fairyland. Their understanding of 'state' is primarily that of bureaucracy, technocracy and overt regulationist intervention. While these positions reflect opposites on the scale of 'transition', albeit shaped by ideological simplifications, common to both is their separate view of society. It is interesting that neither sees society as an integral, active part, but rather outcomes, or subjects, of the actions of state and market. The latter two seem to be seen as acting largely independently, subjecting society to their actions and priorities. Unsurprisingly, the perception of what constitutes 'society' and social actors differs between the two conceptual strands. While the 'statists' see society as a complex system or entity that needs managerial directions and 'taking care of', the market-centric arguments view society as a more or less incidental accumulation of individuals, each in pursuit of solely economic satisfaction as the key to happiness. They are essentially objects in a market process, seeking to fit into the process and thus benefiting from it. They seem to be granted an ambition other than economic profit maximisation (neo-classical). In any case, society is seen by both as a separate, detached outcome of the processes initiated by the respective actors – market or state administration. As part of the transition process, 'society', however defined, can be seen as having shifted from being subordinated to the state (or, more accurately, the state apparatus controlled by, here, the communist machinery) to being subordinated to the market and its principles. In both cases, society is projected as subject to these two dominant mechanisms and ideologies: communitarian and 'integrated' in the first case (with little emphasis on the individual but rather on community), and individualist and rationally opportunity-driven, in the latter.

The economy-centred perspective of 'transition': 'shock therapy' or 'gradualism'?

'Shock therapy' (also referred to as the 'big bang' approach; see Holmes, 1997, p. 206) was originally developed in the 1980s to tackle hyper-inflationary crises in developing countries, especially in Latin America. The collapse of the socialist states at the end of the 1980s coincided with the popularity of New Right policies, as propagated by Margaret Thatcher and Ronald Reagan. They focused on economic liberalism and a minimal role for the state as part of a shift from the Keynesian to the workfare state (Peck, 2001). Popular tax reductions, reduced state responsibility through privatisation of state enterprises and public sector functions and, at the same time, better economic performance, were the main conceptual pillars, drawing on modernisation theory. Unsurprisingly, there was little concern with the possibility of a new, specifically post-socialist form of capitalism emerging, such as 'Market Leninism'. The socialist experience was simply written off as a loser. Pushed by the main global institutions (e.g. World Bank, European Union), and impatience in the ex-socialist countries (e.g. Poland, Czech Republic, East Germany) to be part of the West, emulating western liberal structures as quickly as possible, seemed paramount. Economists at the World Bank developed a 'liberal-isation index' (Bradshaw and Stenning 2000) based on the degree of privatisation and liberalisation, and the opening up of domestic markets, key ingredients of shock therapy (de Melo *et al.*, 1996). The alternative softer option of 'gradualism' in transition, with a continued stronger state role, attracted much less attention (e.g. in Hungary and most FSU states). It seemed less committed to full transition.

Shock therapy consists of three main elements: liberalisation of prices and opening up the economy to competition, reducing subsidies to (loss-making) state businesses and, finally, privatisation of large parts of the public sector economy (see also Holmes, 1997, pp. 206 ff; Lavigne, 1999, pp. 116–118). A crucial requirement for all this to work is a strong state to push through the changes (Skidelsky, 1995), but this was not always the case (e.g. in the former Soviet states; Kaminski, 1996b). Speed and extent, and particular legacies of the socialist state and society, have given post-socialist 'shock therapy' its own character.

Balcerowicz is one of the main advocates of 'shock therapy' to transfer the state-socialist economies of CEE countries into liberal market economies in one rapid sweep. As the Polish finance minister in the early 1990s, he was thus in a rare position as an academic to be able to put theory into practice. Much of his argumentation is shaped by his background in economic and market theory, and this is reflected in his theoretical accounts (Balcerowicz, 1995). They focus first and foremost on the technical, functional aspects of transition. Marketisation is the main focus, with everything else following from there, almost automatically, it seems. Accordingly, 'transition' is essentially reduced to a top-down managed process of macroeconomic stabilisation and microeconomic liberalisation, both to minimise constraints on effective business operation, and facilitate fundamental institutional restructuring to underpin a competitive, advantageous business environment. This reflects the belief in the economic dimension as the crucial driver of the socialist countries' transition towards an approximation of the western countries, especially the United States. And everything else was expected to follow from there. Indeed, he confirms his views by challenging comments about the possibly more successful 'gradual model' pursued by China by emphasising China's different structure as a less developed, more agricultural economy. What may work there was not considered an appropriate 'medicine' for the situation of the CEE countries (ibid., p. 250). He also points to the much lower indebtedness of China, in contrast with most CEE countries.

Of particular interest, however, is the more general argument about the relative sequence of transition: should marketisation follow democratisation, or vice versa? He argues that there is evidence 'that political democracy is not in itself a factor in economic development' (ibid., p. 249), and points to the various forms of strong and weak autocracies with a pro-capitalist (Taiwan, South Korea) or populist (Latin America) approach. Democracy, he thus suggests, is not a prerogative of a functioning (liberal) market economy. By implication, this means that the Chinese model, autocracy cum marketisation, would follow in those countries' footsteps. Indeed, the successful economies of the largely autocratic 'Asian Tiger' countries (South Korea, Taiwan, the Philippines, Malaysia) serve as development models closely studied by the Chinese leadership (see Chapter 6). But these changes need not all follow the same pace, as individual national and sub-national factors shape the progress of marketisation. The main focus is entirely on the economy and its position within the state systems. Society is seen as an integral and dependent part of economy and market, and democratisation as a desirable but not strictly necessary part of the society side of the equation. Thus, while likely negative effects such as job losses are seen as a possible side effect of transformation, this is seen as a necessary price worth paying in the interest of the expected longer-term benefits.

Indeed, any impact 'cushioning' policies, such as make-work schemes or other government intervention, are rejected as effectively 'unemployment on the job', with an associated loss in work skills (ibid., p. 181). But it is difficult to imagine that people would prefer no job at all to one that is subsidised. Again, the presumption seems to be that the market will generate all those replacement jobs, if only it is 'let loose'. Strong traits of neo-liberal, New Right thinking are evident here, as by far the dominant credo among politicians and academics in the early 1990s, including the World Bank and the International Monetary Fund (IMF).

But discussions have not been entirely dichotomic, where a liberal market economy with minimal state involvement is pitted against an all-state-controlled planned economy as pursued by communist ideologists. In fact, several of the early popular movements pushing for the end of 'their' communist systems in the CEE countries had their eyes on a middle way between the two, a 'third way' of 'Market Leninism' or socialist market economy. The latter is perhaps not entirely unlike the western German model of a social market economy, albeit with a distinct emphasis on the 'social'. Inevitably, this would mean manipulating market mechanisms through state intervention, thus making the system inherently ineffective and contradictive in the neo-liberalists' eyes, and not really altogether a capitalist model. It is not difficult to see how those advocating more state-managed reforms found themselves frequently labelled 'socialists' by advocates of neo-liberalism. But statist economic developments need not necessarily be 'socialist' or 'communist', nor need they be struggling, as the successful example of the Southeast Asian economic 'tigers' has demonstrated. Evidence makes clear that 'an East Asian pro-capitalist autocracy ... is capable of a better growth performance than any type of democracy acting under comparable conditions' (ibid., p. 139).

But economic and political systems do not usually emerge in their perfect incarnations, especially when they need to combine somehow, and thus require some degree of accommodation to each other's characteristics. There are various combinations of degrees of 'free' market and democratisation, stretching from an improbable combination of planned economy and a high degree of democratisation (although under extreme conditions, such as warfare, it may be possible) and full marketisation under an authoritarian regime, something tried by China. Figure 3.1 sketches the possible combinations between 'marketisation' and 'democratisation', and also shows indicatively the relative position of the different forms of transitions between countries. Essentially, there are infinite combinations possible, as both variables are effectively continua, although inherent dynamisms will make it difficult to just keep 'sitting on the fence' between the available options in an equidistant position. The balance is likely to be tipped either in favour of generally 'more state' or 'more market'. It is against this background that the different 'global regions' of post-communist transition will be discussed here, also drawing on the concept of 'path dependency' in this transition process, and seeking to evaluate its relevance for the emergence of different forms of post-communist conditions in the respective countries and regions. Are there influential *individual* factors, or is the particular *mix* more important?

High

Liberalised Communist regime Free market *cum* authoritarian regime (e.g. *China, Vietnam*)	Market economies *cum* varying degrees of democratisation (e.g. *FSU, post-socialist Africa, across CEE*)	**Popular democracy** Free market *cum* full democracy (Anglo-American model, e.g. in *Poland, Czech Republic*)
Unreformed Communist regime: Authoritarianism *cum* planned economy (no market), i.e. *communist states*	Variably controlled market with part-democratisation (e.g. *Central Asia*)	**Plebiscitary autocracy** Full democracy *cum* planned economy (exceptional, e.g. warfare)

Degree of marketisation

Low   Degree of democratisation   High

Figure 3.1 Intersection of democratisation and marketisation in post-communist transition

Society-focused arguments

Maurice Glasman (1994) asks why the west German model of 'social market economy' has not been adopted more widely across Central and Eastern Europe, instead of the neo-liberal model. As part of that, he argues, the British (Thatcherite) model was adopted, establishing a free market in labour. As a result, he sees post-socialist transition largely as a shift towards societal commodification, put in place by paternalist authoritarianism (ibid., p. 192). He thus refers to Polanyi's work of the late 1940s, in which he argues the paradox of modernity in that technological advancement facilitates the self-liquidation of society in the face of a centralising state and decentralising economy. Effectively, society is atomised by the proliferated economic interests of each individual member of society. Ultimately, no society as a community of shared interests and responsibilities is left (ibid., p. 193). Margaret Thatcher's infamous claim that there is no such thing as society very well reflects this neo-liberal ideology of individualist striving for economic reward for the personal advantage only.

There is thus an inherent contradiction here between 'free market' and the notion of civil society as the underpinning of 'true' democracy. A truly free market would, by definition, sit very uneasily with such organised force that could interfere with its 'optimal' operation. This, however, Glasman (1994) argues, can clash with other sources of societal factors, such as, in the case of Poland, the Catholic church with its emphasis on solidarity and cooperation. The political marginalisation of the Solidarity movement during the years of implementing an as free as possible market economy would attest to the operation of such mutual exclusion. This obviously contradicts the beliefs and aspirations held by the self-organising democratic movements that emerged as the bedrock of anti-communist opposition in those formerly communist countries where transformation came about through 'bottom-up' pressures. But corporate interests and their close interconnection with political power stood either sidelined by these more independent movements outside the market-driven environment, or simply 'gobbled them up', as in eastern Germany. There, the autonomous popular movement Democratic Forum was absorbed by the established western German parties as they 'conquered' the eastern part. Glasman thus refers to the free market ideology as 'market utopianism', idealising the functioning and rationale of the market *vis-à-vis* societal interests and needs. It was 'the moral attractiveness of the market as a foundation for freedom and prosperity – [that formed] a crucial part of its appeal in Eastern Europe at present' (ibid., p. 200), and it is not surprising that the generally less than perfect results have somewhat disillusioned and disenchanted many, especially those imagining the 'third way' of a socialist market economy through the fusion of the 'best of both worlds', Western and Eastern, as it were.

Instead, in the real world, society was largely reduced to being an onlooker on the sidelines, whether in the shaping of new government mechanisms or economic practices. Sure, privatisation was meant to provide for an element of democratising economic processes, but that remained largely an idealistic ambition, with reality favouring the creation of new, or (quite often) the strengthening of old, elites. The problem with the transition process as it occurred in post-socialist Central and Eastern Europe is that 'in the shift from Marxist to market Leninism, civil associations and cooperation – society in short – have still been ignored'. Thus, for instance, the so-called Balcerowicz Plan of turning Poland into a market economy 'overnight' was not developed with representatives of society, but was imposed by the state in the name of the market (not so much 'society' or the people). 'State and market were the only two institutions that mattered' (ibid., p. 202). And, as the Polish example of the grassroots movement of Solidarity demonstrates, the imposition of martial law during the 1980s demobilised the population and thus reduced pressure for *political* change and the development of civil society. This, in turn, allowed the political actors to change their minds, moving away from political reform and democratisation towards neo-liberal economics as the most effective way of delivering economic prosperity, which had become the main focus of envisaged systemic change. This made popular democratic movements like Solidarity appear increasingly out of the loop and, ultimately, superfluous in the pursuit of the neo-liberal goods. Given their continued propagation as the only way forward to the desired western quality of life, less operational considerations like 'democracy' appeared less important and their advocates therefore superfluous. 'Market subordination and facilitation through the institutionalization of industrial democracy, subsidiary and subsidy were unknown in Eastern Europe', and 'Market Leninism' seemed increasingly too much stuck in 'socialism' and thus, by definition, incapable of delivering the largely materialistic goods.

Post-communist transition and
post-authoritarianism

The concept of 'transition' has not only been developed in response to the unprecedented events of the collapse of state-socialism, but has been discussed earlier in the more general context of a shift from authoritarian rule to a democratic system. Interestingly, and as a reflection of the political-ideological pre-eminence of western paradigms, such transition has been discussed only as a legitimately one-way process. Any reverse processes have been presented as a systemic failure, and unlawful ('putsch'). There is thus the notion of democratic systems being 'normality' and everything else abnormal or aberrant and thus, inevitably, temporary (see O'Donnell and Schmitter, 1986). Their definition of 'transition' as 'the interval between one political regime and another' (ibid., p. 6) offers an attempt at a more generic, neutral definition. The 'transition' discussions gained in momentum during the 'first wave' of post-authoritarian democratisation in southern Europe and Latin America (O'Donnell and Schmitter, 1986), and comparisons have indeed been drawn between that period of change and that 10–15 years later (Linz and Stepan, 1996, p. 199).

O'Donnell and Schmitter (1986) view 'transition' very much as an active, actor-driven process 'from within', where the incumbent ruler begins to modify the existing rules by providing more secured rights for the population, thus encouraging participation in the political process. Crucial in this model is an evolutionary process, with institutional structures largely remaining in place, and just working to different rules. In any case, democratisation is seen as the 'natural' aim of such actor-driven liberalising transition. However, as stated earlier, economic liberalisation may, of course, exist without democratisation, as in China, for instance. As such, liberalisation and democratisation are essentially two processes of change in their own right. They are connected, but (economic) liberalisation may not lead to democratisation. The latter is the more uncertain process, as it depends much more on people's participation – that is, 'socialisation' – and is thus more difficult to manage from outside. The endpoints of the intersection between liberalisation (marketisation) and democratisation are diagrammatically shown in Figure 3.2, and its application to the post-socialist condition in Table 3.1 (p. 61 this chapter).

An important observation is that the factors that brought about the end of autocracy and triggered a transition process may not, in their own right, be enough to facilitate the development of a new system, such as a political democracy. Other factors, institutions and individual inputs are necessary (ibid., p. 65), as a result of which 'political democracy then, usually emerges from a non-linear, highly uncertain, and imminently reversible process involving the cautious definition of certain spaces and moves on a multilayered [chess] board' (ibid., p. 70), with an unspecified number of players. Important here is the non-linearity, suggesting the absence of predictability and simple transferability of processes, instruments and, most of all, results.

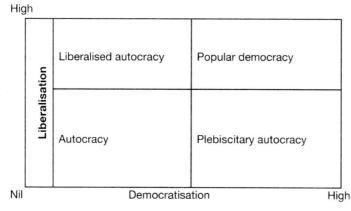

High

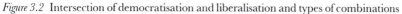

Liberalised autocracy	Popular democracy
Autocracy	Plebiscitary autocracy

Liberalisation

Nil   Democratisation   High

Figure 3.2 Intersection of democratisation and liberalisation and types of combinations

Democratisation as automatic outcome?

The Washington Consensus doctrine viewed democratisation according to western understandings as the natural and only available goal for post-socialist development. The focus on Eastern Europe and Russia made the western European regimes the obvious models. Experiences in other transitional economies and/or regimes, mainly in developing countries, seemed not very relevant, as we were dealing with European, i.e. *developed*, countries. Yet changes during the 1990s have demonstrated that democratisation according to western understanding is not necessarily the default outcome, whether facilitated through sticks and carrots or not. Authoritarian structures may well remain, albeit somewhat obscured by a formal mantle of democratic appearance. Ottaway (2003) speaks in this context of semi-authoritarianism – whether post-socialist or not – as the best description of the pseudo-democratic arrangements following the end of totalitarian state socialism. In her eyes, *socialist* authoritarianism is thus seen as one variety of such regime changes the world over.

Semi-authoritarian regimes are political hybrids. While allowing little real competition for power, they leave enough space for political parties and civil society to form (ibid., p. 3). Many of the former Soviet states, albeit to different degrees, fit the bill of such semi-authoritarianism, e.g. Belarus in eastern Europe, and Azerbaijan and Kazakhstan in central Asia. There, former Communist Party bosses have transformed themselves into elected presidents, but in reality remain strongmen whose power is barely checked by weak democratic institutions (ibid, p. 3). Similarly elsewhere, in particular in sub-Saharan Africa, Arab countries, the Balkan states and Asia, semi-authoritarian regimes rather than democratic structures of western mould have established themselves. Such forms of government are not new, but in the absence of the Cold War divisions pressure has increased on these states to at least pretend adherence to democratic principles to make dealings with the West publicly more acceptable. As a result, semi-authoritarian regimes have become more numerous and are likely to increase in number even further (ibid, p. 4).

An optimistic assessment views such regimes as transitory in a longer-lasting process of proper democratisation. Accordingly, semi-authoritarian regimes sit on a scale somewhere between totalitarianism and western-style democracy, thus challenging in their diversity the rather deterministic, end of history (Fukuyama, 1989) argument propagated by the main western policy makers at the time of the collapse of communist states. Such deterministic interpretation, focusing on set and inescapable developmental pathways, practically applies a Marxist analysis of history by adding democratisation to the developmental path of state–society relationships. Accordingly, semi-authoritarian regimes may be seen less as a stop-gap in the transition process, and more as a deliberate strategy to be *seen* to fit in with the general wave of democratisation out of political expediency, without requiring the relinquishment of *de facto* power by the political elite. These regimes thus challenge the widespread notion, held particularly in the late 1980s and early 1990s, that liberalisation is an effective instrument in producing democratic structures through unleashing dormant democratic forces.

Particular national circumstances, including economic problems, historic structures and social inequality, may limit the appeal of a fully democratic arrangement. In fact, the western perspective of democratisation implies a broad popular consensus and participation in establishing the democratic system, when in truth much of the presumed re-/emerging civil society is often merely a narrow (urban) elite, rather than a broad grassroots movement sweeping a whole nation (Ottaway, 2003). The degree of economic development and prosperity is an important factor, albeit perhaps not quite as dominant as implied by Rustow (1970). Comparing a sample of 75 developing countries in the period from 1962 to 1992, Feng and Zak (1999) conclude that more affluent societies/individuals are more likely to support democratisation, although there are other factors involved as well, such as the distribution of wealth, educational levels, and the strength of preferences for political rights and civil liberties. It remains to be seen whether, as the authors claim, sufficient growth will inevitably push non-democratic governments towards democracy. While, generally, democratisation seems to benefit from higher income, there is no 'magic number' beyond which democracy becomes unstoppable, but democratic transitions are less likely when the level of development is low, income inequality is high, and citizens are poorly educated (ibid.). The authors thus conclude from their international comparison, that Africa's main obstacle to democratisation is economic paucity; among the oil- and cash-rich Arab countries it is the lack of a democratic culture and ambition; in Latin America, it is the relative weakness of the 'middle class' as standard-bearer of democratic practice; and the east Asian countries are currently in an exceptional state of political and economic development, as demonstrated by China, for instance.

State, society and post-socialism

A key feature of post-socialist transition has been the role of society versus the state, whereby the end of communism has been very much seen as a victory of society over a suppressive and distant state machinery (Grzymala-Busse and Jones Luong,

2002). In other words, transition from communism to democracy implied a transition from a state separated from, and dominating, society, to the ideal of a democratic arrangement, where the state is controlled by society. Public discourses created both in the West and, albeit from quite a different experience, in the East, of the communist state was that of an out-of-control, while control-obsessed, behemoth that had *per definitionem* to be reduced in size and capacity. The focus was thus generally on less state following the New Right doctrine of the minimalist state (while dominant market), rather than discussing the rebuilding of a new state altogether.

Grzymala-Busse and Jones Luong (2002) identify three main features typical of the post-communist state-building process: the 'fast forward' pace of creating a new state structure and rationale; the importance of informal next to formal relationships and negotiations between actors; and external pressures strongly influencing the process, such as global financial institutions and the accession conditions for EU membership. The big challenge has been to establish, rapidly, the necessary new legal framework, change the ethics of the bureaucracy from agent of the Party to civil servant, and restructure the networks of security, redistribution, and regulation as the essentials of the advocated liberal market-democratic state. But the preparations for EU membership, in particular, have raised the profile of the state-society interface and the notion of 'state in society' as the antidote to the opposite relationship under communism: society within an all-embracing, ever-present state. Post-socialist developments have brought to attention a more complex picture than that, where state-society relations are seen as bi-directional, reflecting a mutual learning and adjustment process, although states and societies remain visibly separate, frequently competing, entities. The degree of perceived separation between state and society varied between countries, reflecting the different outcomes, or success, of the communist regimes at bridging the practical-ideological gap between developing a ruling state apparatus controlling the people, while also emphasising the Marxist–Leninist vision of society and state – society as lived state – with blurred boundaries between them, giving way to the idealised communist society. Achievements here have been relatively more 'successful' under Soviet rule where it was attempted to overwrite national boundaries, ethnic differences and cultural by the new state ideology. In CEE countries, however, there were strong legacies of the state as a reflection of national identity and sovereignty, and these remained under communist rule. The rapid development of new nation states in the Baltic Republics after independence from the Soviet Union, for instance, reflects a pre-existing legacy of the notion of 'statehood'. The state is not just an oppressive, anonymous administrative control system, but is seen as a reflection of national identity and sovereignty. The state thus became synonymous with nation state and, eventually, independence and self-determination. The other post-Soviet states had a less strong legacy of statehood, and the population was less insistent on actively forming, and participating in, 'their' nation state. It has thus been much easier for sitting communist leaders and apparatus to retain control and societal domination. The recently evident rather unenthusiastic electoral engagement by the public reflects a degree of disenchantment with the apparatus representing 'the state', rather than with the state *per se*.

The difference in the notion of 'state' between Soviet and CEE has been in the amount of space reserved for society to organise independently. Institutional and practical legacies in policy-making as an expression of state operation thus matter for the shaping of the post-socialist state. There was no *tabula rasa* to build on, but existing structures and their bureaucrats were the basis and object of any remodelling of the states. This includes the nature and role of informal arrangements, especially personal links within the bureaucracy. They provided important markers in the emergence of new institutions and administrative-governmental practices, and often provided continuity between 'old' and 'new' structures. But, as Grzymala-Busse and Jones Luong (2002, p. 534) argue, 'the rapid nature of post-communist state-building also serves to privilege elites participating in the initial stages of the transition and their actions'. They set the framework and the condition for their own subsequent operation. But any such changes are embedded in wider economic processes. It is through these that the post-socialist countries gained the completely new experience of globalisation, and its strong impact on national economies, which challenges their newly gained autonomy. This involves new rules and practices of economic competitiveness, but also propagated 'good practices' of governance, democratisation and liberalisation. Particularly with respect to the accession conditions for EU membership, many of the relevant countries felt they had swapped Moscow for Brussels as their effective ruler. But a deeply engrained apprehension towards Russia, with its perceived inherent hegemonic tendencies, meant that they saw effectively no other choice than 'Moscow or Brussels' for their future statehood.

Transformation processes after the end of communism have also highlighted the multi-nodal nature of the state. The state is not just one big homogenous entity acting inherently as one. Instead, it encompasses many competing actors, and centres of power and their networks, operating within and without each state and defining policies. 'Oligarchs, political parties, and presidents on one hand, and international financial institutions or regional trade associations on the other, all have access to the nascent state structures and exert considerable pressures on the processes of state formation' (ibid., p. 532). There are thus different centres of emerging new authorities claimed by the emerging 'new actors on the block', with those representing economic capacity and political power being the main 'sparring partners'. The current competition between the Kremlin and some of the new oligarchs (Osborn, 2005) for political control of the state illustrates these internal divisions within 'the state'. The resulting dynamics have been crucial for the shaping of the nature of the post-communist state. Acknowledging the heterogeneous nature of the state, shaped and driven by competing elites' interests, challenges the projection of the state as one single entity and thus the notion of a state versus society relationship, with clear demarcation lines between them. This 'avoids a return to the debate over the appropriateness of state-centred versus society-centred approaches' in seeking to understand the post-communist state (ibid., p. 545).

Not one, but many post-communist transitions, and more ...

The transition from communist to post-communist (democratic) regimes can be seen as a specific version of transition from an authoritarian to a non-authoritarian regime. Balcerowicz (1995) distinguishes four main types of transition:

1   Classical transition towards democracy in advanced capitalist societies in the late nineteenth and early twentieth centuries.
2   Neoclassical transition after 1945, such as in Germany and Italy, and, in the 1970s and 80s: Greece, Spain and Portugal, some Latin American countries, and Taiwan and South Korea.
3   Market oriented reform in former communist countries of Central and Eastern Europe and the FSU.
4   Asian post-communist transition (China since the 1970s and Vietnam since the 1980s), with its particular duality of continued communist authoritarianism and liberalised economies.

Post-communist transitions show particular characteristics in their scope, complexity and speed, particularly those in CEE countries. There, not only were governmental principles, institutional structures and essential ways of policy-making transformed in a very short period of time, but also economic principles of location, investment and connectivity with other national economies were completely changed. Given the different traditions, geographic contexts and experimental legacies, these processes varied in speed, depth and relationship to each other. Still, with very few exceptions, there was no violence in connection with the (negotiated) revolutionary changes. This allowed for a considerable degree of political and economic administrative continuity, and thus the use of established institutional expertise, albeit under the impact of privatisation. In many instances, public property was transferred to the private control of previous state-appointed bureaucrat-managers. In fact, many of the political elite re-launched themselves as reformed communists, now labelled market-socialists or social-democrats, and as such competed successfully in national elections.

Not surprisingly, variations in the nature and constellation of political and economic actors yielded quite different outcomes for the transition towards post-communist conditions. In correspondence to the typology suggested by Ottaway (2003) for capitalism under authoritarian conditions, there are two main avenues for liberalising the state-planned economies of the communist regimes: a destroyed communist economy and a distorted (modified) communist economy. While the former reflects a complete abandonment of the existing system and the completely new construction of a market-based economy, echoing the notion of the shock therapy paradigm discussed elsewhere in this book, the second path mirrors the basic features of gradualism. Here, elements of the old system remain in place, albeit suitably modified to incorporate elements of a liberal market economy, operating within a framework defined by many checks and balances in order to protect society from the unbridled impact of a wholesale systemic change. Perhaps uniquely, compared with the other transitions from authoritarian regimes during the last quarter of the twentieth century, the (western) media has played an important role by instantly widely publicising early signs of discontent in any one place, and any indications of a weakening in the resolve of a communist regime *vis-à-vis* public criticisms, transcending all boundaries set up by the controlling governments to protect their power. This publicity provided a significant boost to opposition groups in their negotiations with their respective governments. The recent democratisation movements in Ukraine (see Chapter 5), clearly demonstrated the importance of the media in attracting international attention, and thus internationalise the negotiations between the defending state elites and the new popular challenges. This matters, because the relative strengths of the two political opponents very much shapes the outcome as either a negotiated compromise, accepted by both parties as a deal, or an imposed outcome, reluctantly accepted and most likely resented by the weaker, inferior party (see also Sutter, 1995; Swaminathan, 1999). Such transition through transaction (Share, 1987) is fundamentally shaped by the role, attitude and negotiating skills of the elites. The importance of the conditions of change thus becomes clear.

Transformation as a dynamic process

Rustow (1970) questioned the, at the time, prevailing notion in academic and public debate that democratisation needs a specific set of supportive factors as prerequisites to establishing itself and keeping a foothold in a society. These included such broad variables as economic prosperity, level of education and widespread adherence to liberal or democratic values. Instead, he turned the focus on agency, process and the bargaining between actors, and not on set structures and institutions, and thus broke new ground in the interpretation and analysis of transitions to democracy. Thirty years later, the outcome of the 1989 to 1991 events in the former communist Eastern bloc have confirmed the crucial role of the way in which change takes place. Thus, Anderson (1999a) emphasises the importance of the *process* of transition from an authoritarian to a democratic state system. In particular, he identifies the way power was transferred as a key factor in shaping the likely path (success) of any move towards democratisation: either 'orderly' through negotiations, allowing the old elite to have an influence on the shaping of democratic structures and thus ensuring a degree of continuity of governance, or abruptly through revolutionary processes. In the latter case, there is a break with the past and a new beginning, with completely different governmental arrangements. This, he argues, would allow for greater consideration of individual country-specific circumstances, and thus a more accurate reflection of underlying dynamics, than could be achieved from simply employing a standardised one-size-fits-all perspective.

But circumstantial particularities will make predictions difficult, as no easy correlations can be drawn between 'factors in place' and the resulting nature of democracy. Indeed, even seemingly 'infertile' conditions can generate democratisation processes, at least on the surface. Often, these are more for external consumption, though – to be seen to be democratising – rather than an expression of genuine internal changes within a state towards democratic principles. The resulting 'democratic appearance' usually serves to placate external observers and satisfy conditions attached to international (western) loan or trade agreements. Often, these projected 'institutional concessions were little more than Potemkin villages' (Anderson, 1999a, p. 9). They have become necessary, because after the end of the Cold War the American model of civil and economic governance is the only show in town. No longer is there an alternative socialist or communist model on offer that would allow access to different sources of funding and geopolitical support (see Chapter 5).

However, as Haggard and Kaufman (1999, p. 73) point out, the emphasis on the political (transitional) bargaining process, despite its central role in regime change, tends to somewhat ignore the importance of 'resources that contending parties bring to the negotiation and even the institutional stakes of the negotiation itself'. There is no blank sheet to start off from. Past experiences, existing networks and own agendas compete for recognition amidst the prevalent economic conditions and societal structures and particularities (Lock, 1994). They include a socialist culture of doing things (Seleny 1994), or a distinct statist tradition (Szarvas 1993). These particularities, Haggard and Kaufman (1999) argue with reference to path dependency theory, encourage political processes to embark on particular avenues. There is also, they continue, the not to be ignored 'contagion effect' of democratisation processes spreading to neighbouring countries and thus 'taking a shortcut' to changing conditions. The overall importance of economic factors is expressed through such popular indicators of improvement as employment and personal income. The quite rapidly varying political fortunes of the ruling parties attest to their close association with good and poor economic performances and the degree to which there seems to be equitable income (Baer, 1993).

A general failure to generate the promised immediate, blossoming economic landscapes, but instead landscapes of economic decline and disinvestment, soon raised questions about the somewhat simplistic and, with hindsight, naive assumptions that had underpinned many of the political decisions. These include the overtly simplistic and naive presumption that market forces *eo ipso* can effect all the imagined and propagated economic miracles all by themselves, if only left undisturbed by state intervention, as the New Right saw it. Instead, the pivotal role of the state as provider of a market-conducive environment has been recognised, and this includes a stable institutional and legal framework to allow strategic investment planning (Eggertsson, 1998). 'For example, variations in historical patterns of state formation between coercive and capital-intensive paths and in their reliance on internal or international dynamics may well create cultural and social structural predispositions toward certain types of political regime' (Anderson, 1999a, p. 11).

The sense of liberation from external pressure and control has added a considerable amount of national assertiveness and identity-building to the politics of post-socialist transition, and added an extra variable affecting transition outcomes in comparison to non-communist transitions to democracy elsewhere, where there has been less experience with controlling domination by an external force. Post-colonial liberation and nation-building movements may offer the closest parallel here. Whether or not there had been such an external control experience is thus likely to shape the post-authoritarian policy agenda of its democratic successors (Haggard and Kaufman, 1999). The political changes in Central and Eastern Europe during the 1990s illustrate the crucial effect of the experience of an economic crisis on the subsequent path and outcome of post-authoritarian transition towards a democratic system. A seeming inability to address effectively some perceived economic malaise may lead to disenchantment with the transformation/ reform process, and encourage cynicism and apathy in the face of obvious weaknesses of the new system. Low voter participation in elections, voting for radical or single-issue parties as an alternative to the 'mainstream' governing parties, or voting for former communists, are all expressions of discontent and disillusionment, while also leading to an emerging eastern identity *vis-à-vis* all-out 'westernisation' strategies. This supports the view that fully functioning institutions and governance are a key determinant of the nature, quality and progress of democratisation. It also reflects the by far dominant focus on economic issues. Everything else was seen as automatically following for the 'better' (Lavigne, 2000).

The particular circumstances at the time of departure 'define the mode of transition in terms of the identity of the actors who drive the transition process and the strategies they employ' for gaining control/influence (Munck and Leff 1999, p. 193). This competition, through its influence on the roles of elites, shapes the nature of emerging post-transitional regimes and politics by affecting the pattern of elite competition, institutional rules and willingness to engage and compromise (ibid., p. 193). The underlying notion of a path dependency is thus much less structurally determined than process based, and thus ultimately haphazard, unpredictable and particular. As such the modes of transition are *per definitionem* non-repetitive and non-transferable. They circumscribe the way in which power was transferred, either with or against the old regime, and this very much shapes the outcome. While confrontational processes help to unify the opposition and help gain greater concessions from a relatively weak old regime, a concordant arrangement between the *ancien régime* and the revolutionaries may prolong transition and thus their continued political influence and consideration. An important criterion of the mode and outcome of transition is whether the main agents of change come from within or outside the political establishment; Russia, China and Romania are examples of the former, Poland and the Czech Republic of the latter. In between, there are a variety of roles for one or the other. Munck and Leff (1999) thus see the *type* of actors driving the transition process as primarily responsible for the way it proceeds, and the outcome it achieves. They thus draw a connection between modes of transition, and form and degree of democratisation, as illustrated diagrammatically in the Table 3.1.

Accordingly, two types of democracies may emerge at opposing ends of a scale: restricted and full. Reforms led by strong incumbent elites, seeking to defend their ground against the challenging counter-elites, tend to result in more restricted,

Table 3.1 Type of transition and the role of the old and new elites as agents of change

Mode of change	Agent of change: old authoritarian elite	Mode of change	Agent of change: new democratic opposition
Confrontational change	*Revolution* from above through political establishment	Disruptive change	*Revolution* from within society (non-establishment)
Protracted change	Process somewhere between reform and revolution	Marked change	Process somewhere between reform and revolution
Negotiated change	*Reform* from above seeking to maintain as much of old elite's influence as possible	Gradual change	*Reform* driven by democratic opposition, seeking to minimise (terminate) role of old elite

Source: loosely based on Munck and Leff, 1999

negotiated forms of democracy, while weak incumbent elites are more likely to be swept away by a clearly dominating counter-elite, too weak to extract any compromises. The likely outcome here is further-reaching democratisation. Between these extremes sit more complicated and protracted transitions as a result of negotiated 'gives and takes' between less asymmetrically empowered new and old elites. The complexities here are, as the authors observe, raised by the likely less clearly focused agendas of both elites, as they feel less threatened than under an asymmetric distribution of influence. In other words, the weaker and more disunited the challenge, the weaker is the resolve among the defendants and the more likely are pacts about transition arrangements (see e.g. Friedheim, 1993).

Ágh (1999) distinguishes two types of democratisation: internal (intra-national) and external (inter-national). While the former relates to the processes discussed here, that is, a shift from (communist) authoritarianism towards some form of post-authoritarianism within a country, the latter refers to the integration of a democratised (democratising) state into wider world politics. He thus, too, focuses on the *process* of democratisation, rather than on a particular situation at any one time. Referring to the transition model presented by Schmitter (Karl and Schmitter, 1992), he highlights the *process* of transition as a, if not the, crucial factor in shaping its outcome. In other words, the 'mode of transition' is largely responsible for the resulting 'type of democracy'. This mode revolves largely around the degree of compromise versus confrontation, between old (communist) and new (post-communist) elites. The outcome is shaped by the relative impact of the different actors, leading to an agreement between the opponents: reform as negotiated compromise. This may lead to a liberal democracy, as advocated by the 'West', or a nationalistically underpinned democracy, often with more formal than practiced democratic principles. Examples exist in many African post-socialist/communist countries, and most post-Soviet states. An important criterion is thus the degree of a democratic tradition. Does democratisation get driven from within, or is it imposed from the outside, because 'it is good for you'?

The main focus of these understandings of transition is thus the actual process of change-over from one regime to another, rather than the quality of the outcome, usually measured against an ideal scenario as desired result. And this process is shaped by a combination of structure, defining the arena and scope of transition, and the actors as initiators and implementers of change. Pickel (2002), for instance, while clearly emphasising the nature of post-communist transition as a political *process* first and foremost, stresses the need to look beyond structure and adopt a more integrated, holistic perspective of both structure and action on the ground. It is here that the usual understanding of transition has its weakest point, as Bönker *et al.* (2002a) point out. The whole concept of transformation and its evolution shifted during the 1990s from an entirely neo-liberal, purely economy-focused under-standing, to a more comprehensive paradigm, involving government and society. This variation reflects the absence of a well-founded theoretical underpinning of transformation research. Much of the existing work is rather less than comprehen-sive in its scope, restricted largely to an economy-centred interpretation, requiring policy makers to learn relevant policy-making on the job.

Process of change and outcome: transition as a disruptive, destructive or constructive process – transformational recession or transformational growth

The close connection between delivering the (economic) goods and attitudes to the post-communist condition has been a particular feature of developments in CEE countries and the former Soviet Union (Baer, 1993), not least because the gap in opportunities for consumption between West and East had become one of the main drivers of disaffection and, ultimately, opposition. The new rulers in those countries thus placed great emphasis on facilitating or engineering visible economic improve-ments to gain legitimacy for their hold on power. In most instances, they followed some form of the propagated liberalisation agenda, and this meant largely breaking with the past and abandoning state support for the existing economic structures. Inevitably, economic contraction followed. In contrast to the Chinese experience, post-communist regimes were immediately associated with conditions even worse than before. This perception has in many parts survived until today, especially as regards the more peripheral, disadvantaged parts of the national economies. In eastern Germany, this has resulted in a view back through rose-tinted glasses to the old system, when life seemed to be not easy, but more predictable and secure, and less harsh and less like a rat race. 'Ostalgie' has become a widespread phenomenon, and found its expression in the resurgence of the political far left, including former East German communists, in the most recent German federal elections.

There are several possible explanations for this transformational recession, as Lavigne (2000) points out. There are particular historical legacies, such as inherent competitive disadvantages, which were exacerbated by the new competitiveness introduced through liberalisation and stabilisation policies, as advised by the Washington Consensus. But there were also particular ways of doing things, especially a bureaucratic, hierarchical thinking, waiting for orders to come from above and thus removing the need to make decisions and take responsibilities. Not only was this a result of the previous communist regime's workings, but also, as has become evident from many interviews on economic policy-making conducted by the author in eastern Germany over the last decade, of an acute sense of insecurity and uncertainty among bureaucrats, especially during the early years of transition. Many were concerned about maintaining their jobs in a climate of post-communist purges of former functionaries, and not 'sticking one's head out', and avoiding any potential for criticism of one's ability to do the job effectively, seemed a good survival strategy. It thus was the propagated and effectively imposed western model of economic development that caused the initial and immediate economic contraction and, subsequently, social cost of 'instant' westernisation (Offe, 1996).

On the other hand, one needs to keep in mind that there is also a relative aspect to the perception of decline, at least in its quantitative aspect. Figures released by the communist states were inherently unreliable, because they also served propagandist goals, to demonstrate the regimes' successes to both an internal and, especially, an external western audience. The experiences with German unification highlighted that, where claims of hundreds of billions of Deutschmarks in asset and productive value made by the then East German government as their contribution to the 'marriage' turned out to bear little reality. Most of the highly valued productive capacity turned out to be well out of date, producing non-marketable goods. This highlighted the problems with a purely quantitatively based, technical and arithmetic assessment of economic conditions. The actual situation 'on the ground' matters at least as much as officially released figures, especially when the releasing source has a monopoly on information. The mixed results of applying the Washington Consensus wholesale and inflexibly attest to the need for developmental models to be placed in the political-economic and societal realities of their target areas. Econometric modelling alone cannot grasp all the many factors influencing economic development, and this has become one of the main criticisms of the largely uncritical, technocratic application of the terms of the Washington Consensus.

The Washington Consensus paradigm and the complexities of post-communist developments

The spectacular changes in Central and Eastern Europe at the beginning of the 1990s spurned a whole new breed of researchers looking into the process of transition and making predictions about its likely course. They also dispensed advice on best practices and gave recommendations for effective policy approaches. Generally, this entailed the by then neo-liberal mantra of market-centred liberalisation. The resulting politics of transformation (Bruszt, 1992) focused almost exclusively on market processes as the responsible engine of societal and political change. The early reform programmes called for structural adjustment policies – stabilisation, liberalisation and privatisation – policies that had been prescribed to Third World countries since the early 1980s. This 'neo-liberal discourse of radical reform' (ibid.,

p. 5) took a strictly economic perspective, and gave little consideration to other factors. These already established concepts of economy-centred democratisation movements seemed confirmed by the events in Central and Eastern Europe. The medicine dispensed to Latin American countries a few years earlier to help tackle their debt problem through good financial governance seemed ready to be taken off the shelves and applied to the former communist countries in Europe straight away. This new, comprehensive ambition was summarised and projected in the Washington Consensus, which was concluded between the main (US-based) bodies of global financial and economic governance, the World Bank and the IMF. The Consensus refers to a set of supposedly uncontroversial theoretical and political assumptions about the composition and sequence of market-oriented reforms for economic growth. This discourse was powerful, because it structured the field of legitimate arguments, linking scientific rationale and (derived) policies, and introducing the politically and morally potent distinction between radical reformers and conservatives (or populists) (Bönker et al., 2002a, p. 5). International trade was considered to be a, if not the, major vehicle of driving the economic transformation process.

The Consensus became the dominant (almost dogmatic) paradigm of politics towards post-socialist countries, specifically eastern Europe (see also Herrschel, entry 'Shock Therapy' in T. Forsyth (2004)). It also became a tool to reaffirm the West's ideological victory, over Russia in particular. Turning the latter into a western clone would be the ultimate symbol of the 'aberrant nature' of socialism/communism, and remove any future danger of possible challenges from there to the western paradigm (Bönker et al., 2002a, p. 5) This was portrayed as further evidence of the superiority of the liberal market principles in the Thatcherite/Reaganite mould.

The Washington Consensus was driven by the institutional power of international (US-led) financial organisations (IMF, World Bank), combined with the prestige of neo-classical economic theory supplied by the transitologists in academia, and an orthodox belief in the efficacy of market forces, which had gained in profile through the rhetoric of the Thatcher and Reagan governments and their dispatched advisers. As a result, orthodox transformation theory and policies of the early 1990s revolved around the mantra of stabilisation, liberalisation and privatisation. All advice to transforming states was based on these three factors in a near-dogmatic belief. This economy-centred orthodoxy, however, has been increasingly challenged by developments in post-socialist Central and Eastern Europe. Interestingly enough, the impact of the experiences in these countries on the institutions responsible for the Washington Consensus has been much greater than had been mustered by the developing countries in Africa and Latin America.

The problem with the simple policy transfer of the Washington Consensus paradigm to the post-communist countries in Europe was that policies did not quite fit the particular challenges posed by the nature of this transition. The policies of the Washington Consensus were tailored to the problems of economic instability and associated indebtedness in developing countries, particularly Latin America. They were thus aimed at economies that were, more or less, following market principles, albeit not always in a very successful way. The Consensus was aimed at distorted *market* economies, and not those with a completely different approach to, and operation of, resource allocation. Kolodko (2000a, b) blames the insufficient knowledge of those involved in 'transitology' of the situation in communist countries as one of the main reasons for the misjudgement of likely transition paths and their outcomes. They simply applied their understanding (however incomplete) of developments in developing countries to the situation in eastern Europe, either ignoring, or being unaware of the significance of, the fundamentally different underlying structures and legacies, although these countries were categorised as 'Second World', rather than 'Third World'. Much of their focus was on post-Soviet rather than eastern European experiences, possibly because of the strong Cold War-era focus on the Soviet Union as the embodiment of the 'East'. There had, of course, been some aborted attempts at establishing communist regimes in Africa and Asia too, but they had been seen largely as peculiarities or, at worst, an irritation, in the development trajectories of Third World countries. So, there was some information available here, but the relevant countries did not feature much on the geopolitical and geo-economic radar screens of the main international financial agencies, and the 'western' governments, leading to limited interest in designing policies specifically for those conditions.

The main tenor of the Washington Consensus thus remained unchallenged, although its policy recommendations had been written for 'clients' in circumstances quite different from those facing the collapse of their communist regimes. Shock therapy was advocated as the one and only 'proper' approach to conducting the shift to a market economy, and curbing inflation was the central part of its implementation. Its basic assumptions and reassessments have thus remained the same as developed for the circumstances of the indebted developing countries in the 1980s. There were only a few minor cosmetic changes. These included the option of 'gradualism' as a possible, if less preferable, alternative to the shock therapy approach. Not surprisingly, Stiglitz (1999, p. 21), a vociferous opponent of the largely uncritical application of the Washington Consensus's *modus operandi* within the World Bank, refers instead to 'incrementalism', thus emphasising the hesitant aspect of this approach. The Washington Consensus of the mid-1980s did not envisage application to post-communist conditions. But when the situation arose, its medicine was prescribed to those particular circumstances wholesale and without reconsideration.

In effect, post-communist changes, including those in Europe, were put on a par with the challenges faced by industrialising developing countries. The immediate interpretation *vis-à-vis* the collapse of communism was 'that it would be sufficient to liberalise prices and trade and then fix the financial fundamentals' (Kolodko, 2000b, p. 1). This view is somewhat surprising, considering the then realisation that 'this is a process of replacing an old system with a new one ... rather than tinkering with the way it [the old system] performs' (ibid.). Privatising state assets was a crucial instrument of the propagated and expected adjustments. Economic growth was expected to follow automatically, without state-institutional reform to underpin the new markets. In fact, driven by economists, the state was considered as the main perpetrator of the problems in eastern Europe, and rolling back the state as much as possible would, therefore, by definition, help markets develop in their most efficient form. There was no consideration of the need for appropriate institutional structures to provide a necessary legal and administrative framework for 'proper', that is predictable and reliable, market operation. This lack turned out to be one of the main weaknesses of the Consensus. Indeed, as early as 1992, Murrell pointed to the classical analysis of *inter alia* Karl Popper to point to the potential pitfalls and social costs of implementing 'utopian blueprints', rather than opting for more 'piecemeal social engineering' (Murrell, 1992, p. 16). In fact:

> catchwords like 'social market economy', 'people's capitalism', 'mass privatization', 'improvement of efficiency and competitiveness', 'fast growth' and 'higher living standards' were used as substitutes for a comprehensive blueprint … Visions about the outcome were fuelled more often by the experiences of somebody else, especially the developed market economies, rather than by critical analyses of one's own abilities and constraints.
>
> (Kolodko, 2000b, p. 43)

Only slowly, with experience, the initial policy rationale was revised to take the particular features of the post-socialist countries into account. As a result, post-communist countries were able to move from an initial 'shock without therapy' to a later, more informed, phase of 'therapy without shock'.

Thatcherism and Reaganomics rejected the state as a separate force outside the economy, potentially interfering with its presumed optimal functioning if left alone. Consequently, the political process was not seen in terms of parliamentary procedure, consultations with stakeholders, and the art of compromise, but in terms of a technology of implementation (Bönker *et al.*, 2002a, p. 11). This technocratic approach focused on marketisation and de-statisation of economic development as the universally applicable and appropriate strategy. These policies were generally applied and monitored from above rather than initiating, and drawing on the outcomes of, debates and discussions from within as part of a democratising process of an emerging civil society. Despite the exclusion of political process, and the fact that the implementation focus was later seen as failing to realise the importance of institutional contexts for the proper functioning of markets (see e.g. Poznanski, 1999), it should not be forgotten that at the time of the collapse of communism, time seemed of the essence to avert the danger of a possible premature end to the opening up of the communist states, without having the right structures put in place. Again, structure was the main focus. Installing western principles quickly to establish irreversible changes to the communist-era structure was seen as a quickly implemented guarantee of embarking on a path of no return. And this was the ambition of people both inside and outside those states. Radical liberalisation appeared as the only effective way of 'detotalizing the regime' (Bönker *et al.*, 2002a, p. 10); economic reform was seen as the centrepiece of rapidly and permanently westernising the East effectively, and neo-liberalism was, by far, the most favoured strategy at that time. So it was not surprising that under post-communist conditions economic liberalism and political liberalism became two sides of the same coin (ibid.). Political liberalism became the presumed automatic outcome of installing a free market economy – but, as the example of China shows, this need not be the case (see Chapter 6).

One of the main drawbacks of the New Right ideology of Thatcherism and Reaganomics was the neglect of institutional reform, together with a clear Euro-centric perspective as far as communist regimes were concerned. The 'minimal state' was relentlessly advocated as *de rigueur* for a successful market economy. The role of the state and institutional reform have been raised by the European Bank for Reconstruction and Development's (EBRD) annual Transition Reports (available from: www.ebrd.com/pubs/econo/series/tr.htm), applying a host of indicators to assess 'progress' in implementing the rules laid down by the Washington Consensus. But what is the role of the state in a transition process that is meant to be driven by liberal market forces (nearly) alone? Is it merely a detached enabler of a liberal market, providing institutions and then standing back? One observation, made by Kolodko (2000a, b), is that transition economies need a more carefully designed set of anti-cyclical policy measures to avoid boom and bust economic cycles. In this way, government can pursue its role, as ultimately expected by most 'ordinary' people, to ensure robust economic growth and a fair distribution of its results. While this may not be following its textbook role envisaged under neo-liberalism, it has been a major consideration by the electorate when placing their votes. In effect, it was attempted to implement a stabilisation package in a market vacuum (see e.g. ibid., p. 24). But the crucial role of government involvement became increasingly obvious. In its absence, market failures are likely to prevail, and informal institution-alisation takes over, possibly leading to 'bandit capitalism' (Kolodko, 2000a, b). Liberalisation and privatisation alone cannot provide the conditions needed for the effective operation of a market economy. Stable legal frameworks, especially regard-ing property rights, are necessary (ibid., p. 74). It is at that point that differences between countries in their degree of implementing socialist practices become apparent. Those with a more liberal approach to socialism (e.g. Hungary's hybrid model; see e.g. McDonald, 1993) and thus some experience with the needs of 'markets' went through the learning process more quickly than those that continued to follow a more technocratic, dogmatic approach (e.g. Romania). So it may not come as a surprise that in 1990 Kornai suggests, for Hungary, a dual economy, split between state-run and private-sector businesses, to be the most likely outcome of post-communist transition and, indeed, for other countries, too (Kornai, 1990).

The hegemonic domination of economic liberalism meant that dissenting, or even cautious, voices were deemed out of step with the reality of events. Many of these concerns focused on the nature of privatisation, the very key plank of liberalisation as the motor of transition (Laski and Bhaduri, 1997). Against the dominance of the neo-liberal dogma at the time, those advocating a less ruthless, more gradualist, state-managed transfer of ownership were deemed to be fantasts, failing to learn their lessons from the communist experiment and hanging on to socialist dreams instead. Nevertheless, the discussions about gradualism or shock therapy for privatisation and transition, together with early observations about the nature of various privatisation processes, encouraged a somewhat more differentiated, less solely market-centred view. The importance of political processes, discussions and representation, rather than mere institutional structures, became increasingly visible as the true backbone of democratisation (Lavigne, 2000).

The absence of a public debate on the nature and direction of post-socialist development, and the instead externally driven establishing of a one-size-fits-all western model of state-societal and economic organisation, raised in itself obvious questions about democratisation, offering little scope for finding more locally based, specific avenues of post-socialist transformation. This, of course, contradicted the very notion of democratisation as a path to building civil society. In effect, it may be argued that the dogmatically advocated abstract liberalism of Western policy advice, even if unwittingly, thus contributed to the emergence of this situation (Bönker et al., 2002a, p. 16) of reinforced inequalities in the outcomes of transition. Events made obvious the much greater complexity of privatisation in the formerly socialist states than had been propagated by the western-based Thatcherite advocates of shock therapy. In particular, powerful interest groups at the interface of state (administration) and economy held more control of the process of asset transfer than had been expected in the rather more simplistic understanding underpinning the Washington Consensus. Thus, for instance, the state did not control all means of production, allowing a simple, clear-cut and immediate transfer of ownership from one hand to another. In reality, various forms of mixed economies existed as syntheses of state socialist and market principles, such as in Hungary and Poland. This meant a more difficult, differentiated picture of control and stakeholding of the means of production, with varying claims to their ownership, which often were overlapping and even contradictive. Thus, large socialist integrated industrial complexes, driven by the ideology of extending the visibility of an industrial proletariat, sat next to traditional forms of a rural peasantry deeply rooted in pre-socialist tradition (for example, in Poland), or next to small workshop-style businesses (as in Hungary). Against this background, by the late 1990s, a more differentiated picture had emerged, taking into account the institutional, political and structural dimensions of privatisation (ibid., p. 17), as evident on the ground.

The disappointment with neo-liberalism in Eastern Europe coincided with similar experiences in other parts of the world. The results of almost two decades of 'structural adjustment' in Latin America have been less than convincing, and its responses to the Asian crisis at the end of the 1990s did little to boost the IMF's reputation as an instigator of successful economic management. Comparing the international policy responses to economic development in those global regions with the ones in formerly socialist countries, will cast some wider light on the policy objectives of 'structural adjustment' in general, whether under post-socialism or not. 'The call for adjustment with transformation' (Bönker et al., 2002a, p. 20) and second-generation reforms emerged in the African and Latin American contexts, where conventional programmes had produced widespread dissatisfaction. It is somewhat surprising, therefore, that few of these considerations made it into the policies towards eastern Europe, but the perceived difference of their developmental state as 'Second World', and their Europeanness, may have contributed to the absence of immediately drawn parallels, and the adoption of a technocratic, ideology-driven approach instead. The economic decline following suit in Central and Eastern Europe was considered a necessary medicine to prepare the field for the propagated growth thereafter, and was seen as an inevitable trade-off of macroeconomic reforms (Bönker *et al.*, 2002a). Just as in the developing countries, income distribution became more unequal, and real wage levels also fell, at least in some parts (Carter and Maik, 1999). This affected, particularly, those regions and countries with less immediate economic advantages from transformation and westernisation, encouraging a backlash against the reforms and favouring a resurgence of reformed former communist parties. They were expected to possess a better understanding of the indigenous difficulties, and thus be in a better position to genuinely defend the relevant populations' interests. It may also be seen as an act of defiance, such as now observed in eastern Germany.

The evident weaknesses of the Washington Consensus, demonising the state as bloated and ineffective *per se*, while ignoring the importance of creating functioning institutions to create a pro-growth economic environment, led to the emergence of a 'New Development Paradigm' (Bönker *et al.*, 2002a, p. 20). This advocates 'modern-ising the state' rather than simply 'rolling back' the state as far as possible. The paradigmatic change developed gradually, with 'old' and 'new' understandings overlapping. Thus, for instance, the 1996 World Development Report, *From Plan to Market* (World Bank, 1996), presents both views in one document: in Part One it contains an orthodox, technocratically directed assessment of economic variables, including a list of performance indicators, and in Part Two it boasts concern with differences and particularities between and within states, suggesting the need for a more differentiated evaluation and policy-making, stressing the importance of long-term institution-building and a strong state. The late 1990s thus produced a more comprehensive, three-dimensional perspective of transformation processes, insisting much less on a 'rigid and dogmatic' one-size-fits-all standard approach. Nevertheless, policies were still clearly Eurocentric in focus.

Research into post-communist transformation needs to take the diversity and complexity of not just structures, but also, and in particular, processes, into account, and this seems to be increasingly the case. But it also needs to look beyond the post-Soviet, CEE perspective, even though that is the most visible, high-profile arena of change. Comparative studies at the wider geographic scale are few and far between, with most work concentrating on a particular global region, mainly the CEE countries, the FSU and China. Beyond that, work becomes thin on the ground. One of the few further-reaching comparative assessments is that by Dryzek and Holmes (2002), stretching from central/eastern Europe, via the former Soviet Union, to China, in an attempt to identify types of transition based on their progress. There is no presence, however, of either Africa or Latin America, which tend to be subsumed under 'Third World development' (Snookes, 1999), rather than versions of post-communist development where relevant. But the more recent perspectives tend to extend beyond the relatively narrow, economy-centric and technocratic argumentation between gradualism and shock therapy, so much loved in the early 1990s. With a generally growing awareness of the principal workings of global processes, limitations to liberalism have become more apparent and recognised. Global forces of free trade do not necessarily favour, and automatically create, 'free' western-style states and societies, however defined. Reality is much too imperfect to generate the outcomes predicted by simplistic, idealistic development models, such as those underpinning the Washington Consensus. There are also opportunities, such as for the Asian transition economies, of looking towards the 'tiger economies' for inspiration (Pomfret, 2001).

Experiences with post-communist development have highlighted the importance of the role of the state in the interplay with society and economy. Transition, so it became apparent, cannot be reduced to economic variables alone, and a mere formal shift from state control to market liberalism. Instead, it needs to be seen in a much more holistic sense, as 'economic society' (Linz and Stepan, 1996), where the economy is an integral, if very important, part of societal structures and processes, rather than a stand-alone feature. Particular attention needs therefore to be given to the interaction between market requirements, state capacity, the democratic openness of the political process and civil society activity. 'The interplay of these elements is the common point of reference for more complex explanations of diverging transition paths, and for different variants of political capitalism' (Bönker et al., 2002a, p. 26). They may distinguish between 'trailblazing', 'halting' or 'late developing' avenues of transition, as suggested by Dryzek and Holmes (2002). For instance, the application of the paradigm underpinning the Washington Consensus first and foremost to Russia/the Soviet Union, and only as a second thought also to eastern Europe, reflecting the somewhat naive, Moscow-centric evaluation of communist Central and Eastern Europe, with little awareness and appreciation of underlying differences between nation states, and between the former Soviet Republics and the CEE countries (see also here Chapters 4 and 5).

But in the absence of an awareness of such diversity, more readily available paradigms, based on conventional scientific (economic) approaches, were adopted, propagating and assuming just one 'right' model. This simplistic and rather prescriptive view encouraged:

> many economists and some political scientists [to feel] well-equipped to advise on political reform. In contrast, sociologists, geographers, and other social scientists, not to mention historians and philosophers, were perceived as not possessing knowledge particularly relevant to the practical problems of transformation.
>
> (Lavigne, 1999, p. 112)

The rationale behind that was, of course, that transformation was entirely, and exclusively, an economic challenge, revolving solely around establishing a liberal market economy, with all other aspects of post-communist transition following from there by default. Experiences since then, however, have triggered a learning process and a realisation that there are variations and, indeed, other factors, such as legacies (see Grabher and Stark, 1997), that matter. This has led to a gradual reinterpretation and redefinition of the nature and process of transformation. Increasingly, with the limitations of an economy-only interpretation more evident, area specialists from within the social sciences, drawing on experiences with transition to democracies elsewhere (southern Europe, Latin America; see e.g. Linz and Stepan, 1996), gained in importance within the debate. They challenged the econometric view and, from the mid-1990s onwards, helped to formulate a re-conceptualisation and new explanation of post-communist transformation (see Bunce, 1995; Karl and Schmitter, 1995).

Alternative models to the Washington Consensus: growth-led versus recession-led marketisation

The organisation of markets is the main difference between post-communist and other post-authoritarian countries. From a Washington Consensus point of view, transformation victory was seen as facilitating 'creative destruction', replacing the inherited structures and starting anew, with immediate growth following. But this implied an *immediate* and *complete* introduction of a fully operational market economy. The promise of immediate economic rewards for everyone, especially greater personal affluence and consumer choice, seemed remote in the early years of the 1990s, raising question marks over the salience of the *modus operandi* of the transition process. Not surprisingly, by the end of the initial period of enthusiasm among the population, the willingness to make sacrifices diminished, and positive results delivering the goods were expected with growing impatience. This was accompanied by clear signs of reform fatigue (Bönker *et al.*, 2002a, p. 14). There were attempts at different approaches to post-communist transformation, usually coupled with independence from global financial support, and thus attached strings. There, a less wholesale, more controlled approach to reform was attempted, largely leaving in place existing (inherited) political-economic structures, and merely amending them with some new policy measures and initiatives. China, followed by Cambodia, is the most prominent example of this carefully reformist approach, leaving much of the existing system in place, and just making it responsive to market principles, albeit in selected territories only.

Pei (1996) seeks to identify general patterns of transition by comparing the Soviet-centred and China-centred regions of transition. While the latter is discussed as exemplifying growth-led transformation (e.g. China, Vietnam), the former is described as recession-led (eastern Europe, Russia). Accordingly, the main difference between the two models of transition is the approach to the privatisation of the economy: an experimental, carefully staged approach in the Chinese model, and rapid replacement of the whole state sector in the Washington Consensus-based approach in eastern Europe/Russia. Thus, in the former, the existing structure stays in place, and private investment is added to that. Under the eastern European model, most, if not all, existing structures and capacities were restructured, which often meant disinvestment and huge losses in institutional capacity and knowledge. The result was economic collapse and an obvious necessity to rebuild the state, society and economy from scratch. This is a massive task by any measure. The initial experience of the affected population is thus not one of growth and development, but rather of problems, unemployment, and an atmosphere of crisis and deterioration. It is no surprise, therefore, that the reform politicians, with their propagated liberalism, were soon voted out of office, in some cases in favour of the former communists. In China and Vietnam, by contrast, private input was merely felt as an add-on, not a replacement of existing structures (see ibid.). This helped maintain the position of the ruling Communist Party, which continues to claim political control. There is no possibility of seeing people's assessment of the changes.

Not until more recently, have non-economic factors been considered in the transition process, especially the impact of institutional legacies and their relationship to the observed paths of economic transition. Relevant legacies include the 'completeness' with which communist structures and principles were established and implemented, in particular the degree of centralism, and thus the permitted existence of (semi-) private businesses (even if only small) outside the state-led economic apparatus, such as in Hungary. 'Such institutional differences, moreover, can be used to explain the different patterns of micro-level response to economic reform and its political consequences' (Pei, 1996, p. 132). China and Vietnam are of particular interest here, as they are the only transitional communist countries with continued economic growth throughout the transition period. The different approach to 'transition' and 'reform' is a key part of this phenomenon. The crux of the evident economic success thus seems to rest in maintaining the basic structures and capacities, while carefully changing the 'rules of the game' in favour of 'more market'. This way, existing institutional knowledge, infrastructure and networks can continue to function where appropriate, rather than everything having to be built up from scratch. Such gradual 'transition' is thus understood as reform, rather than revolution. This can be selective, as in China, restricting changes to just one dimension of the process, such as *economic* as against *political-governmental* conditions.

In addition, of course, the different legacies matter when embarking on one route of transformation or the other, and China and the former Soviet Union certainly have different legacies, especially the degree of industrialisation (McIntyre, 1992). Much of China's communist-era development effort focused on industrialising the rural economy, inspired by Soviet strategies, but, in contrast to there, policies allowed and supported smaller-scale, village-based projects. Thus, some decentralised decision-making skills were facilitated, contrasting with the Soviet Union's rigid centralism, which allowed few, if any, such skills to develop. This difference proved important for China when introducing market principles. There was at least some basic grasp of, and partial experience with, making decisions and judgements regarding opportunities. In China, therefore, economic development did not depend entirely on the performance of new, massive industrialisation schemes, but gained its capacity through the sum of dispersed small village enterprises, in combination with large (usually state-owned) firms. The main engine of economic growth thus emerged as rural industrialisation through developing businesses in towns and villages, providing a much broader and more 'advanced' starting point for economic transformation towards marketisation than was the case with the Soviet system.

Given these differences, but also the unreliability of communist-era economic figures, Pei (1996), emphasises the need to look beyond the debated arguments for or against shock therapy and gradualism, and instead to examine the particular economic and political circumstances of each country (or region).

Much of the disappointment about the lack of instant progress and success after the change of the system was also due to the overoptimistic, unrealistic expectations in the private sector's response. This deflation of expectations affected, in particular, the shock therapy approach, where state-owned firms were believed to be responding quickly, released of their 'shackles'. But that did not happen, owing to a number of impediments such as lack of experience or understanding of market mechanisms, long-established ways of non-market thinking, and a close interconnection between political and economic elites. It is not surprising then, that countries that had experimented with modernising socialist economies through the injection of some form of private enterprise, such as Hungary (Bozóki et al., 1992), provided a more fertile bedrock for post-communist economic development than those that clung to dogma.

The main difference is the form of institutionalisation of state-socialism, that is, its thoroughness and 'purity'. The highest concentration of state power, and thus the lowest scope for entrepreneurialism, can be found in the more orthodox communist states, whereas the less orthodox systems show somewhat greater scope for entrepreneurial response. These differences in legacy seem to reflect on the speed with which market principles could establish themselves in practice, permitting individual entrepreneurialism or relying more on formally imposed arrangements. In those cases, where there was some form of a mixed economy, enterprises seem to have developed the necessary skills in circumnavigating governmental red tape and administrative obstructionism, whether intended or not. This is also a useful skill under market conditions. Other key differences affecting the shape of post-communist transition processes are varying mechanisms to deal with property rights and social welfare, and thus the way of democratising the state-run system. The two types of transformation processes – recession-led transformation (Soviet-centred) and growth-led transformation (China-centred) – also seem to have produced different elites. While the former facilitated a new business elite, often through insider-informed privatisation schemes, allowing state apparatchiks to turn into entrepreneurs, the latter system used a more decentralised system of 'entrepreneurialisation', although there, too, close links between government officials and entrepreneurs exist. The lack of an effective state, providing clear rules and supervision for the privatisation process, was one of the main reasons for the quite particular way this process unwound in the former Soviet Union. How these differences impact on the development of democratisation and the emergence and operation of a civil society, and thus a 'successful' transition by the criteria of the Washington Consensus, remains to be seen.

Conclusion

The notion of 'transition' has gained enormously in currency in public discourse over the last 15 years, since the high-profile end of communism. Yet it has not become much clearer in its meaning. This chapter has sought to outline some of the key arguments revolving around 'transition', whether seen as prescriptive in relation to the Washigton Consensus or, in its strategy, as either 'growth led' or 'recession led'. In this context, has post-communist transition been a particular version of post-authoritarian democratisation à la Southern Europe or Latin America of the late 1970s/early 80s? What has become evident is the dominance – by far – of the economic dimension, that is liberalisation and marketisation, in tandem with democratisation as 'automatic' outcome. While there has been a clear hegemony of the Washington Consensus, its appropriateness to the conditions of post-communist countries has been less clear. The costs, especially social (unemployment) and investive (loss of productive installations) have been enormous, and the subsequently rapidly deepening inequality between the 'winners' and 'losers' of the regime change is becoming increasingly problematic, and could become destabilising if not addressed. 'Democracy' and 'market' will look less desirable if living conditions turn out to be harsher than under the previous regime; pragmatism will surpass wider philosophical notions of 'freedom' and 'legitimation of government', if life is a struggle for survival.

The outcome may be different forms of post-communist regimes than are presumed by the comfortable western advisers who descended from their institutional worlds on the late communist countries to turn models and theories into reality – that is, forming reality according to theory. The last 15 years have demonstrated that such linear strategies and predictions are not necessarily matched by developments 'on the ground', and there is a growing awareness of new forms of regimes emerging that fuse the particular legacies of the communist countries, in all their underlying diversity. This means that, yes, there has been a shift away from the old communist regimes, but whether it is a 'transition', in the sense of a clearly defined route towards a visible destination, is not so clear. Different starting points and different destinations emerge, with the journeys to them following a variety of routes at different paces. This diversity, and the inherent 'fuzziness' of the notion of 'transition', thus suggests, at least, the use of the plural 'transitions', as there are so many variations in their nature. It will take time for the likely shapes of their destinations to become a little clearer.

4

MARKETISATION, DEMOCRATISATION AND INEQUALITY IN CENTRAL AND EASTERN EUROPE

Craig Young

Introduction

This chapter provides an overview of the nature of post-socialist (post-communist) change in Central and Eastern Europe (CEE). It sketches out the key characteristics of this change, particularly drawing attention to the marked differentiation within the region. Post-communist change in CEE has been distinctive compared to that in the other global regions of post-socialism considered in this book, although there are perhaps some similarities with the nature of change in parts of the former Soviet Union.

What characterises post-communist change in this region is the strong influence of external actors, ideologies (especially the Washington Consensus; see Chapter 3) and processes on the nature of change. The focus on the former Eastern Europe and FSU made the Western European experiences obvious models for transition and the moral appeal of the market as the basis for freedom and prosperity enhanced its appeal in CEE. The Washington Consensus model and the activities of international economic organisations and consultants shaping aid and trade policies has implanted the notion of the 'correct' path to development being the adoption of market liberalism. This has been further enshrined in the process of accession to the EU for some CEE states. Domestic politics across the region emphasised the need to become accepted members of the international community by joining organisations such as the EU, NATO and the OSCE and to gain international political recognition of sovereignty and economic credibility. The characteristics of post-communist change here are thus strongly marked by the region's particular relationship with Europe, both in terms of geographical proximity and historical and cultural links with 'Europe' as an idea, and also the existence of neighbouring 'models' of democracy and statehood. Thus there have been powerful political projects to 'Europeanise' CEE states. The most focused form of this Europeanisation has been the process of accession to the EU, which was achieved by eight of the CEE states in 2004. This process exerted a powerful external force on these nations to conform to a particular idea and form of European statehood. This process has also impacted more widely within the region, including in the next CEE accession states (Bulgaria and Romania in 2007) and future potential applicants, most notably Croatia in the short term. However, even for those CEE states not currently part of the accession process 'Europeanisation' has been important in attempts at democratisation and creating a new national image.

However, this chapter will be concerned with examining how the playing out of these powerful external forces, which have often been internalised through the policies of post-communist governments, has produced a complex range of changes within CEE. Post-communist change in the region has not been a simplistic, linear 'transition to capitalism', as was suggested by approaches such as the Washington Consensus. Indeed, some processes are producing a fragmented pattern of development within the region. EU accession has created a new divide within the region, with non-EU CEE states pursuing a range of other approaches to regional integration and development. States that are more geographically remote from Europe but nearer to the Russian Federation or the Mediterranean countries have pursued different policies to international integration, albeit with an eye on developments in Europe. The marketisation of the economies of these countries has proceeded unevenly with processes central to the Washington Consensus view, such as foreign direct investment and privatisation (see Chapter 3), actually producing highly uneven geographies of economic change. Domestic economic and political strategies intertwine with these external forces at a range of scales to produce great variation within CEE. In addition, there is the key process that the Washington Consensus and related approaches to post-communist change really underestimated: the legacies of state socialism in the region and their impact on economic and political processes. Rather than just being swept away at the 'end of history' to leave a *tabula rasa* upon which Western models of democracy and capitalism could be instantly established, the networks and socio-economic relations of state socialism have continued to impact on post-communist change in the region.

Thus although there are certain characteristics that may mark out the nature of change in CEE compared to the other global regions considered in this book, great care must be exercised when discussing the impact of these processes on the nature of change in the region. Rather than a simplistic 'transition to capitalism' in the region, post-communist change has been marked by highly uneven and fragmented patterns of development, which are strongly influenced by the legacies of communism in the region. This chapter thus outlines some of the key characteristics of change in CEE to identify the nature of post-communist change in the region. After briefly reconsidering the key policies that were developed to shape change in the region post-1989, the chapter then illustrates the resulting geographical complexity in key areas. Patterns of economic change are examined at a range of geographical scales to outline the production of a fragmented space-economy as a result of the marketisation of centrally planned economies. 'Europeanisation' as an important process in redefining the nation state, and in particular the EU accession process, is then outlined, an analysis that also brings out the contested nature of this process.

The process of democratisation across the region is then analysed to illustrate the complex political development paths that have emerged since 1989.

The fall of communism in Central and Eastern Europe

The reasons for the end of Communist Party rule in the former Eastern Europe in 1989, and in the USSR in 1991, are complex and long term. Although the end of communist rule is often portrayed as unexpected, in fact pressure for change had existed within the communist system since the 1970s and had been predicted within Eastern Europe itself in some quarters. Communist-shaped societies and economies, such as those in CEE countries, contained tensions and contradictions that ultimately produced their own downfall. Poor economic performance in combination with a series of social and cultural changes eventually led to the political fall of these regimes. A key issue was that the Communist regimes failed to solve the issue of the need to legitimate one-party rule among their populations. Discontent found various outlets, such as culture, religion and rock music, and provoked tensions surrounding ethnicity and class (Ramet, 1995). Over time, communist systems were unable to create consensus. Recurring demands for democracy indicated a failure of the policies of socialisation and a failure of efforts to inculcate the values of communism in the population.

Ramet (1995) provides a summary of the key processes that underlay the eventual fall of Eastern European state socialism (though these varied from country to country). First, long-term economic deterioration was a factor also related to the Marxist–Leninist systems' declining legitimacy. The rapid economic growth demonstrated in the 1950s and 1960s was replaced in the late 1970s by economic decline. Increased borrowing from the Western capitalist nations in the 1970s led to problems with servicing debt repayments, which in 1987 took between 20 and 67 per cent of annual export earnings as interest rates rose rapidly. The average citizen thus faced harder working conditions, lower real wages, shortages of consumer goods (as these were marketed abroad), and energy shortages to support the paying off of national debts (Romania in particular suffered from harsh austerity measures). This led to an increase in strikes in key industrial sectors and also contributed to the sense of alienation from the communist regime. In response, new groups emerged in the political system that proved to be influential in political change. The propaganda of the workers' democracy gave the working class a sense of political entitlement (for example, the Solidarity trade union movement in Poland). Urbanisation, secularisation, education and changing sexual ethics contributed to new movements among young people, women, and religious groups. These processes were strengthened by the defection of intellectuals. Academics, writers, artists and other intellectuals were, to varying degrees and despite censorship, able to voice dissent with the regime, in some cases forging alliances with workers' movements, urging political and economic reform.

These factors were exacerbated by state inefficiency and corruption, which added to the loss of state credibility among the population. Corruption, which the bulk of the population engaged with, became a necessary part of life. In several countries in the 1980s, the authorities admitted corruption within government, but the attempts to deal with it further heightened the regimes' loss of credibility. Despite the Party's control of the media and production of propaganda, surveys in the 1980s revealed a high degree of disaffection with the regimes among the public. Control by Communist parties was further weakened by factionalisation among the political elite, e.g. between reform-minded and other Communists. In the 1980s, several Communist parties admitted to past mistakes and harsh policies, which shook their credibility and self-confidence. These problems for Communist rule were not helped by the often inept use of force by the state, as the 1980s saw many instances of the use of force by the state police and/or armed forces to maintain the communist regime. All of these factors contributed to a longer-term expectancy of change. In the late 1980s, most of Eastern Europe witnessed increased levels of belief in the possibility of change, and the continuation of strikes (as in Poland in 1988) demonstrated that people were still prepared to demand it.

A key process in this longer-term demise of state socialism in Eastern Europe was the attempts by the Communist parties to solve the problem of poor economic performance without major internal structural reforms. The experience of each country varied, but a general pattern was that in the 1970s the economic perform- ance of Eastern European societies began to fall, especially relative to Western capitalist economies, and by the late 1980s they were experiencing economic crisis. Individual countries were failing to meet the production targets specified in five- year plans. In Czechoslovakia, for example, 31 per cent of enterprises failed to meet their production targets in 1987 (Ramet, 1995). Qualitative deficiencies in produc- tion were also an issue, such as the poor quality of goods, which limited sales on domestic and foreign markets. The standard of living within countries was affected as export markets became increasingly important for generating hard currency. The response of most communist regimes to falling industrial output in the 1970s was to borrow capital from the Western industrialised nations to purchase new technologies to support modernisation and exports to the West (although Romania and Czechoslovakia pursued more isolationist policies, which in the long term produced technological decay and dependence on the West).

However, the resulting debt servicing caused further problems for these regimes. Their goods lacked competitiveness in export markets, adding to the problems of debt servicing that increasingly took a large proportion of what export earnings were generated. Policies were introduced to reduce imports and introduce austerity programmes to service debt. Shortages of consumer goods eroded confidence in the political system run by the Communist Party. This represented a double crisis for the state, in production and consumption, especially given the ideological and propaganda aims of socialism. The resulting material hardships further undermined the legitimacy of Communist rule.

Managing the economic problems inherent in centrally planned economies through external borrowing introduced a fundamental shift in the position of these economies with respect to the rest of the world economy. Despite their ideological intentions, Eastern bloc economies were opened to Western capital. Borrowing from Western economies may have offered a solution, but it resulted in these economies becoming intertwined with rapidly changing global capitalism at a particular time in its evolution (Verdery, 1996). Borrowing and debt servicing increased communism's interlinkage with increasingly globalised and newly flexible capitalist systems. Marxist–Leninist economies thus had to cope with becoming increasingly integrated with global capitalist systems at the very time that those global systems were undergoing fundamental change.

These changes exposed the limitations of communist economies. In addition, this favoured the reformist elements in Communist parties arguing for structural reforms in state socialism. Gorbachev's efforts at reforming the USSR from 1985 through *perestroika* ('restructuring') and *glasnost* ('openness') also created political space for attempts at structural change (see Chapter 5). An initial response was the formation of 'political capitalism', a dual, hybrid economy partly state-controlled and partly open to capitalism. Over a long period of time, these changes in attitude within the Communist Party combined with the social, cultural and political changes discussed above to produce a profoundly destabilised political structure which ended spectacularly in 1989–91.

While the end of state socialism in CEE and the FSU may be a clearly identi-fiable historical event, what has happened subsequently is less clear. As this chapter will go on to discuss, post-communist transformation in CEE is an ongoing, complex process. Some key processes that are common to the region can be identified as a starting point (following Light and Phinnemore, 1998; Sakwa, 1999). A common characteristic was the end of Communist Party dominance over politics, economics and society (although the variation in the nature of state socialism in each country must not be forgotten), and subsequent attempts at democratisation, that is, the establishment of a multi-party democracy and civil society. However, establishing a genuine multi-party system, and the creation of a society with clearly understood rights for, and responsibilities of, citizens, has proved difficult and there is great variety in the kinds of states that have emerged. A radical reorientation of foreign policy also marks the post-communist transformation, especially with regard to membership of Western organisations for economic (EU) and security (NATO, OSCE) reasons. A pervasive belief has also been established that the only possible option for development is reintegration into the global economy. This has been accompanied by attempts to introduce market economies to replace centrally planned economies, although this should be thought of as a process of uneven marketisation of these economies rather than a simple replacement of communist economies. The key elements of market economies that have been introduced (albeit unevenly) are:

- the transfer of economic units (state owned factories and collectivised farms, land and buildings) from state ownership to private ownership through privat-isation programmes;
- the abolishment of fixed prices and exchange rates of currencies to allow the market to determine prices (price liberalisation) and exchange rates (currency convertibility);

- and attempts to attract private investment, particularly foreign investors, to inject capital and 'know-how' into post-communist economies.

These economic changes have been accompanied by rapid changes in the class structure and the employment structure and increased social polarisation and poverty. There has also been rapid change in the area of identity politics, including the rise of nationalism and ethnic tension, and a complex working out of a multitude of cultural processes including a contested relationship with the communist and pre-communist past. Indeed a key feature of transformation are the strong institutional, cultural and social legacies of the communist period in the post-communist order. In elite politics, for example, many former Communist politicians continue to hold power. There are many new institutions and practices as a part of post-communism, but they are often hybrids of old and new practices, personnel and ways of operating. Continuity from the communist period is a key characteristic.

The 'Washington Consensus' and discourses of post-communist change in Central and Eastern Europe

The main development discourse guiding the general processes outlined above was that advanced in the name of the 'Washington Consensus' and applied in different ways by the governments of most of the newly emergent states in CEE post-1989. The Washington Consensus has already been described in Chapter 3.

The early conceptualisations and applications of the Washington Consensus, which very much revolved around the four credos of liberalisation, stabilisation, privatisation and internationalisation, have received numerous criticisms, pointing *inter alia* to the importance of regulative institutional structures (e.g. Smith and Pickles, 1998). Crucial elements were neglected in early versions of the Washington Consensus, notably the redesign of the state, institution-building and the improvement of corporate governance of the state sector prior to privatisation (Kolodko, 1999). As Bradshaw and Stenning (2004, p. 14) summarise, 'without the supporting institutional infrastructure, such as a functioning fiscal and legal system, markets cannot function. Marketisation itself cannot generate these institutions; the state must create the infrastructure necessary for the market to function.'

This recognition of the need for the appropriate development of the state and associated institutions has become more accepted over time and has been incorporated into the 'post-Washington Consensus' (see Kolodko, 1999; Lavigne, 2000). The need for liberal markets and open economies persists, but the role of the state and market organisations and the links between them for sustained economic growth is more acknowledged as, it is claimed, are the social costs and responsibilities of economic growth (Kolodko, 1999). However, the post-Washington Consensus continues to reproduce discourses prioritising the implementation of 'Western capitalism' as the solution for CEE. The sections below go on to outline the development of CEE under these policy prescriptions, contrasting countries' experiences of post-communist change with the simplistic view of a 'transition to capitalism' advanced under the (post-)Washington Consensus.

The 'Europeanisation' of Central and Eastern Europe

A key characteristic of post-communist change in CEE, which distinguishes it from the other global post-communist regions, is the drive towards the 'Europeanisation' of CEE states. This relates perhaps most obviously to those CEE states that have gone through the process of accession to EU membership. However, even for the other CEE states, which may or may not immediately aspire to EU membership, other discourses of 'Europe' have been significant. As mentioned above, 'Europeanisation' refers to processes that are in a variety of ways generated externally to these nations (e.g. as a model that aspirant 'European' nations should adhere to) but also internally (e.g. as a part of the formation of national identities that draws on notions of a 'return to Europe'; e.g. see Âgh, 1998). This section considers 'Europeanisation' in two ways. First, the formal process of EU accession is outlined to show how it is an important, externally imposed process focused on creating states that conform to 'Western' concepts of statehood. Second, the importance of the idea of 'Europe' is explored more generally, and other discourses are also considered.

The fall of Communism in Eastern Europe in 1989 was greeted with great optimism in Western Europe. A great deal of rhetoric was expounded by Europe-wide and international organisations about the need to incorporate the newly independent CEE states into Europe. Early EU initiatives included trade agreements (the 'Association Agreements' and subsequently the 'Europe Agreements'), aid (such as the PHARE programme) and involvement in financial institutions (e.g. the European Bank for Reconstruction and Development (EBRD)). However, the efforts of the EU in the first years of transition were rather uncoordinated and lacked a comprehensive approach to what to do with CEE, and certainly fell short of a quick entry to the EU (Gibb and Michalak, 1993; Mayhew, 1998). The process was further stalled by objections from existing EU members to enlargement. However, beginning with its 'Agenda 2000', the EU began to identify a more comprehensive approach to tackling the possibility of enlargement into CEE. In 1993, the European Council meeting in Copenhagen identified three criteria against which the progress of CEE countries towards accession to the EU could be measured. These criteria, known as the 'Copenhagen Criteria', covered three key areas in which prospective member states had to comply with definitions of statehood as defined by the EU:

- the creation of a functioning market economy that could compete within the EU;
- the development of democratic institutions incorporating recognised legal systems and respect for human rights;
- the capability to adopt the *acquis communautaire*, i.e. the ability to assume the obligations of membership of the EU.

Accession Partnerships were established between the EU and applicant CEE nations to negotiate and implement the changes required to achieve EU membership through 'negotiating chapters'. Applicant countries have to produce Regular Reports that lay out how the requirements of the Accession Partnerships are progressing relative to the Copenhagen Criteria. This approach to integrating the CEE countries into 'Europe' clearly follows the Washington Consensus view of post-communist change as a process of linear change from communism to the favoured end-point of neo-liberal capitalism, European style. Those CEE states involved in the process were in a new context, as the process relied on these states driving through these political conditions internally (Pridham, 2002). The outcome for states that achieved membership was a dramatic realigning of their economic and political goals and institutional infrastructure to EU demands (which were very much inspired by a neo-liberal agenda) (see Vintrová, 2004).

In May 2004 eight CEE states joined the EU: Estonia, Lithuania, Latvia, Poland, the Czech Republic, Hungary, Slovakia and Slovenia – the so-called 'accession eight' Two other states – Bulgaria and Romania – were judged not to have made sufficient progress and were set likely accession dates of 2007. In 2005 negotiations have continued over the prospect of Croatia being the next CEE state to be invited to join the accession process. How other CEE states who have as yet not engaged with the accession process will be drawn in is not clear, but Ash (2005) provocatively suggests that the October 2005 offer to Turkey to engage in accession talks, however far away membership for Turkey might be, has decisively shifted the map of future accession and will result in much more pressure for membership from those 'European' states lying between the EU and Turkey. The outcome of such developments remains to be seen, but the prospect of further EU accessions among the CEE states holds out the prospect of further post-communist states in CEE being driven along this development path, with even countries such as Ukraine having paid some attention to relations with the EU and possible EU membership (Protsyk, 2003).

EU accession is thus a powerful process in defining post-communist development paths in CEE, and one that does much to distinguish the nature of post-communist change in this region from that in other areas of transition. However, this is not to portray it as a monolithic, totalising process. Support for joining the EU also varied between the candidate countries during accession. Although care must be taken when interpreting Eurobarometer results, polls taken during the accession process demonstrated that while overall 61 per cent of people in the accession-eight countries thought that EU membership was a 'good thing', this varied from 78 per cent in Romania to 32 per cent in Estonia. During the accession process, the support of the Polish public for accession remained high but did decrease, a pattern that Szczerbiak (2001) interpreted as suggesting that Poles had become cynical about the accession process and consented to the idea of membership rather than being enthusiastic about it. Further, qualitative studies of public perception and attitudes revealed considerable national differences, and that even those who felt accession was 'a good thing' worried about the process, particularly on the basis of their relative economic underdevelopment (Kucia, 1999; EC 2001). While accession maintains a strong influence over economic development trajectories, the influence of the EU on sovereignty, citizenship and national identity in CEE has varied. The

EU's accession negotiations with Slovakia, for example over the treatment of minorities, progressed very differently from the negotiations with Romania. Romania's persistent economic problems and a weak state capacity hampered its ability to harness the EU's political considerations over minorities as a mechanism for democratic consolidation (Pridham, 2002). In the event, the populations of the new member countries voted in support of accession, but this does not signal total acceptance of membership.

Despite EU efforts to Europeanise national identity in CEE, to construe the nation state as part of an evolving supra-national community (Banchoff, 1999), notions of citizenship and identity at different scales are contested in the relationship between national and 'European' identity. Even within the 'old' EU countries the notion of a 'European' identity is variable and not generally considered to supersede national identity. EU policy also celebrates diversity and reinforces national and sub-national identities. Throughout CEE, there is considerable national and sub-national variation in the extent to which citizens see themselves as 'Europeans', and this source of identity is secondary to national or regional identities (Kucia 1999; Pridham 2002; Šabič and Brglez, 2002). Different occupational groups or age groups, for example, see the future in Europe very differently. Opposition to EU membership can be part of an overtly nationalist stance. In the early years of transformation, Vladimir Tudor's far-right Greater Romania Party was able to mobilise support through an explicitly anti-EU campaign that was linked to powerful notions of Romanian nationalism. In recent years, this support has been interpreted more as a form of protest vote, and the party has developed a more open stance on the EU as it attempts to remain in mainstream politics. Slovenia offers another example of a country in which different political parties' discourses on Europe and national identity offer contested versions of the role of 'Europeanisation' in national identity formation, with many of them advocating the development of a Slovenian identity as a part of a European identity and culture (Šabič and Brglez, 2002).

Further geographical variation is produced by those countries that currently lie well outside the accession process politically and geographically, such as Belarus, Moldova and Ukraine (although following reforms in 2005, Poland is now actively advocating that Ukraine be allowed to start the accession process). The populations of these countries exhibit a general sense of 'Europeanness' in terms of their geographical location, but less clarity over what it means to be 'European'; few people think of themselves specifically as 'Europeans' but possess a stronger orientation towards Russia as an important future influence (White *et al.*, 2002).

For both new-EU and non-EU CEE states, a further aspect of 'Europeanisation' has been the production of discourses centred on the symbolic construction of belonging to Europe as a geopolitical space (Kostovicova and Young, 2003). EU accession was shaped by the invention of the spatial narratives about the 'new Europe' and the accession countries' central place in it (Moisio, 2002). This symbolic spatial repositioning within Europe emphasised the European character of their national political, historical and cultural heritage. One element that was drawn on in this process was the use of maps of the post-communist states that, through their design, challenged the existence of Eastern Europe as a geopolitical concept by offering a visual re-regionalisation of Europe. Hence, to correct the particular Cold War geo-vision that had been constructed for them, the post-1991 Baltic states are presented at the centre of northern Europe, or in the case of the Czech Republic in the geographical centre of the European continent (Ziegler, 2002; Kostovicova and Young, 2003). Such cartographic representations can be commonly seen in many official discourses produced about the state throughout CEE, for example on the websites of government inward investment agencies seeking to market the CEE nations as suitable sites for foreign direct investment.

After the fall of socialism, the idea of the 'return to Europe' was primarily a territorial policy presented in national terms, as a solution that was in the national interest of the prospective accession countries. The policies of the ruling elites towards 'a return to Europe' were premised on the redefinition of the nation's spatiality – the nation's place was in Europe, and its politics was a European politics (Kostovicova and Young, 2003). Representations of Europe have moulded the political dynamics between the post-communist countries and Europe. As Ágh (1999, p. 266) suggests, Western Europe has developed an extensive 'European political architecture' since 1945 and:

> the basic regulations of these institutions have become mandatory for all European states, and their provisions, vital for democratisation, have actually been 'forced on' … CEE … as preconditions for their own 'Europeanisation'. The historical events of the nineties have witnessed the increasing … activity of all-European institutions as quasi supreme organisations in the … region.

However, Batt (2001) has pointed out the discrepancy between the idealisation of Europe prior to the fall of state socialism and the reality of the actual negotiations with the EU and its perceived heavy-handedness. In reality, the process has been an interactive one in which national and sub-national actors in CEE react to and shape the integration process (both in the accession states and the 'old' EU states), and thus domestic pressures and national frameworks also play a role in shaping post-communist change in these countries (Friis and Murphy, 1999; Ferry, 2003). In these processes, 'Europe' has been understood in different ways – as a geographic entity, as an experience, and as an institution (i.e. as the EU) (Paasi, 2001).

An appreciation of 'Europe' as discourse has been, and is, important in the CEE countries, where the image of Europe was vital in the reconstruction of national, and the construction of European, identities (Kostovicova and Young, 2003). Observers note that rather than conceiving of the enlarged Europe as seeking to be a Westphalian federal 'superstate' with clear-cut borders and a given territory, perhaps thinking of it as a 'commonwealth' (Ash, 2005) or a 'neo-medieval European empire' with overlapping authorities and multiple cultural identities (Zielonka, 2001) would provide a clearer understanding of how the EU's eastward expansion will progress. As negotiations continue over potential further EU enlargement, and the future of non-EU CEE countries is discussed, it is likely that it will remain a powerful idea in the future development of post-communist CEE. However much the EU accession process has shaped the economic, political and social development of those countries, it is not a totally homogenising process and national imperatives and domestic policies have still impacted on the development of CEE.

The fragmentation of economic space in post-communist Central and Eastern Europe

The fall of state socialism in CEE in 1989 was accompanied by a widely held belief, certainly among political and economic elites, that the introduction of capitalism, particularly along the lines promoted by the Washington Consensus, would achieve a rapid convergence between the economies of the former communist countries and those of Western Europe. It did not take many years for it to become apparent that economic transition in the region was a much more complex process than these rather optimistic policy prescriptions had suggested. Economic change in the first decade of transition, and in some countries still today, has been a much more protracted and complex transformation. This section outlines the key characteristics of economic change in the region to bring out this complexity.

The initial years of post-communist transformation in CEE were characterised by region-wide economic recession, particularly associated with de-industrialisation (detailed by Smith, 1997; Kolodko, 1999; Sokol, 2001; Dunford and Smith, 2004) characterised by Kornai (1994) as a 'transformational recession'. Smith (1997, pp. 334–5) summarises the negative impacts of this recession by commentating that 'the experience of "transition" [up to 1995–6] has been one of economic collapse, labour shedding, rationalisation, an onslaught on labour, and social and political disorientation [accompanied by] a deep-seated social and psychological crisis', a situation that Sokol (2001) illustrated not to have changed a great deal by 2000. It is difficult, for a number of reasons, to analyse meaningful changes in even basic economic indicators such as gross domestic product (GDP) for this time period. However, Dunford and Smith (2004, pp. 43–5) outline the dramatic economic decline experienced throughout the region from 1989. Table 4.1 gives some indication of the dramatic downward trends in GDP and industrial output in the early years of the transition, but also that most countries had achieved positive growth in industrial output by 2000. However, it has taken at least until 2000 for a significant number of CEE countries to regain gross national product (GNP) levels equivalent to those achieved in 1988, and (though the calculations are based on particular assumptions) a number of countries will take even longer to achieve net gain. That these countries, on the basis of this calculation, include the Czech Republic and Poland (forecast as achieving net gain by 2006–7), which were admitted to the EU in 2004, gives an indication of how deeply this recession has impacted on the CEE states. Romania, whose EU accession date will be 2007, is estimated not to achieve a net gain in GNP until 2025 (the detailed calculations are presented by Dunford and Smith, 2004, pp. 44–5). This data suggests an immediate qualification of the optimistic forecasts of the Washington Consensus regarding economic transition in the region.

Table 4.1 GDP per capita, GDP change and change in industrial output in East and Central Europe, 1990–2002

Region/country	Percentage change in GDP 1990–1991	Percentage change in industrial production 1990–1991	Percentage change in industrial production 1999–2000	GDP per capita in US$ 2002
Baltic Europe				
Estonia	−13.6	−9.5	9.1	12,260
Latvia	−10.4	−2.1	3.2	9,210
Lithuania	−5.7	−26.4	−9.9	10,320
Central Europe				
Czech Republic	−11.5	−21.2	5.7	15,780
Hungary	−11.9	−16.6	18.2	13,400
Poland	−7.0	−8.0	4.3	10,560
Slovakia	−14.6	−19.4	9.1	12,840
Slovenia	−8.9	−12.4	6.2	18,540
Eastern Borderlands				
Ukraine	−8.7	−4.8	12.9	4,870
Balkan Europe				
Bulgaria	−11.7	−20.2	2.3	7,130
Croatia	−21.1	−28.5	1.7	10,240
Romania	−12.9	−22.8	8.7	6,560

Source: Business Central Europe online: www.bcemag.com/statsdb; UNDP, 2004

The depth and extent of this initial recession surprised those who had adopted the rhetoric and policy prescriptions of the Washington Consensus. Kolodko (1999) identifies that a crucial assumption of the policy prescription offered – that is, that the privatisation of property and transfer of the allocative mechanism from the state to the free market would enhance savings rates, capital formation and allocative efficiency – was flawed. This early experience of economic transition highlighted how the lack of mature institutional structures impacted on post-communist economic change in CEE. Rather than the simplistic introduction of Western models of capitalism, or the naive assumption that there was one version of capitalism that could be established, what emerged in CEE was a complex mix of hybrid economic arrangements that synthesised in complex ways the new influences of the emergent globalised, flexible capitalism with the legacies of communist systems. The emerging economic relations were in fact deeply 'embedded' in the pre-existing socio-economic relations and networks of communism (Smith, 1997).

Thus despite the efforts of political and economic elites to establish a new neo-liberal order, 'the reality is that there are a variety of strategies pursued at local, national and cross-national scales to consolidate particular regulatory dynamics, none of which is becoming dominant. The results are divergent realities throughout the region and a mixing of "old" and "new"' (ibid., p. 332) in patterns of profoundly geographically and socially uneven development. There has been a fragmentation of political and economic spaces, which are currently being subjected to intense processes of re-regionalisation (see Dingsdale, 1999). As Smith and Pickles (1998) note, marked regional differentiation is emerging in CEE between sets of regional economies experiencing dramatic restructuring and globalised growth related to their ability to mobilise their global connectedness particularly through their ability to attract foreign direct investment (FDI), and sets of marginalised economies increasingly left behind in the process of capitalist restructuring.

This diversity of local responses in post-communist transformation arises out of the inter-relation of previous sets of socio-economic relations with new forms of regulation and accumulation, rendering transformation a path-dependent process (ibid.). Thus instead of the straightforward implantation of a neo-liberal model of capitalism, the new forms of economic activity that emerged in the early years of transition were what Smith (1997, p. 336) terms 'speculative economies', that is, 'a diverse set of semi-formalised, institutionalised structures through which capital is recycled through semi-legal and legal activities revolving around the basic tenet of "fast money"'. Smith (1997) thus identifies the emergence of new core-periphery patterns of economic development and one type of regional development that is characterised by the end of the 'old':

- regional industrial structures, based on large state-run enterprises that dominated local economies, have suffered economic decline and closure because of 'the intersection of "old" local dependencies with the "new" law of value ... ' (ibid., p. 336);

and four 'new' major forms of 'speculative' economic activity:

- formalised capital and money markets that have particularly provided a dynamic for metropolitan growth but may be poorly embedded in the CEE economies and unsustainable in the long term;
- 'political' capitalism, that is, a hybrid economic form comprising new productive relations based on an intersection of existing socio-economic networks of production relations and new dynamics of power;
- 'kiosk economies' and new consumption spaces;
- the rise of mafia, protection and illegal speculation – these are partially linked to the inheritance of social relations and paternalism from communist times and examples from within Western capitalism (see also Dunford and Smith, 2004).

The emphasis is thus on the variety of forms of 'capitalism' that have emerged in CEE and on the complex ways that new influences (policies, investment, capital, enterprise strategies and key economic actors) interact with the pre-existing socio-economic relations of economic activity under state socialism to produce new hybrid or recombinant forms of economic activity. Underlying the production of this differentiated and uneven geography of hybrid economic forms is an alternative transition model that questions the neo-liberal model that proposes a straightforward change from state socialism to liberal capitalism. Sokol (2001) instead proposes a 'vicious circle' scenario in which economic decline, regional fragmentation, socio-political polarisation and instability, and crime and corruption underpin a move to forms of 'hybrid' and 'bandit' capitalism. The result is a 'deadlock of transition' (Sokol, 2001), with no way back to communism and liberal-capitalism a distant reality, which demonstrates that economic and political transformations do not necessarily support each other but can in fact undermine development.

Ironically, one of the key examples of how new policy prescriptions are combining in complex ways with the legacies of communism and even pre-communist development patterns to produce hybrid and diverse forms of capitalism, is the impact of FDI on economic change. FDI is seen as a key element in the Washington Consensus prescription for development. It is seen as a key source of capital that is missing in domestic markets to provide the investment necessary to achieve restructuring, and in particular it has been highly significant in privatisation processes. However, although FDI has proved important in the restructuring of some regions, it has also driven the increasing fragmentation of the space-economy. FDI has been sectorally biased in terms of where investment has been focused. It has also been geographically biased, focusing mainly on those countries closest to the EU and the CEE accession states and the major metropolitan areas or regions in the west of CEE countries that previously bordered the EU (Turnock, 2001). It has also been far from clear whether the expected benefits of FDI (in terms of upgrading local economies, importing capital, know-how and expertise, and establishing good linkages with local economies) have actually been achieved. Some forms of FDI have formed 'island-type' developments, which are isolated or disembedded from the surrounding local economies and that fail to receive the benefits that investment was supposed to bring (see e.g. Hardy, 1998; Pickles and Smith, 2005).

Smith (1995) illustrates this through the example of Slovakia and shows how FDI combines in complex ways with economic legacies from the communist period (regional patterns of industrialisation) to produce a fragmented geography of uneven development at the regional scale. The Communist state in what was then Czechoslovakia aimed at reducing regional inequalities within the country through spreading industrialisation more widely. Regional convergence was achieved under state socialism but, despite the dispersal of development, regional inequalities persisted into the 1970s, with significant concentrations of industry. Industrialisation in the 1950s and 1960s created a concentration of industry around the major metropolitan centres and a western core of industrialised regions. The 1970s and 1980s saw branch plant industrialisation in the more peripheral regions, but the rural peripheries in the south and east remained largely agricultural. Despite its ideological goals of equality and overcoming rural–urban differences, communist-controlled development was itself an uneven process producing divergent development pathways.

The end of state socialism led to national economic collapse during 1990–95 and Slovakia's recovery has been marked by considerable regional variation. The emerging fragmented space-economy reflects the ability of existing regional structures to respond to globalisation and marketisation. The metropolitan regional economies have generally diversified into tertiary functions with strong SME development and high levels of FDI (e.g. Bratislava and western Slovakia). Those regions that have become dominated by FDI, enterprise restructuring and export-led growth are mainly those that developed most under state-communist industrialisation. By contrast, the peripheral late-industrialising regions under communism are struggling to restructure. Thus, these divergent pathways of post-communist economic development depend on the interaction of the regional economic structures established by state socialism with the individual restructuring strategies of new and privatised firms.

How post-communist economic transition is conceptualised can therefore have a major impact on how what is really happening is understood. This is illustrated further by the widely observed emphasis in the new economy on a reliance on 'household survival strategies', particularly those based on a range of 'non-official' economic exchanges – self-sufficiency (particularly in the home production of food), barter and exchange (for accounts, see Meurs and Djankov, 1998; Smith, 2000; Smith, 2002). As Smith (2002) notes, reliance on these economic forms is often interpreted as a survival strategy in times of economic austerity that the collapse of state socialism and attempts to marketise economies have produced. However, Smith (ibid., p. 238) undertakes a different reading of such economic activity, suggesting that it needs to be understood as 'practices with long-standing cultural and economic significance in the (re-)production of household economies' that draw on networks of socio-economic and cultural relations dating from state socialism and even before it. The point is that they illustrate how post-communist economic activity comprises a 'diverse economy of variously constituted practices' derived from capitalism, but also sources such as 'feudal' household processes. Thus, they represent 'forms of community partially outside of capitalist market relations' (ibid.). What is emerging in post-communist CEE is thus a vast range of economic practices that combine old and new resources in a multitude of ways to produce various hybrid forms of capitalist and non-capitalist economic systems. Such a situation cannot be simply conceptualised as a 'transition to capitalism'; we need to view economic practices through a different conceptual lens and this in turn helps to explain the vast differential in economic performance that can be observed across the region.

These differential economic impacts have been felt at the individual level through a dramatic polarisation of people's life chances. While some people have become rich in the new economic context of post-socialism, for the majority of the population their economic circumstances have declined, particularly with the withdrawing of state provision of social welfare. Just a few statistics provide sobering evidence of the level of development that some of the CEE countries are experiencing. Life expectancy is a key indicator of people's life chances. The probability at birth of not surviving to age 60 (projections for 2000–5 expressed as a percentage of the birth cohort) exceeds 20 per cent in Estonia, Hungary, Lithuania, Latvia, Belarus, Romania, Ukraine and Moldova. While it is problematic to take income data for this region at face value (incomes are under-declared to avoid taxation and/or people derive 'incomes' from a variety of sources), the percentage of the population living below US$4/day (1996–9) was very high in a number of countries – Lithuania (17 per cent), Estonia (18 per cent), Bulgaria (22 per cent), Romania (23 per cent), Ukraine (25 per cent), Latvia (28 per cent) – and in Moldova encompassed a massive 82 per cent of the population (data from UNDP, 2004). That many of the countries listed here are among the first wave of EU accession states indicates the consider-able gap between living standards in Western and 'Eastern' Europe that still remains. Evidence shows that social inequalities have developed very quickly during the post-communist transition (Duke and Grime, 1997).

These inequalities are also repeated at the regional and national level (for an extended account, see Dunford and Smith, 2004). A great deal of development has focused on capital cities and key metropolitan regions. Table 4.1 demonstrates the continuing disparities in wealth creation at the national scale, with GDP per capita varying significantly across the region. The differentiation in development at a national scale is illustrated even more clearly in Table 4.2. The UNDP Human Development Index ranks the level of development of countries based on a number of indicators measuring economic factors, health and education (for the same analysis based on 2000 data, see Bradshaw and Stenning, 2004, p. 27). All of the CEE countries are classified as at least 'Medium human development', distinguish-ing them from the poorest countries in the world, which, as Bradshaw and Stenning (2004) point out, is indicative of the legacy of development under communism and pre-communist development. Nine CEE countries fall into the top category of 'High human development', and it is indicative of the different development paths experienced in CEE that these include the eight new EU members plus the country that is probably next to join the accession process, Croatia. Slovenia is the most developed, ranked at twenty-seventh place, which puts it below all the Western European countries (although on a par with Portugal) and equivalent to other nations, such as the Republic of Korea. Twenty-three places separate the top (Slovenia) and bottom (Latvia) ranked countries in this group. Since 2000 these countries have gradually improved their rankings, although even new EU mem-bers, e.g. Hungary and Slovakia, have dropped in the rankings.

However, the variation becomes more acute with the remainder of the CEE countries that fall into the 'Medium human development' classification. Both Bulgaria (the highest ranked of this group, at 56) and Romania (69) are included here, indicating that their development levels are far behind even the new EU members from CEE, which has implications for their accession to the EU in 2007. Fifty-seven places separate the top and bottom countries in this group, indicating the variation in development. To place these countries in context, all are ranked well below any Western European country. Bulgaria (56) is on an equivalent level to the Russian Federation (ranked 57) and Malaysia (59); Romania (69) and Ukraine (70) are bracketed by Venezuela and Brazil; while the lowest-ranked Moldova (113) is ranked with Indonesia, Vietnam, Bolivia and Honduras. Thus within CEE there are countries that are experiencing convergence with the 'old' EU members (though 'real convergence' is still to be attempted; see Vintrová, 2004), but others that lag behind this, and some that share levels of development with middle-income

Table 4.2 The UNDP Human Development Index 2004 ranking of Central and Eastern European countries

Country (rank order out of 177 countries)	HDR rank 2004	HDR rank 2000	Change in rank 2000–4	HDI value 2000	HDI value 2004
High human development 2004					
Slovenia	27	29	+2	0.879	0.895
Czech Republic	32	33	+1	0.849	0.868
Estonia	36	42	+6	0.826	0.853
Poland	37	37	No change	0.833	0.850
Hungary	38	35	−3	0.835	0.848
Lithuania	41	49	+8	0.808	0.842
Slovakia	42	36	−6	0.835	0.842
Croatia	48	48	No change	0.809	0.830
Latvia	50	53	+3	0.800	0.823
Medium human development 2004					
Bulgaria	56	62	+6	0.779	0.796
Macedonia, TFYR	60	65	+5	0.772	0.793
Belarus	62	56	+6	0.788	0.790
Albania	65	92	+27	0.733	0.781
Bosnia and Herzegovina	66	—	—	—	0.781
Romania	69	63	−3	0.775	0.778
Ukraine	70	80	+10	0.748	0.777
Moldova	113	105	−8	0.701	0.681

Source: UNDP, 2004: 139–42; Bradshaw and Stenning, 2004: 27

Note
The UNDP Human Development Index is based on a number of indicators that measure a country's achievements in providing its citizens with a long and healthy life, knowledge and a decent standard of living.

and/or developing countries around the world. There are many reasons for this national variation in development, responding to particular histories of the pre-communist and communist-era developments, and their interaction with today's domestic economic policies in the new regional and global context. However, the key point is that this variation again challenges notions that neo-liberal capitalism can be simply introduced to solve the development problems of all of CEE.

These patterns of national difference are echoed at the European scale where the countries of CEE are peripheral relative to a Western European 'core' or 'pentagon' of wealth and dynamism centred on London, Paris, Milan, Munich and Hamburg (see Sokol, 2001; Dingsdale, 2002). To the east of this core, wealth creation (measured as GDP as a percentage of the EU average) falls away steeply. Analyses suggest that, even allowing for the existence of some advanced city-regions, CEE as a whole still lags behind this Western European core and the EU. Indeed, Sokol (2001) conceptualises CEE at this scale as part of a 'super-periphery', which he divides between super-periphery A (East-Central European countries, bordering the EU, plus the Baltic states) and super-periphery B (the FSU minus the Baltic states). The scale may obscure some urban and regional differences but the overall pattern is clear. CEE lags considerably behind Western Europe and the EU; the further east travelled, the greater this discrepancy, and there is growing disparity within the 'super-periphery' itself. As Dunford and Smith (2004, p. 55) suggest:

> There is little evidence of convergence on the levels of development of economies in the EU. There is much evidence to suggest that liberalised markets and integration into the European and globalised economies have created and further deepened already existing territorial and employment inequalities.

Gorzelak and Jałowiecki (2002) predict that EU accession will increase divergence rather than convergence in Europe.

Interrogating the nature of economic change in CEE thus offers a great deal of empirical evidence that questions the accounts of post-communist change offered by linear 'transition to capitalism' models. As Sokol (2001) suggests, it also calls for the need to closely examine the political economy and market effects that were supposed to achieve convergence between 'West' and 'East'.

Post-communist democratisation in Central and Eastern Europe

The fall of state socialism in CEE was clearly linked to a desire to establish multi-party democracies in the former communist countries. The dominant discourses about how this was to be achieved centred on the adoption in the 'East' of Western paradigms of state-societal organisation, especially those based on democratic market economies. Across CEE, the adoption of democracy was seen in most countries as an important part of rejecting the communist past and establishing these countries as modern independent states that belonged in Europe and as part of the international community. This involved most states in CEE making efforts to develop democracy and reinvigorate civil society. Significantly, the process was also externally driven as such efforts at democratisation were also central to attempts to join international institutions, such as the EU or NATO, but were also important in attracting foreign direct investment and development aid, such as from the European Bank for Reconstruction and Development (EBRD). However, there has not been a straightforward move to establishing fully functioning democracies across CEE. Some countries (e.g. Poland, Hungary) have undergone a rapid adoption of 'Western neo-liberal development' paths with relatively advanced 'democratic state-building'. However, others have experienced relatively slower changes (e.g. Romania), have had to re-establish democracy after a protracted period of conflict (the countries emerging from the former Yugoslavia), or have experimented with market reform while preserving authoritarian forms of governance (e.g. Belarus or, until 2005, Ukraine).

There are a variety of processes shaping the different experiences of democratisation in CEE. Post-communist state-building is taking place within a context in which international actors, with their own particular experiences of, and models for, building democracy, are able to exert a significant influence over the nature of democratisation (Grzymała-Busse and Luong, 2002). These countries have been characterised as 'penetrated societies' whose external environments play an important role in political developments (Ágh, 1998, 1999). Post-communist state builders have thus been influenced by international aid, and financial organisations and other institutions, and in particular in CEE the EU has been an important actor in defining the nature of democratisation through the EU accession process, as outlined above. However, the form and degree of international pressure has varied over space and over time, as has the response of national governments to those pressures (Friis and Murphy, 1999; Ferry, 2003). The approach of different governments to adopting these international processes of state-building has been influenced by their ability to accede to, or resist, international pressure, their geopolitical location and their economic performance.

Post-communist state-building is often influenced by the nature of economic transition and the electoral response of the public. The institutional arrangements that have arisen during democratisation are shaped by the experience of transition as growth-driven or shrinking economic performance (Pei, 1996; Grzymała-Busse and Luong, 2002). In countries that experience economic failure or recession-led transformation, support has at times swung to the centre left and former Communist politicians, while centre-right parties emphasising liberalism may hold power in countries with positive growth. However, even this posited relationship does not always hold true, as in the case of Poland with its return to 'centre-left' governments from 1993.

Grzymała-Busse and Luong (2002) further note that post-communist state formation is a dynamic process, which they conceptualise as elite competition over the authority to create policy and policy-making institutions. The type of actors involved in transition thus effect its progress and outcome. Again, this can vary from country to country as domestic politics encounters different international pressures, a process further complicated by the nature of institutional legacies from state socialism or even pre-communist political traditions. The domestic situation inherited from the communist period can thus impact on democratisation processes. 'Incomplete' communist systems may leave a legacy of a more flexible political climate that is more capable of responding to reforms (again favouring right-of-centre parties), while those that had 'complete' communist systems are often less responsive to reform opportunities (Pei, 1996). Two extreme forms of democracies can thus emerge. Reforms led by strong incumbent elites seeking to preserve their position may lead to 'restricted' democracies, while weaker incumbent elites may allow more thorough change and wider political representation in 'full' democracies (Munck and Leff, 1999). The emergence of 'semi-authoritarian' regimes, which restrict any real competition for power, while allowing a seemingly independent press and political opposition, has been typical in some areas of CEE (Ottaway, 2003). Semi-authoritarian states have emerged through former Communist Party bosses transforming themselves into 'elected' presidents who remain authoritarian rulers with their power relatively unchecked by the weak democratic institutions. While these semi-authoritarian regimes may have been attempts by incumbent elites to retain power while satisfying external demands for democracy, widespread discontent among their citizens often produces further change, suggesting that they may represent a stage in a longer-term democratisation processes (ibid.).

All of these points raise questions about the nature of democratisation in CEE and suggest that the process is far from the Washington Consensus view of a straightforward introduction of Western models of democracy, aided and underpinned by the introduction of market economies. The diversity in democratisation outcomes challenges simplistic 'end of history' accounts of transformation and also the idea that liberalisation necessarily or effectively produces democratic regimes (ibid.). The post-communist state is not a unitary actor but is characterised by multiple actors at a variety of scales (inside and outside of the state) forming multiple centres of authority-building (Grzymała-Busse and Luong, 2002). These actors in turn operate in different political and societal contexts that are influenced by the previous experience of state socialism. The nature of democracy that emerges may also depend on what was present in these countries under state socialism, such as whether there were any pre-existing democratic ideals and organisations (Ottaway, 2003). Thus the nature of legacies from state socialism are significant in the different paths of democratisation (Barany and Volgyes, 1995). Formal and informal legacies act as constraints on post-communist state formation because they are among the primary resources available to elites competing for authority (Grzymała-Busse and Luong, 2002). Post-communist change in CEE, with the exception of the former Yugoslavian states, was a negotiated change generally lacking in violence, allowing more continuity in elites and administration. This point is demonstrated by the post-1989 electoral success of former Communist politicians and parties in many CEE states, which was related to their ability to break with the communist past while simultaneously reshaping those elite resources ('portable skills' such as political expertise and administrative experience) that could be used to continue their grip on political power (Grzymała-Busse, 2002).

There is not space here to provide a detailed outline of the different experiences of democratisation in CEE. Neither is it easy to produce simple classifications of groups of states that fit into a 'stage' in democratisation. If we accept the definition of a democratically legitimate state as one that has a 'democratically elected, popularly accepted and widely supported government, [and] which exercises actual control over its full territory and manages to direct the crucial aspects of socio-economic development' within its boundaries (Âgh, 1999), then the full diversity of the democratisation process in CEE becomes apparent. According to Âgh (1999) some states have managed effectively to control their whole territory while maintaining political unity and shaping economic development, particularly those in 'central Europe' such as Poland, Hungary, the Czech Republic, and Slovakia, but the same could be said for Slovenia. Romania and Bulgaria, who fell into the second wave of accession countries, have managed to some extent to sustain political control over the whole country, but have struggled to manage socio-economic development successfully. Bojkov (2004) highlights how this leaves Bulgaria and Romania in a difficult position relative to the EU and southeastern Europe. As part of an advanced EU accession process, they are decontextualised from their geographical location in southeast Europe. On the other hand, delay-ing their accession until 2007 has disconnected their achievements in economic and political transformation from those of the accession eight.

Other states, and particularly those emerging from the conflict in the former Yugoslavia, are only just beginning to achieve a measure of control over the whole of their territory, e.g. Bosnia and Serbia (see Jackson, 2004), and also Albania. Further diversity is illustrated by those states that have experimented with market reform while preserving authoritarian forms of governance. Ukraine represents an example of a CEE state that did not emerge as a fully authoritarian regime but until quite recently had not achieved significant levels of democratisation (Dyczok, 2000; Jackson, 2004; Kuzio, 2005). Ukraine maintained a hybrid fusion of the former Soviet system combined with an emerging reformed polity and economy, which Kuzio (2005) labels as 'competitive authoritarianism' in which multi-party demo-cracy was instead represented by 'superpresidentialism' and opposition 'pseudo parties' (Ishiyama and Kennedy, 2001). As a 'failed authoritarian regime' Ukraine remained fairly stable under Kuchma, but with the opposition's claim that the 2004 elections were rigged came popular protest that overcame the semi-authoritarian regime during the 'Orange Revolution' of 2005. The emergence of a more democratic and Western-oriented leadership will potentially revive relations with the EU and offer the potential for further democratisation (Kubicek, 2005). In Belarus, severe economic and social difficulties are linked to the authoritarian nature of the regime. Belarus has largely dropped out of the broader processes of marketisation and democratisation at work in the rest of CEE, and it is marked out by its official policy of incorporation into a union-based relationship with the Russian Federation (Eke and Kuzio, 2000).

Thus establishing political democracy is non-linear, uncertain and even poten-tially reversible and is a process lacking predictability and the simple transference of processes. Democracy can be achieved by more than one route and the political circumstances of each country or group of countries in CEE require examination (Pei, 1996).

Conclusion

This chapter has provided an overview of post-communist transformation in CEE in order to outline the distinguishing characteristics of post-socialism in this region. While it is impossible to cover all aspects of post-communist transformation, three key areas – Europeanisation, economic restructuring and democratisation – have been outlined to illustrate the nature of these processes and how they are playing out in the CEE context. The experience of post-communist transformation in CEE has been distinctive in relation to other regions (though with similarities in the case of the Russian Federation) principally through the intensive application of the neo-liberal policy agenda defined by the (post-)Washington Consensus. Thus a key characteristic of post-communist change in the region is the external pressure exerted by the international political and economic community. This includes international financial organisations, such as the World Bank and International

Monetary Fund, but it has become particularly enshrined in the shaping of CEE states that have gone through the process of accession to the EU. The dominant discourse of transition in the region has been to assist (most of) CEE in a 'return' to Europe as its legitimate home and to guide CEE states along a linear path of development to fully functioning liberal market economies using Western notions of statehood as a model to be aspired to.

However, as the sections above have illustrated, while this is a view of change in the region that dominates domestic and international political and economic thinking, and is often reproduced in the media, post-communist change in the region is much more complex, subtle and above all heterogeneous. Indeed, the analysis above has stressed the difficulties in even trying to write about or conceptualise CEE as a coherent region, or trying to categorise its member states in simplistic classifications. The Washington Consensus and processes of Europeanisation are powerful external influences on the nature of post-socialism in the region but they are not monolithic or simply imposed on these states; domestic strategies towards the EU are also significant, for example. Analysis of the economic geography of the region reveals neither a homogeneous convergence with the rest of Europe, nor a simplistic imposition of a standard model of a functioning market economy.

Consideration of the three key processes impacting on CEE and guided by ideologies such as the Washington Consensus, reveal the deeply contested and uneven forms of post-communist development that have emerged since 1989:

- Europeanisation – the accession of eight CEE states into the EU in 2004 has created a new division within CEE and Europe as a whole. The 'accession eight' have had to dramatically realign their political, economic and social systems to achieve EU membership – but they still retain unique characteristics and have a long way to go to achieve the expected convergence with existing EU states. The division between CEE EU and non-EU states further divides CEE and presents a challenge for those currently outside of the EU as to their future development.

- Marketisation – the model presented to the CEE as the ideal path for change was a linear transition to market capitalism based on the 'Western' model of capitalism (ignoring the great variations even within that category). While the new EU members from the CEE are considered to have achieved satisfactory progress in that respect, the overall pattern is one of economic fragmentation. While some regions have demonstrated growth, many others have suffered decline and stagnation. A variety of hybrid economic forms have emerged, many of which combine the new influences of marketisation with the legacies of state socialism. Overall CEE lags behind Western Europe in terms of economic development and is deeply differentiated internally. That some of the CEE countries can be equated with middle-income developing countries illustrates the great diversity of development and the lack of impact of the neo-liberal policy prescriptions.

- Democratisation – democratisation was seen as central to post-communist change in CEE and was expected to develop along the lines suggested by existing Western European models and to follow from the establishment of liberal market economies. Again the pattern within CEE is one of considerable variation. EU accession has marked out those states that are considered to have achieved the transition to functioning democracies, but even those states that are part of the next accession wave are still considered to have problems in establishing fully functioning democracy. The rest of CEE demonstrates considerable variety, from states only beginning the establish democratic control as they emerge from conflict to those only emerging from or even retaining 'presidential' or 'semi-authoritarian' regimes. Establishing democracy in the region has equally proved to be non-linear and unpredictable.

What marks out the nature of post-communist change in CEE is thus its relation to Europe as a geographical entity, a political organisation and an idea, while also bearing in mind the influence of the Russian Federation and now also the role that countries such as Turkey may play in European integration. External influences and domestic economic and political responses are key. Simultaneously, however, post-socialism in the region is marked by its heterogeneity, and its lack of any simplistic linear track 'to capitalism' and economic and political convergence with Europe. Post-communist change has proved to be much more complex and uneven than expected. In particular, change has been typified by the continuity of the socio-economic-political relations of state socialism and their modification in combination with the new influences brought by marketisation (Swain and Hardy, 1998). Thus, new forms of political-economic organisation combine in highly complex and new ways with the 'old' practices, resources and personnel of state socialism in a process of 'recombinance' (Stark, 1996) in which differences in Communist regimes lead to varieties of 'post-communist capitalism'. Hybrid economic forms arise, combining elements of central planning and market economies, and this interaction underpins dramatically uneven development (Smith, 1997). Post-communist transformation in CEE is thus 'embedded' in the existing locally institutionalised practices and networks of economic life, while processes of 'asset conversion' allow those well placed under one regime to remain so. Specific state–society–economy relations constrain choices for development. Foreign technology and capital combine with local networks of knowledge and resources in a variety of ways that lead to development that is 'path dependent' (Hausner *et al.*, 1997). Thus the post-communist transformation in CEE has not been as expected and its future development is questionable.

5

THE SOVIET UNION AND AFTER

'Incidental transition towards a *formal* democracy'

Introduction: why distinguish between post-communist transition in Central and Eastern Europe and in the former Soviet Union?

Post-Soviet, and, because of its particularly pre-eminent role in the Soviet Union, Russia's, post-communist transition contrasts significantly with the process of change in Central and Eastern Europe (see previous chapter) in several ways. First, there is the legacy of the communist experience *per se*. While in Russia, the implementation of the communist regime was essentially a domestic affair, in the CEE states it was imposed after the war by a victorious occupation force, resulting in many negative associations with the very notion of 'communism'. Second, there was the experience of an economic system designed for a feudal, pre-industrial economy, that was simply extended to, and imposed on, the more advanced economies of the CEE states. In addition to this structural mismatch, there was the disconnection of these economies from their established connections to the rest of Europe's space economy and polity, and forced re-orientation towards Russia's and the rest of the Soviet Union's economic requirements. Third, the nature of the end of communism differed significantly between the Soviet and the CEE countries – essentially 'bottom-up' with popular engagement and grassroots involvement, versus an elitist, top-down instigated reform course 'gone wrong', with little popular involvement, in the Soviet Union. The fourth point of difference is the varying outcomes of the collapse of the communist regimes in CEE and the FSU – active democratisation with the involvement of an emerging civil society, against an essentially merely formal democracy disguising an ever more autocratic regime. Lastly, there is the role of territoriality and nationhood, which generally were of lesser importance in the CEE countries, as national territories clearly existed and were unchallenged. The former Yugoslavia and Czechoslovakia were the only exceptions. This contrasted with the territorial struggles between newly independent nations in the aftermath of the break-up of the multi-ethnic Soviet Union. Territoriality these had been largely defined by political considerations and 'divide and rule' policies under Stalin. While within the Soviet Union, boundaries between technocratically defined territories were of limited relevance for people's movements, they mattered more as policy areas, including limited autonomy for particular ethnic groupings. Raised to the status of international border, these divisions had suddenly gained in importance and separating effect, and their location become more of a concern

Despite these evident differences, all countries shared the abrupt confrontation with the necessity and opportunity to develop new structures, and principles of governance and economic development, without any preparation and 'warning'. There simply was no time for learning best practices. Instead, 'learning on the job' was needed. This was the case in China, too (Chapter 6), but there a clearer understanding about the outcome and direction of reforms avoided the rather protracted nature of post-Soviet development.

Given these complexities, it had become evident by the late 1990s that the initially widely projected view of a linear trajectory of post-Soviet transition towards an emulated western market democracy had been overly simplistic, ignoring the past and its impact on current values, ways of doing things and ambitions, but also (and especially) the engagement of the people and their ownership of the whole transformation process (see e.g. Smith, 1999). In the Soviet Union, changes had been initiated 'from above' by a political elite, as previously in history. Struggles within the elites about these reforms, and their likely impact on the future of the Soviet Union, effectively led to the rather tumultuous and protracted events of 1991. Rather than halting the disintegration of the Soviet Union, the hardliners' intervention effectively accelerated it, and brought about the independence of Russia and 14 other states. Not unlike CEE countries, some of the new states maintained their established political elites, if 're-badged' as non-communist and, first and foremost, newly nationalist. Of course, the Soviet Union was by no means a homogeneous state construct, and its collapse led to quite different paths of post-communist (or post-Soviet) development in terms of democratisation, economic development and cultural emphasis, as will be explored in this chapter.

Difference and similarity between Eastern European and post-Soviet transition

Political-geographical proximities often seem to suggest common features of post-socialist transition in the Central and Eastern European countries and those of the former Soviet Union. Artisien-Maksimenko (2000), for instance, subsumes CEE and the FSU under 'Eastern Europe' in his edited overview of country-specific experiences with multinationals in the privatisation process. By including examples from Transcaucasia and Central Asia, he goes beyond the usual focus on those countries on the western end of the former Soviet Union, bordering Central and Eastern European countries, when post-communist developments in the FSU are being discussed. This includes in particular Russia (again, with a distinct Euro-centric perspective), Belorussia (now Belarus), Ukraine, Moldova, Georgia and, especially, the Baltic States, Estonia, Latvia and Lithuania. Russia, because of its paramount importance in the Soviet Union, and its internal ethnic and geographic divisions, has attracted particular interest, to the point that post-Soviet developments have often been implicitly seen as post-communist Russian developments. Other countries, apart from the Baltic States and, perhaps, Ukraine, have attracted much less attention, with the new states of Eurasia even less in the centre of post-communist studies (exceptions here include the series 'The International Politics of Eurasia', edited by K. Dawisha and B. Parrott of 1994 onwards, including eight volumes on *Russia and the New States of Eurasia*). This chapter, too, will concentrate to a considerable degree on Russia as the evident main player among the former Soviet Union states, which is also reflected in the fact that Russia is the legal successor to the Soviet Union as far as international commitments are concerned. Nevertheless, the developments in the other parts of the FSU will also be discussed with the aim of characterising their transformation paths after formal independence.

'The collapse of the communist system was everywhere a long process of erosion. However, it fell apart in quite different ways, each of which have important consequences for the process of post-socialist transformation' (Mendell and Nielsen, 1995a, p. 10). But despite the differences, stretching from Poland's 'shock therapy' approach, moving directly 'from protectionist authoritarianism to atomism' (ibid., p. 7), to the much more gradualist approach in Hungary or Slovakia, for instance (see e.g. Carter and Maik, 1999), there have been some underlying commonalities. These include many varying speeds and degrees of moving towards a liberal market society, with its commodification of the labour force, focus on privatisation (Frydman *et al.*, 1993), and growing role of multinationals (Artisien-Maksimenko, 2000), while largely ignoring civil organisations, especially those grassroots initiatives that were instrumental in bringing about the collapse of the communist regimes. Thus, for instance, 'the Balcerowitz Plan was not worked out with representatives of society, but was imposed by the State in the name of the market. State and market were the only two institutions that matter [sic]' (Glasman, 1994, p. 192).

This embracing of 'market' and marginalisation of the 'revolutionary grass roots' was typical of the transition process in CEE countries and the former Soviet Union. It may also be seen as the second phase of revolutions, where the original revolutionary interests and subsequent political forces get separated, leaving the revolutionary elite behind, replaced by a post-revolutionary agenda and its propagators (Mendel and Nielsen, 1995a, p. 10) This contrasts with the other model, as found, for instance, in many of the Central Asian republics of the FSU, China or, indeed, Cuba, where the revolutionary forces have retained power.

The almost paramount surrender to the 'temptation' of the liberal market approach was caused by a combination of external and internal forces. For once, market forces embodied the antidote to the communist regime and thus, by implication, were very attractive to symbolise the departure from that era. The more market, the bigger the distance to communism and its legacies. In addition, market liberalism, propagated by the leading New Right advocates Margaret Thatcher and Ronald Reagan, had become the new 'trendy' credo of the 1980s in response to globalisation. This applies also to the anti-communist paradigm emerging within the affected countries, where 'free market' was seen as anti-communist (while anyone expressing reservations against all-out liberalism was branded 'communist'), and therefore the only 'real' alternative. But there was also an opportunistic element to it. When the ruling *nomenklatura* realised that 'free market' was the new currency, they happily swung behind the new paradigm, anxious to retain their positions. 'This certainly reflects the reality behind the predominance of the language of the free market in the East: a new yes man's code replacing the similarly misleading Soviet rhetoric of "peace, friendship and solidarity"' (Wainright, 1995, p. 33).

It could come as no surprise, therefore, that 'trendy' government advisers and policy think tanks, and through them, also the World Bank and IMF, offered this model as the 'only show in town', although some minor variations were available. The problem with pursuing this approach was, however, that it meant the death knell for the carefully constructed high-industrial society of the communist countries. If Britain's industrial structures had seemed out of date in the late 1970s, leading to its rapid restructuring during the early 1980s, the gap between the existing communist 'old industrial' society and the 'post-industrial' (or late modern) global economy was even greater. The result was a rapid and fundamental collapse of the old, now well out-of-date regime of accumulation, and the associated social structure.

Once unleashed, the changes proved difficult to manage. Where there was no managerial constraint, as in Poland or the Czech Republic or Russia, many of the anti-communist dissident groups found themselves overtaken by events and literally swept aside – their social civic aspirations, looking for a 'third way' along the lines of the German model of a 'social market economy', swept away by the rationale of the free market. Neither the Hungarian movement of the Free Democrats, nor the Czech Civic Forum, nor the East German Democratic Forum (see also Herrschel and Forsyth, 2001) could gain access to the new political elites and government. 'However, nowhere was the suicidal inclination in the post-communist transformations as evident and dramatic as in Poland, where the workers of Solidarity were among the great losers and where the alliance of workers and intellectuals rapidly crumbled' (Mendell and Nielsen, 1995a, p. 12). Each had to face the new realities of free market reforms and thus the obsolescence of much of the inherited structures, societal arrangements, privileges and values.

The intellectuals, having played an important role in articulating opposition and thus helping to focus the anti-communist movements as part of the highly respected 'intelligentsia', lost their influence and privileges. Under post-communist market conditions, many academics had (and are still having) to supplement their frozen incomes through second jobs, for example as taxi drivers. Similarly, the industrial working classes, that is the majority of the working population, effectively voted for their own redundancy by supporting marketisation and liberalisation. This is one of the reasons why the social costs of the transition process have been so high. The existing societies literally had the economic rug pulled from underneath their feet. So it cannot come as a surprise that the new governments, which gained their power through the grassroots' pressure on the old regimes, began:

> post-communist life as a very strange interest group indeed: arguing that
> the interests of its members were best served by accepting deep sacrifices

on behalf of a class that did not yet exist [new entrepreneurial middle class], in return for benefits that it was hoped – and only hoped – would accrue in the future.

(Ost, 1992, p. 12)

The extent of the adjustment costs and challenges were not unlike those in the industrial northeast of England in the early 1980s.In fact, they were even more fundamental, extensive and long-lasting, uniquely combining both a 'delayed and accelerated process of restructuring' (Bachtler *et al.*, 2000a, p. 2). This is illustrated by the continued structural economic problems with dependency on on-going western subsidies, even after years of massive cash injections. The continuing massive inequalities within the different countries, with highly localised 'winners' and a much more widespread distribution of 'losers' of the transformation process, attest to the underlying problems and challenges, especially the formation of a firmly embedded civil society. The collapse of communism also triggered the need to identify strategies and options for further development through new procedures and unaccustomed political negotiation.

Economic transformation and inequality

Van Brabant (1990) summarisingly observes that market economies had to be constructed in Eastern Europe essentially from scratch; they do not simply emerge in full fettle upon the retreat of central planning. 'Among others they require a complex infrastructure of laws, financial systems … and certain habits of economic behaviour … ' (ibid., p. 196). Against this background, the author questions the salience of the arguments about 'shock therapy' as a realistic model or interpretation of post-communist economic transformation. 'Since changes in structure and economic behaviour can materialize only with some delay, the arguments presented point to a more evolutionary approach to the reform process than the rapid, shock treatment suggested by some observers' (ibid., p. 196). This, van Brabant sees as being caused by an over optimistic expectation of transformational progress. But that is also caused by at times somewhat blurred distinctions between micro- and macro-economic considerations. Stern (1997) also points to the false nature of the division between 'shock therapy' and 'gradualism' and the fact that it depends on the types of initiatives that decide the speed of their introduction. 'Some things can and should be done quickly, others take longer' (ibid., p. 53). In any case, crucial for successful economic transformation appears to be the operational framework as set by the respective governments, especially generating credibility of their programmes. Without realistic certainties new investment will be hesitant in coming forward. Admittedly, this is difficult, as many countries moved into uncharted waters. 'Once a reform programme gains credibility, the question of gradualist versus shock or "big bang" approaches becomes secondary' (van Brabant, 1990, p. 197). However, as long as sufficient progress with reform and economic development is made to satisfy the people's expectations, the question about the speed is of lesser importance, as experience from elsewhere, such as China, seems to suggest.

In addition, as evident from several years of experience, there are many more variations of transitions between countries, mainly owing to their different circumstances when embarking on the process of change (Zecchini, 1997a; Bastian, 1998; Bachtler *et al.*, 2000a). In some countries, by the mid-1990s, there is evidence of 'transition fatigue' (Zecchini, 1997a, p. 2) as a result of the ongoing changes facing people and policy makers. At the centre of the difficulties is seen to be the extent of necessary changes, with four concurrent tasks: deciding on speed, sequence, depth and relation of demand management (ibid, p. 2). In addition, the lack of information available on the economic structure and situation at the starting point of change (e.g. hidden unemployment through 'hoarding' of people by firms) adds to the difficulties in assessing 'achievement', 'progress', and so on. Nevertheless, the common challenges/tasks faced by all included establishing market structures, developing market participants (households, entrepreneurs), shaping a new financial system and reducing the size of the public sector as service provider, while developing tools of indirect economic management.

Another commonality was the general top-down nature of implementing the new system and principles, overnight, with instant price liberalisation, opening up to external markets and establishing a new legal framework as 'legislative shock' (Zecchini, 1997a, p. 9). Subsequent policy actions, at times easing competitive pressures by temporarily establishing higher customs import duties, were the answer to the enormous political pressures resulting from rapidly deteriorating economic prospects, although the picture varies between countries (Table 5.1). The differences between countries' policy approaches rested primarily in the order and relative emphasis on the different elements of transition to a market economy.

There are thus common, but also many distinguishing, features of policy responses to the ever more evident inequalities (Bradshaw and Stenning, 2000) across space and society. For instance, Poland and the then Czechoslovakia went through rapid privatisation and liberalisation, although the relatively debt-free national economy of Czechoslovakia made subsequent liberalisation easier, as inflationary pressures were lower. Romania and Hungary both pursued a much more gradual, step-by-step approach, albeit for different reasons. Hungary had already begun to reform its economy under the late communist government and so felt less compelled to race forward (Szamuely and Csaba, 1998), while Romania saw a political continuity in the sense that many of the old policy makers stayed in office as 'reformed communists' and pursued change in a more haphazard way (Kornai, 2000). Thus, for instance, market mechanisms were allowed to a limited extent only.

Overall, Zecchini (1997a) observes that one of the main challenges and difficulties of economic transition has been the lack of continuity in policy initiatives, suffering from frequent discontinuities, changes, even contradictions and reversals (ibid., p. 16). This has made conditions less predictable and so kept investors and businesses cautious. He also challenges the notion of 'shock therapy' as, in fact, no country implemented the changes in such an abrupt way 'all out'. Instead, reforms were introduced and then modified in response to observed outcomes, especially in terms of unemployment and domestic economic performance (affecting political acceptability of changes). In that way, changes were 'gradual' despite being called 'shock'.

Table 5.1 Differing paths of economic transition in CEE and the former SU

Country	Real GDP 1990 % change	Real GDP 1992 % change	Real GDP 1994 % change	Private sector share of GDP in % in 1995
Central and Eastern Europe				
Albania	−10	−10	7	60
Bulgaria	−9	−7	1	45
Croatia	−9	−9	1	45
Czech Repubic	0	−6	3	70
Former Yugoslavia/Macedonia	−10	−21	−4	40
Hungary	−4	−3	3	60
Poland	−12	3	6	60
Romania	−6	−9	4	40
Slovak Republic	0	−6	5	60
Slovenia	−5	−5	6	45
Baltic States				
Estonia	−8	−14	−7	65
Latvia	3	−35	−15	60
Lithuania	−5	−38	−24	55
Commonwealth of Independent States (FSU)				
Armenia	−7	−52	−15	45
Azerbaijan	−12	−23	−23	25
Belarus	−3	−10	−12	15
Georgia	−12	−40	−39	30
Kazakstan	0	−13	−12	25
Kyrgyzstan	3	−25	−16	40
Moldova	−2	−29	−1	30
Russia	−4	−15	−9	55
Tajikistan	−2	−29	−11	15
Turkmenistan	2	−5	−10	15
Ukraine	−3	−17	−17	35
Uzbekistan	2	−11	−2	30

Source: Stern, N., 1997

In all instances it was important that the old system was seen to be abandoned, opening up new opportunities and expectations. Psychology was just as important as pure economics.

Nevertheless, by the mid-1990s, there still were considerable structural obstacles to 'free' business development, not least through bureaucracy and resistance to change by communist-era *nomenklatura* who continued to populate the civil service. 'Because of these impediments, even in countries where privatisation has been rapid, a strong drive towards good corporate governance and investment in fixed capital has not yet been produced' (ibid., p. 17). Not surprisingly, perhaps, many new business ventures went semi-legal and operated in a shadow economy or 'grey market'. For many new would-be entrepreneurs, therefore, not much changed in the reality of setting up business ventures, with them still being required to by-pass, circumnavigate and avoid state institutions and their ('old style') representatives/bureaucrats. It has not been a rare occurrence to find the same communist-era *apparatchik* behind a desk in a town hall, for instance, expected to further democratic and market-based principles only a short while after the collapse of the old regime that had propagated just the opposite values. These obstacles, of course, varied between countries, depending on the extent to which they had seen more or less wholesale changes in the political climate and attitude to change.

It is difficult to change established practices, ways of doing things and attitudes in the public sector after at least 40 years of preaching good socialist practices, but the extent to which old political elites had been swept away and replaced *en masse* by new political masters affected the workings of the apparatus. Effectively, therefore, 'the non-reformability of socialism was institutionally based: any reforms under-taken were blocked' (Hausner, 1995, p. 57). This author learnt through comments during interviews with economic policy makers in eastern Germany in the mid and later 1990s that one of the main problem was uncertainty. People in public adminis-tration were worried about their jobs, about being identified as 'socialist' through their practices, and thus resorted to following any instructions to the letter, rather than using good common sense and judgement when making decisions. Inevitably, the outcome was bureaucratic procedures that were anything but supportive of new business ideas or facilitating economic development in a rapidly changing environ-ment. Privatisation and changing the public sector service provision alone could not automatically yield 'success'. Just as important is the credibility of policies and their implementation, and the avoidance of confusion and uncertainty.

Transition in Central and Eastern Europe and the former Soviet Union

Post-Soviet transition from a communist to a market-based, democratically struc-tured arrangement has shown some obvious parallels with the events and processes in Central and Eastern Europe (though even there considerable variations have emerged), but there are distinct and important differences that justify placing it into a category of its own. Sakwa (2003) refers in particular to the much bigger task in Russia, for instance, in creating civil society and entrepreneurialism from scratch, because of the much longer time spent under communist rule and its attempt at inhibiting independent civil engagement, whether political or otherwise.

Four key features that have shaped post-Soviet transition and its outcome, contrasting it with the developments in CEE countries, may thus be identified:

1 The length of time spent under communist rule stretches over three gener-ations, the longest anywhere, making recollections of pre-communist con-ditions nearly impossible as 'living memory'.
2 In contrast to the grassroots-driven revolutions in Central and Eastern Europe, the transition ending the Soviet Union was set in train, from above, by an elitist approach to reform first and democratisation later.
3 The Soviet Union was a multi-ethnic, multi-national state, where many boun-daries between nations and areas of habitation did not coincide with pre-Soviet

divisions and ethnic-based territorial identities, particularly as a result of Stalin's policy of uprooting strong national groups (e.g. Chechens) to reduce their potential political threat.

4   Underneath the Soviet mantle rested many different nationalities with their roots in European, Central Asian, Christian-Orthodox and Muslim traditions. These re-emerged as newly ascertained identities after the loss of the common Soviet reference point. Russia's definition of its new identity has been particularly difficult, because of its close intertwining with the Soviet identity – for many, including Russians, Soviet equalled Russian, and thus the lost Soviet empire effectively meant the loss of its status as superpower for Russia.

As a result, post-Soviet and, indeed, Russia's, transition produced particular dynamics and patterns of change and outcomes, with effectively several transitions overlapping (ethnic, territorial, historic; see e.g. McFaul, 2001). Against this background, Offe (1996) sees the challenges of post-communist (post-Soviet) transition as threefold:

1   definition of territory and citizenship (who is in/out the nation state);
2   constitution of polity (system of governance), including civil society; and
3   redefinition of a welfare state in the face of rapidly growing inequality.

The legacy of Soviet Communism

The year 1917 marks the origins of communism as applied socialist ideology. The 1917 Bolshevik October Revolution, under Lenin's leadership, was effectively four revolutions interacting and overlapping, reflecting the diverse interests of participating groups, such as the economically struggling peasants or the disaffected intelligentsia, and various ethnic-national groupings. Suppressed under the subsequent rule of the Communist party, these diversities re-emerged when the weakness of the communist state had become apparent (Sakwa, 2003). The underlying national differences and dormant identities affected the course of events in the run-up to the collapse of the Soviet Union, and shaped the process and outcome of the subsequent transition away from communism. Just as the attempts at reform towards the end of the Soviet Union were instigated 'top down', so also was the establishment of the communist state in the first place. Lenin's revolution was largely driven by a relatively small elite, manipulating and utilising the political weakness of the interim government following the Tsar's resignation. The somewhat alien and far-fetched nature of Lenin's repeated references to the 'proletarian masses' becomes obvious from the fact that these industrial workers made up only 1–2 per cent of the total Russian population (Pipes, 2001). In reality, rather than empowering the people, Lenin imposed a draconian rule and suppressed any dissenting ideas and objectives through his 'ban on factions'. Simultaneously, he pressed everyone to follow his official discourse of social and economic development under the leadership of the Communist Party, and thus, with him as the Party's general secretary, his leadership.

Economically, the ultimate goal was to establish a command economy with no individual property rights, that is to nationalise all human and material resources. Central control would cover all aspects of public (and many of private) life through an intricate network of informers, secret police and surveillance. Sowing distrust among people was an effective policy of 'divide and rule', as it made grouping into political organisations, or any organisations at all, so much more difficult. And in the end, whatever non-political associations were to be established, the Party's eyes and ears were also part of these through informers. As a result, the new political leaders, and the Communist Party, became distinctly separate from the people 'on the ground', controlling them at will with little immediate concern for public opinion.

Just as importantly, because it, too, has fundamentally shaped the outcome of post-communist transition to this day, was the particular form of ethno-federalism that underpinned the creation of the Soviet Union in 1922. Individual republics (Ukraine, Byelorussia and the Russian Soviet Federation, as founding members) established the USSR, with the Central Asian and Caucasian Republics joining shortly thereafter. Ethnic divisions thus underpinned the internal territorial structure of the USSR, always presenting a latent challenge to its integrity – a challenge suppressed by Stalin with brutal force during his leadership from 1923 to the 1950s. For instance, he relocated whole peoples, such as the Chechens, away from their historic homeland, to weaken their ethnic identity and thus their potential to challenge Soviet domination. This history is an important factor in their drive for independence from the Russian Federation today. Stalin's other main agenda was modernisation through industrialisation, seeking to transform rural, feudal Russia into an industrial society of the late nineteenth-century type, paid for by the domestic agricultural sector. State and Communist Party control extended into all aspects of public (and individual) life, seeking to eradicate 'alternative' (subversive) political ambitions, and thus effectively destroying civil society.

Deprived of the exercise of independent, grassroots politics for some 75 years, or three generations, and given a historic absence of any experience of meaningful democratic principles, there was no pre-communist tradition to fall back on when the Soviet Union collapsed. This differs from the situation in many CEE countries, especially in Central Europe, where the basic principles of popular democracy had been forgotten and were to be learned 'from scratch' with post-communist democratisation.

> The distortions of the Stalinist command economy, the destruction of the most active people in the countryside, the neglect of the service sector, the reduction of money to an internal accounting unit and the relative isolation of the Soviet economy from world development, all left the post-communist Russian economy with severe structural problems.
>
> (Sakwa, 2003, p. 6)

After Stalin's death, there was a sense of relief, and a desire to go back to the basics of the original Leninist objectives and achieve a time of 'normality' (Pearson,

2002). But there were several competing, often conflicting understandings of what true Leninist socialism actually meant, a problem also discovered by Gorbachev when, in the mid-1980s, he attempted to redefine the purpose of socialism anew. There was no single 'right' form of Marxist–Leninist socialism. Khrushchev's attempts at improving rural productivity in the 1950s and early 1960s through regional economic councils and developing the consumer industry were part of a post-Stalinist softer approach. 'Peaceful co-existence' with the West, limited marketisation of agriculture and production, and expansion of influence into Third World countries all had brought many changes.

It is against this backdrop of continuous change that Brezhnev's long rule as general secretary, for nearly 20 years (1964–1982), needs to be viewed. His maxim was to ensure the 'stability of the cadres', thus giving the bureaucrats automatic jobs for life with no incentive to achieve goals and 'perform'. Patronage for personal appointments in the *nomenklatura* became widespread. Administering with the least effort and for maximum personal benefit was the guiding principle. The bureaucrats (*nomenklatura*) thus ruled supremely, thwarting any new initiatives that might mean change. The inevitable outcome was, in Gorbachev's terms, a period of stagnation (White *et al.*, 1993; Sandle, 1999; Pearson, 2002). In 1991, there were some three-quarters of a million *nomenklatura*, many of whom ensured the continuation of their privileged lifestyles by using their positions and connections to their advantage during the privatisation process in the early 1990s (Boycko *et al.*, 1996). Many of the current so-called oligarchs come from that background, mutating from a political to an economic elite (Baev, 1996). These *oligarchs* utilised already existing networks between the production units, as well as extremely low asset valuation. Many of the largest factories were valued at just 0.5 per cent of the value that a comparably sized western unit would realise (Boycko *et al.*, 1996). But stagnation turned into effective regression, when compared to the increasingly technologically more advanced world outside the Eastern bloc. A comparison with the newly *post-industrial* 'West', for instance, made the Soviet Union look increasingly outdated and its system inferior. It was this relative falling behind of the Soviet economy in all its aspects compared with western technology-driven development, especially in the 1970s and 1980s, that spurred the Communist Party's Politburo to appoint Mikhail Gorbachev as General Secretary, to break with stagnation and reform the system to bring the Soviet Union's development forward, and at least halt the relative decline. It was this that triggered Gorbachev's policy of *perestroika* and *glasnost* after he became general secretary in 1985 (Sandle, 1999). Incentivising innovation and productivity were the main goals of (economic) 'perestroika', that is (economic) restructuring.

A central element of Gorbachev's reform efforts was to establish a 'socialist law-governed state', defined through legal statute (White *et al.*, 1993, p. 212), rather than personal networks, patronage and privilege. 'Perestroika exposed the contra-dictions between the attempt to transcend the market and the realities of the command economy in which informal economic activity and corruption were rife' (Burawoy, 1994, p. 426; also Mandel, 1992). However, being implemented through decree from 'above' (Mandel, 1992) meant that there was no public debate or pressure supporting this initiative, allowing the members of the *nomenklatura* to undermine the efforts wherever they could.

The Brezhnev era had made the ambition to maintain the status quo particularly obvious and, with discrepancies between living conditions within and outside the Soviet bloc becoming ever more accentuated and visible to everyone, effectively hastened the system's demise. The tensions between actual and promised conditions – conditions that were seemingly readily available in the West, there and then – once tentatively set free by Gorbachev, generated their own momentum and exceeded by far the narrower agenda set under *perestroika*. The new Russian president of 1991, Boris Yeltsin, realised the underlying dynamics and decided to run faster than the flood following him. He thus went for a complete abandonment of the communist system in favour of full liberalisation and marketisation. This went far beyond Gorbachev's ideas of reform. He believed the old system was viable and just needed some fine tuning. *Perestroika* (restructuring), together with *glasnost* (openness) about political-administrative incompetence and lethargy, was to kick-start new development.

This 'reform communism' contained three main elements. First, modernisation of the economy in terms of both mode of production and sectoral structure, would inject new energy into the stagnant system. Given the limited availability of labour, technological improvements were seen as the main drivers of change. Second, decentralisation of control of economic activity was intended to give regions, and factories and farms, greater say in production, an approach also adopted by China, for instance. The third element of reform involved the introduction of a limited market discipline to establish a mixed economy, not dissimilar to the Chinese model, albeit more moderate and conservative towards private (inward) investment. In order to make the changes appear ideologically acceptable, Lenin was cited as implicitly approving these changes. Cooperative ownership structures were preferred, as they complied better with Marxist doctrine. With no independent access to resources, and facing hostility by established state business and popular sentiment, any new business initiative under this scheme was soon brought to an end, leading to the abandonment of the project not long after its inauguration. The second policy, *glasnost*, aimed at exposing a lethargic bureaucracy to public scrutiny by making administrative processes more transparent. This was seen as a means of introducing democratic control of bureaucrats, and ending inefficient and corrupt practices. *Glasnost* thus set out to get society involved with the aims and processes of reform, but soon stimulated a much wider discussion of the system *per se*, not just aspects of its 'performance'.

Gorbachev's belief was that it was the inadequate *implementation*, rather than systemic flaws, that caused the lagging development process of the communist system. But in exposing incompetence and corruption in the civil service and in government, the state's and the Party's authority were effectively eroded even further. It became increasingly obvious that the Party had to be seen to be separate from the state (and government) if it was to carry on as an independent political force in a reformed regime. This meant ending the close identification of the state with the Party, and with that bringing to an end one of the main hallmarks of

Lenin's implementation of Marxism. Gorbachev sought to steer a middle course between an all-out change to the system on the one hand, and maintaining key elements of the status quo, especially the central role of the state apparatus, on the other. But centrist 'sitting on the fence' pleased nobody: neither the advocates of systemic change towards democracy, nor the 'old guard' seeking to maintain as much of the status quo as possible.

By 1990, however, 'reform communism' had run its course, promising no clear direction and progress in transition. The then Moscow mayor, Boris Yeltsin, seized upon the opportunity presented by the then more radicalised public opinion, seeking all-out change. The August 1991 coup ended the moderate reformists' course, broadening the agenda to the question of sovereignty for the republics, and thus the dissolution of the Soviet Union. 'Independence for many ... became a higher immediate political priority than democracy' (Sakwa, 2003, p. 437). The question of statehood was seized upon by many of the Soviet Socialist Republics' political elites who decided to don the nationalistic hat and present themselves as the creators of national independence and statehood. Having the Soviet Union as a perceived, and projected, threat to that new ambition gave extra impetus to the nationalistic cause.

The nature and outcomes of this process differ between Russia and the surrounding borderland states of the FSU. Russia held a particular, dominant position within the FSU, 'and more than any other nationality, Russians were encouraged during the Soviet period to think of their homeland as synonymous with the spatial expanse of the Soviet Union' (Smith, 1999, p. 8). Since the tsarist empire, 'Russia' has been related to a much larger territory than the ethnic homeland, adding to the difficulty of redefining their new identity as Russia 'proper', 'that is its "effective national territory"', also conceptualised and discussed as the 'Russian Heartland' (Bradshaw and Prendergrast, 2005, p. 83) after the collapse of the Union. The sense of post-imperial inferiority adds to the difficulty in adjusting to the 'lesser' Russia, following the perceived loss in stature and standing in the world, and this continues to define its foreign relations and policies, especially *vis-à-vis* the 'West'. It also affects those Russians based in the other republics, who find themselves 'alienated' – seen as unwanted 'aliens' – rather than feeling dominant (Pilkington, 1998). The attempt at marginalising the large Russian minorities in the Baltic States, especially in Latvia (Dawson, 2001), and turning them into second-class citizens by placing hurdles in front of any naturalisation, certainly raised eyebrows in the run-up to European Union membership (Pilkington, 1998).

> While for many Russians decolonisation is about focusing on the creation of a new sovereign and democratic Russia, for others the idea of re-establishing, in whole or in part, an empire abroad and recolonising the former Soviet borderlands, is inextricably bound up with Russian national identity.
>
> (Smith, 1999, p. 9)

The shift to a post-communist society involved, in particular, the development of links between the state, the government and the general public, which had largely been a bystander to the battle fought out among the elites about the future direction of political developments. While there were some signs of an emerging civil society in the form of grassroots initiatives, such as the miners' strike in 1990, this did not mature into a widespread general development of civil society, especially not in the borderland states of the FSU. The Baltic States are in this, as in many other respects, an exception, because of their histories and only late integration into the Soviet Union. Following the example of many of the CEE countries, and encouraged by the World Bank, IMF and other western advisers at the beginning of the 1990s, Russia and Ukraine embarked on a 'rapid transformation' path but, with the social and economic costs of such immediate marketisation becoming evident, they slowed down the process and sought to maintain some protectionist measures for struggling domestic industry.

Inequality and divisions of Marxist–Leninist modernisation

The Stalinist state developed an urban-centred industrial economy geared primarily to heavy industry and military hardware, concentrated in newly developed old industrial districts modelled on those in the West. These urban industrial centres remained the main objective and focus of investment, way ahead of those for social consumption. Developed in the 1930s, this approach remained in place over the subsequent decades, with little regard for the economic and technological changes affecting the West. The economic structure was thus very one-sided, and effectively remained stuck in the 1930s-style heavy industry structure. The cities acted as designated centres of the modernisation drive, with large numbers of the rural population either rehoused in the new estates around the existing cities, or migrating to the cities (as in western industrialisation 100 years earlier). Overcrowding and housing shortages in the main cities led to restrictions being imposed on resettlement, similar to China's control scheme, with residence permits required for registering at an urban address. Such controlled access to the relative privileges of urban life is not dissimilar in effect to the mechanisms responsible for 'shanty towns' in developing countries (see also Chapters 6 and 7).

These divisions have translated into the different attitudes to communism and its reform, with the urban population being mostly enthusiastically pro change, and the rural population more likely to be against it. But it was the growing awareness of the stagnation in their relative privileges that caused the urban elites to question the salience of dogmatic communism, and thus the need for new goals and principles of economic development. Lewin (1991) suggests two key stages in the modernisation project of the Soviet Union: first, the 'ruralisation of the cities' under Stalin, bringing rural labourers to the new urban industrial complexes to serve as regime-loyal bureaucrats to control and administer the cities and their distrusted elites; second, the general 'urbanisation of society' after 1960. 'Soviet society was therefore undergoing a vital urban transformation at precisely the moment the Brezhnev administration was shying away from any engagement with economic or political reform' (ibid., p. 23). It was the increasingly more qualified urban elite, and more educated Party membership, too, that became more and more aware of, and dissatisfied with, the apparent stagnation in the quality of life and general economic and technological development. 'In short, this new urban world ... sat uneasily with a communist system which stifled economic and political change' (ibid., p. 25). And it was this urban world that drove and supported the modernisation attempts for communism. They saw the danger of continued stagnation *vis-à-vis* the aspirations of a population becoming increasingly aware of the possibilities offered by the western system, and comparing that with the offerings available under communism.

Protracted transformation in post-communist Russia

Russia's post-communist transition is not one but a sequence of transitions, reflecting the somewhat haphazard and unfinished nature of change (McFaul, 2001). The first was the failed revolutionary attempt in 1991, the culmination of the changes initiated by Gorbachev in 1988, but resisted by 'hardliners' within the Party, in a coup attempt while he was by the Black Sea on vacation. The second, failed attempt at moving towards a democratic state, in the fashion of Eastern Europe at the end of the 1980s, took place in 1993. It was another military-backed attempt at repudiating the planned changes to Russia's constitution by Boris Yeltsin, the new Russian president. This second challenge was the starting point of installing new institutions as a framework for a post-communist *Russia*, and no longer the Soviet Union. In 1993, Russians approved the new constitution in a national referendum, but there were no immediate elections to maximise the political legitimation of these institutions by giving them public approval. Thus, they remained, in essence, installed 'from above'. Nevertheless, these arrangements have remained in place to this day.

Comparatively speaking, Russia has thus been through a series of post-communist regimes along its path of transition, each with a different territorial and/or political focus and underlying rationale. The inevitably ensuing uncertainty was exacerbated by the inability to develop and establish new structures in tandem with the collapse of the old, leaving the country and political actors in limbo. Having had the political-ideological rug pulled from underneath the existing state-societal arrangements, the result was an inability to establish a comprehensive new structure, or set of actors and policies, to maintain a continuity of effective governance. This meant a lack of 'guidance' when it came to the privatisation of state assets, and the delivery of public services. This void, together with the somewhat detached introduction of 'democracy' without much popular involvement, and thus no real popular ownership of the process and outcome, are key distinguishing factors between Russia's and CEE's versions of post-communist transformation. Other important factors include the personalities of the key actors, especially the Soviet Union's General Secretary Gorbachev, the ascendance of Boris Yeltsin from provincial office to Moscow's mayoral office at the end of the 1980s, and the multi-ethnic composition of the Soviet Union and Russia.

The political choices and resulting actions of the key individuals need, of course, to be viewed against the legacies and historic factors underpinning the emergence of the Soviet Union, including the relationship between Russia, as the lead nation, and the other nationalities (see also McFaul, 2001). Circumstances and political objectives do, of course, vary over time and between places, and so do responses to perceived challenges. Gorbachev, for instance, responded to the perceived terminal decline of the Soviet economy by initiating liberalisation to breathe life into the ailing state-controlled system. This initiative was inspired by his insights into the developmental gap between the communist and western market-based systems, during a visit to Canada in the mid-1980s. This made him question the wisdom of the conventional communist mantra of the system's inherent superiority, when the 'real world' suggested otherwise. His questioning of the fundamentals of the communist ideology, not surprisingly, alienated hardliners within the system, who feared the collapse of the Soviet Union, and thus loss of their privileges. Out of desperation, they sought to pull the emergency brakes on the reforms through their coup against Gorbachev in 1991. This, however, achieved the opposite effect, accelerating change by discrediting the hardliners and what they represented.

After a brief period of regrouping within the political establishment, with no clear direction towards an institutionalisation of democratic principles, another confrontation between the contradicting elites resulted in 1993. It brought to a head the inability of the elites to overcome their fundamentally opposed views about the future of the Soviet Union, the role of Russia, the establishment of a planned market economy, and, especially, democracy. With no experience of political bargaining and negotiating, compromise was not considered an acceptable way forward. The different phases in the dismantling of the communist state reflect the shifting power relationships between those seeking change and those wanting to retain the status quo. After the 1993 'battle' of the Russian Parliament, with President Yeltsin pictured in the global media on top of a tank outside the building, symbolising the defence of the changes against the old guard, the modernisers clearly had the upper hand. This allowed Yeltsin to effectively dictate the new democratic principles of government and related institutions. 'Imposition, however, rarely produces liberal democratic outcomes. The mode or path of transition influences the kind of institutional arrangement or regime that eventually emerges' (McFaul, 2001, p. 22). The more confrontational the transition process, the less likely the outcome is to be genuinely democratic (popularly supported), especially if all the arguments remain entirely within the political elites, excluding the public realm from any meaningful involvement.

It has become increasingly clear over the following years that:

> 'democracy from above' and the move to a law-based state through legislative enactment cannot succeed without the development of an appropriate culture within the society as a whole. Not only must the old formal structures be eliminated, but the old patterns of thought and the informal

> structures of power must be superseded by a new culture of politics. Such
> a culture must recognise the legitimacy of certain sorts of governmental
> activity on the part of independent political actors. Such a culture cannot
> emerge through legislative *fiat* ... Such a culture can only develop as
> a powerful entity, if it does so through its own means. It must be self-
> propelled ....
>
> (ibid., p. 229)

But there was no precedent, nor time, to develop such a culture during that initial period, and the jury is still out about the scope for it to develop in the immediate future. The inability to negotiate a compromise in the competition between the paradigm of a Soviet statehood and an emerging Russian national awareness meant that only one of the two could survive. During the early 1990s stand-off, power, including public support, shifted from the Soviet to the Russian leader, as did the strategic policy focus with its move from a concern for maintaining the Soviet Union towards securing Russia's newfound independence.

The nature of the Soviet Union as an assemblage of nations, ethnic groups and formerly independent states, held an inherent volatility, which those seeking to maintain the Soviet Union's integrity could only see being achieved through the old order. Once this essential principle had been abandoned, the 'actual' trans-formation process of economic, political and institutional reform began, and new identities and territorial reference points were created. It was not until the second coup attempt in 1993, however, that the urgency of establishing formal institutions and principles to accompany the bandwagon of 'liberalisation and marketisation' had become evident. Actors had clashed first over the principal issues of sovereignty and territorial power and control, and then over economic issues, while concerns about political reform were shunted to third position. When it was sought to put them into practice, however, much of the initial euphoria, political engagement and democratic interest by the population had cooled down again, against the backdrop of the experience that changes did not bring about immediately the imagined (especially economic) rewards expected from them.

This almost inverse sequence in the reform process, together with the absence of a clear majority among the political elite in favour of reform, marks a major difference from Central and Eastern European countries. A roughly evenly sized conservative and reformist camp within the administrative-governmental *elite* made negotiations lengthy and more difficult, as neither was able to effectively dictate the terms to the other, weaker side. In most CEE states, by contrast, the democracy movements, driven by popular support and voice, were clearly domi-nant. Based on the variations in the relative importance of 'reformers' versus 'conservatives' in Central and Eastern Europe, and Russia/the former Soviet Union, three main patterns of democratisation can be distinguished (see McFaul, 2001, pp. 20–1):

1   A clear dominance of the reforming, democratic forces within society (not just the elite) leads to a swift and wholesale shift towards democratic government

structures and politics, and liberal market economy, without contestations, as was the case, generally, throughout CEE and in the Baltic States.

2   At the other end of the scale, the incumbent authoritarian forces retain their strong position and are thus able to defend the status quo, albeit usually with a drop of the term 'communist' and adoption of nationalistic language. The outcome is a post-communist, or rather post-Soviet, authoritarian regime. Many of the central Asian republics of the former Soviet Union fall under this category, such as Uzbekistan, Kazakhstan and Turkmenistan.

3   In between those two uneven distributions of power sit the countries with a much less clear situation. Small differences in the standing of reformers against 'the old guard' lead to contestations and, if inflexible, confrontations. Russia, Ukraine, Tajikistan, Bulgaria and Moldova exemplify this 'messier' transitional path.

Does the eventual 1993 arrangement, and thus the installation of the Russian state as an autonomous entity, mark the end of transition, asks McFaul (2001). That, he points out, depends on the goals set and the yardstick applied. *Technically* speaking, as far as the installation of a democratic system and market-based economy are concerned, transition seems to have reached its destination. The situation is less clear when including the practice of democracy. Has it made it to the hearts and minds of all actors, including the general public, and embedded itself as the 'natural' form of the political system? *Civil society* has taken an important position in the discussions on democratisation, the state, democracy, and so forth. Important is not just the existence of individual, independent groups of interests and representation, but also their scope to make an impact on actual policy. This implies the existence of accepted rules of behaviour, circumscribing the state's sphere of competence. This also means sufficient visibility of grassroots organisations, local NGO representations, and external funding for explicitly democracy-building organisations and initiatives. In the absence of much of this, it is not surprising that there is rather limited public participation in essential democratic activities (Crotty, 2003), including voting, and thus limited scope for the development of an active civil society. The weak position of political parties, especially on the left (Christensen, 1998) in Russia, contributes to the general lack of mass-based interest groups and representation. Their limited visibility, even on the Internet, where their websites are not advertised and are difficult to find for web search engines (Semetko and Krasnoboka, 2003), is evidence of the limitations to a functioning civil society. 'Self-interested motivations for adhering to democratic rules have not translated into normative commitments to democracy' (McFaul, 2001, p. 4), and the pursuit of individual, egoistic objectives is still paramount.

Russia's own transition as half-finished democratisation and continued elitist rule in a 'formal democracy'

The particular feature of Russia's path of 'transition' is its failure to 'go all the way' to being a fully democratic polity. Instead, it got stuck between different transitions, and the process of change has been protracted, conflictual and imposed from the top. McFaul (2001) argues that the particular nature of Russia's transition from communist rule has impeded the consolidation of liberal democratic institutions and values, but it could also be said that Russia's particular history and legacies shaped the very nature of this process of change, thus reproducing and ingraining the underlying structural and ideological differences between Central and Eastern Europe and the former Soviet Union. One such difference is the varying degree of involvement by the public in the actual transition process. While public engagement was instrumental in setting in train the processes ending communist rule in the CEE countries, it did not become involved in Russia until the final stage of building an independent state. This difference in the nature of transition may be seen in conjunction with the varying quality of democratisation between the CEE and FSU countries, with an emphasis on formal structures over practised democratic principles in the latter. Russia's historic authoritarian legacies may contribute to this difference. Evidence from other post-authoritarian, *formally* democratic regimes suggests that not necessarily fully practised democracy may follow (see McFaul, 2001). The current outcome of Russia's post-communist transition appears to show few signs of progress towards liberal consolidation.

One visible indication of Russia's 'in between' status on the notional route towards democracy is the weak, largely marginal status of political parties as independent political actors, rendering them rather ineffective as a counterbalance to growing presidential power (Christensen, 1998). This may be viewed as demonstrating the second differentiating factor of Russia's transition – a weak development of civil society. Obviously, this is in direct correlation with the rather limited participation of the public in bringing about the end of the communist regime in the first place. Much of the political representation, and dealings with the executive, bypass the parties, and work instead through networks of established political elites of the state apparatus. This includes the new business elites in a form of state corporatism, an arrangement that allowed the emergence of the oligarchs in the wake of the privatisation process. Effectively, therefore, politics is being made over the heads of the political parties and, by implication, of the general public. Political parties appear more a decorative element, populating the Parliament (Duma) as a 'must have' feature of a formal democracy, rather than as an instrument of effectual democratic policy-making. This is in considerable contrast to the CEE countries. Aside from the Baltic States, Ukraine is so far the only former Soviet republic where such a formal democratic arrangement has, just now, been democratised *a posteriori* through grassroots movements. Yet to what extent this leads to the firm establishment of a democratic polity in the medium and long term remains to be seen.

There are several reasons for this weak position of political parties (McFaul, 2001). For once, after 70 years of Communist Party control of all aspects of life, there was strong resentment of 'parties' and party politics, and getting involved with parties seemed the last thing to aspire to. The absence, in contrast to many CEE countries, of a pre-communist legacy of democratic parties, that could have been used as a 'bridge' over the communist years, meant that there were no positive role models and experiences with democratic principles, including political parties.

Instead, with no more compulsory attendances at political events and party meetings, withdrawing from political engagement altogether was a frequent, immediate response to the end of the communist regime's control of people's lives. The second factor concerns the extent of transformation of Russian society after the end of the communist regime's control over the assemblage of nationalities that constitute the state. Many new divisions emerged and old certainties (social standing, economic security) were lost, all creating a shifting and unclear arena of diverse, often unfamiliar, political issues. While some of the new parties focused on specific 'popular' issues, most adopted a more general programme about outcomes of the transition in general, to increase their chances of appealing to a larger part of the electorate. A simplistic contrasting of conditions before and after the changes was a favoured format, because it was easy to politicise and for the electorate to relate to. But such simplicity tends to obscure underlying trends and individual outcomes.

The strong corporatist element that developed in the newly formed Russian state under its first president, Boris Yeltsin, was the third key factor undermining the role of political parties as independent arbiters of popular political interests. Privatisation was dominated by a few large actors able to accumulate big chunks of the state's assets, often through existing insider knowledge and close connections to the political system (Blanchard, 1997). The outcome was that a relatively small elite shaped the nature and course of transition both economically and politically, while leaving much of the Russian public effectively in little more than a spectator's role. Furthermore, economically, this capital concentration process has had a limiting effect on the role of small to medium-sized businesses (SMEs) and, as a consequence, the relevance, both economically and politically, of the emerging new entrepreneurial middle class. Remarkably, there are few signs of the nature of transition – 'shock' versus 'gradualism' – having had a significant impact on business productivity (Brown and Earle, 2003), and the relative under-representation of the traditionally more entrepreneurial and innovative SMEs may well be attributed to that. The narrow source of entrepreneurialism becomes apparent when making comparison with CEE countries such as Poland, which, with a quarter of Russia's population, boasted some two million non-agricultural businesses in 1996, compared with Russia's 900,000; this represents an eighth of Poland's density of business formation.

In Russia, 'exorbitant taxes, inflation, lack of liberalisation at the local level, the mafia, and monopoly-controlled markets, have combined to create a very difficult environment for market entry' (McFaul, 2001, p. 319). The main reasons for this sluggish development of SMEs in Russia lie *inter alia* in a lack of an entrepreneurial tradition, a weak and fractured government that could easily be influenced by powerful interest groups (e.g. large businesses), strong income inequalities, corruption and business-unfriendly legislation with a lack of clear rules and legal certainties (Kihlgren, 2003). An important, traditional backbone of civil society with a keen interest in the representation of democratic interest, is thus largely missing. Instead, politics tends to be dominated by political dealings 'at the top', including the oligarchs, who control much of the country's capital assets, and it is heavily Moscow-centric. Feelings among the general (business) public of not really

being 'part of the game' contribute to more lethargic attitudes towards participating in politics: 'What does it matter?' may be the conclusion drawn by the public
when assessing whether it's worth the effort to vote, for instance.

Taking these factors together, therefore, the slow development of civil society in
Russia is not surprising. For one thing, the long period of Soviet control has done its
best to excise any independent, non-organised, genuinely bottom-up, let alone
spontaneous, expression of political interest. There was thus no civil movement as
seedbed of a popular understanding, pushing for change. In Poland, by comparison,
the strong role of the Catholic Church was instrumental in forging the formation
and articulation of resistance to the communist regime as early as 1981, through
the Solidarity movement. In Russia and the former Soviet Union, the extra length
of communist attempts at suppressing traditional religion as a potential rival in
capturing people's loyalties, had done its best to remove such a point of reference
for possible 'counter-revolutionary activities', as any expression of criticism was
immediately branded. There has thus been much more to be re-/built as far as an
actively engaging civil society and a practised democracy are concerned. Although
many personal networks existed, frequently, as a means of self-help to bypass institutionalised obstacles to the conduct of everyday life, translating that into active
democratic political engagement is still a relatively big step to make. Under
communism, official organisations were installed to control society, not to act as its
agent, but society needs to realise and internalise the different roles such interest
representational organisations are meant to take; they need not be instruments of
state control *per definitionem*.

The biggest challenge to the development of civil society has been the cost of
economic restructuring, of the 'recessionary transition' approach (see Chapter 3)
excluding many from the newly emerging opportunities. The oligarchs are an
illustration of the elitist outcome of economic change, while imposing costs on the
many through job losses and much reduced welfare. Privatisation they witnessed as
a scramble for the state's assets by a few well-connected individuals, making official
commitment to democratisation look rather less convincing. Many felt like mere
bystanders at events that affected their livelihoods (as under communist rule), without permitting them, whether effectively or perceptively, much scope for having an
effective influence.

Nearly 15 years after the formal end of the Soviet Union and the communist
regime, Russia and most of the former republics still have some way to go to achieve
full democratisation, if, indeed, that is the target they set out to achieve. The degree
to which changes towards democratisation have been achieved, vary, of course.
Indeed, there are many signs that Russia, for instance, is not moving towards more
'real' democracy with any undue haste. Vladimir Putin, since coming to office in
2000, has sought to increase his power as president, while weakening the representational institutions, as well as press freedoms. Rather than moving towards a
stronger civil society, a shift that seems rather illusionary at the moment, scope for
its development has been steadily curtailed. So far, there has been no sign of public
pressure, as in Ukraine, for instance – having been at first sight, a less likelier candidate for democratisation than Russia. The centralising tendencies of the Russian

presidency include 'cracking down' on the oligarchs, too, if they are deemed to be potential political challengers, as the Khodorkovsky case has demonstrated. A similar rationale is likely to lie behind the constitutional change to making regional governors Kremlin appointees, rather than democratically elected. Apart from thus reducing the 'risk' of them forming their own power bases, it also disconnects the public even more from the state and its political machinery.

The unclear distribution of powers and responsibilities among the centre and the regions and municipalities has added to a general sense of uncertainty, and has made public participation more difficult. The creation of a federal order out of a highly centralised state has been difficult, too, as areas of responsibility had to be negotiated. Rather than being statist, like the communist state, the federal arrangement contains an inbuilt dynamism between government tiers about responsibilities. But this can work both ways – centralising or decentralising. For Russia, the former seems now to be the case. 'In fact, the state has begun to be "re-nationalised", with the influence of the oligarchs curbed and state functions restored to the state' (Sakwa, 2003, p. 462). This seems to invoke a considerable danger of shifting further away from civil society, rather than towards it.

The resulting political order is thus a hybrid of different traditions, or parts thereof, a varying mix of old structures and practices, and new conditions. The outcome is an unstable balance, depending on the relative importance of the different factors at particular times. One important legacy is the fact that 'the democratic state-building slogans of the early years were trampled on in the rush for power and privileges of a narrow political and social elite' (ibid., p. 465). Democratisation in Russia has thus been a result of varying 'mediated outcomes of the asymmetries in access to power and weaknesses in the accountability of that power to society's representatives' (ibid). With civil society still in its infancy, largely bypassed by a network of a powerful elite at the top of the national economy and government, the difference to many CEE countries, especially those in Central Europe, becomes obvious. The unclear Russian nationhood stimulated an exaggerated notion of statehood and accumulation of power, and thus the acceptance of the perceived predominance of state over society, raising questions about the likely future scope for democratisation.

Effectively, the 'party-state' that had emerged under Soviet communism has transmuted into Russia's regime state, following the principles of a 'delegative', rather than 'representative', democracy (O'Donnell, 1994), as was envisaged by the West. This entails a particularly pre-eminent governmental authority of the presidency, closely linked to the personality of the incumbent of the time. Personalities and interpersonal connections and networks thus emerged as the channels of power and policy-making, rather than the formally established institutions. This new form of regime looked both ways for inspiration: back to the command-style policy-making of old, with its bureaucratically regulated economy; and forward to a genuine separation of powers, and of the economy from the state, as is the case under 'western' liberal market democracies. The outcome is a synthesis drawing on old established ways of doing things, if with a new way of interpretation and rationalisation, while seeking to incorporate new forms of policy-making and structures, as well as actors, to move towards democracy. This ambivalent 'in between' approach has been followed until today, with a varying emphasis on 'old' and 'new', depending on the personal political agendas of the president and his immediate political elite, and political expediency at the time. Currently, under Putin, the state seeks to expand its capacity and control at the expense of the corporatist 'regime elite', comprising economic leaders, especially the 'oligarchs'.

The oligarchy that emerged from the particular way of privatisation, favouring those with insider knowledge, represents a fusion of financial and industrial capital with direct access to government. By their very nature, they undermined the separation between the market and the state as the backbone of a liberal market economy, placing in their stead informal (and thus unaccountable) lobbying and network connections. In effect, non-elected and non-accountable actors gained quasi-governmental status. But in exchange for their favourable position, loyalty to the political leadership is expected. Otherwise, loss of status, influence, and, ultimately, freedom, are the price to pay. Inevitably, tensions emerged between system and (actual) regime, especially between the institutionalised and personalised expressions of political authority. There is a danger that the growing importance of personal links renders institutions rather irrelevant, reducing them to mere institutional façades of democracy.

Post-communist Russian politics has thus been shaped by struggles between the competing and contrasting policies of opening up internationally, especially economically, while at the same time seeking to define and assert a new sense of identity within the new, post-imperial, international setting. Sakwa (2003) speaks here of a struggle between globalisation and nativisation. As a result, Russian policy has often looked confused, torn between security concerns, democratisation and defining its new identity. In Russia and most other FSU states, the legacy of a long-established patriarchal community and form of governance, combined with a strong egalitarian paradigm under communism, has made this balance particularly challenging. It is not just the institutions that need to be shaped, established, and rooted, but also the mindset and way of doing things. In many ways, this marks the divide between a mere formal and a practised, popular democracy. The resulting uncertainty in the mid and late 1990s, however, detrimentally affected the perception of Russia as a place to invest, thus hampering economic development, and with it, demonstrable 'success' of transition in the public's eyes.

The elitist, formal and delegative nature of democratisation in Russia may encourage comparisons with such processes in Latin America, characterised by growing class divisions, semi-privatised and state-owned businesses, and the absence of a strong middle class. However, the main difference in Russia was the absence of the military as a key player in the process of change. The absence of private property also meant the lack of a well-established and connected oligarchic property-owning class at the time. There was thus some scope for the political elite to battle their differences out independently of other groupings, although this separateness meant that society at large did not gain much influence or ownership of the democratisation process. Thus, 'Russia generated its own synthesis of tradition and modernity, of old and new elites, of legal-rationality and charismatic rule' (ibid., p. 442). But, as a direct legacy of its multi-ethnic nature, the detached, top-down dispensed democratisation also allowed old, previously suppressed differences to re-emerge between Slavophiles and Westernisers, and between nationalists seeking to reaffirm national independence and autonomy, and liberals wanting to engage with globalisation and internationalisation. When it came to economic change, however, the differences between the various social groups were much smaller, which limited the interest in engaging in political arguments, weakening the role of political institutions in the democratisation process (Ahl, 1999).

The competing ideologies and 'recipes' advocated for Russia's future during the 1990s revolved around the question of the nature and quality of democratisation. This aims effectively at 're-civilising' the country by establishing a civil society, and 'correcting' communist-era 'mis-developments', especially dead-end old-style industrialisation. The latter, inevitably, would lead to enormous replacement costs, usually in conjunction with a new geography of new investment. As a result, to the public's eyes, destruction and democratisation went hand in hand, offering a very different image and experience from that followed by the Chinese. There, 'reform', rather than 'transition', and economic growth went hand in hand, at least for the majority of China's urban centres, especially those on the southeastern coast (see Chapter 6). The particular challenge has been, as in the CEE countries, the complete restructuring, not just of the political-institutional system, but also of social *and* economic structures, making it particularly difficult in all that flux to find a reasonable 'fit' between the two. The way this 'fit' is arrived at, and its quality, are fundamentally shaped by the self-perceptions of Russian society, its legacies and interpretations and, especially, its new identity as 'Russia' in a post-Soviet, post-imperial, setting.

The rapid and tumultuous end of the Soviet Union led to an immediate disintegration of the existing structures and linkages that had developed across the Union. These include cultural, economic, political and social-ethnic differences, which had been kept under the seemingly unifying mantle of Soviet statehood. With the removal of that cover, underlying competing ambitions and variations in historic and cultural legacies, as well as geopolitical and economic geographies, came to the fore, leading to a much more differentiated and heterogeneous picture than had been visible previously. The following sections look at some of these diversities in cultural, economic and political respects.

Competing post-Soviet identities of Russia

Russia's and the other former Soviet republics' paths of post-communist, post-Soviet, development, have been fundamentally shaped by the 75 years of Soviet policies and attempts to forge a new Soviet identity and statehood, on top of the many existing national identities across the Union. Russia's search for a new identity is of particular interest, not only because it is the largest state, and legal successor to the Soviet Union, but also because of its special role and position within the Union framework. Russia was the clearly dominant Republic, and its language, territory, culture and interests became extended across the Union. As a result, Russian identity became intermingled with Soviet identity, not only from an outsider's perspective, but also from within Russia. Loss of the superpower status, and of territorial control, had affected the Russians' sense of pride and self-percep-tion. 'For Russians, adjusting to their new [smaller, more peripheral] homeland status has therefore been disorientating and painful' (Smith, 1999, p. 48). At the same time, many Russians considered the loss of the borderland republics, especially in the south and in central Asia, with their continued need for economic assistance, as beneficial to Russia's own economic development. By the same token, there is also concern about a possible continuation of resurgent nationalism within the multi-ethnic Russian state itself, something vehemently resisted, as the ongoing battle with Chechnya has illustrated.

There are three main competing discourses on Russian identity (see Smith, 1999) that have gained importance with the collapse of the Soviet Union, whose territory and geopolitical standing as a 'superpower' had been closely interwoven with Russian identity.

1   *The 'westernising perspective'*: Russia, as part of 'the common house of Europe', in Gorbachev's words, is seeking to 'join' Europe, thus overcoming its geographic peripherality. It is an ambition most starkly reflected in Peter the Great's building of St Petersburg as 'Russia's Window to Europe'. Adjusting to 'western' values and ways of doing things is an integral part of that view.

2   *The historic Slavic perspective*: Russia as having a particular culture, distinctly different from that of 'Europe' and 'Asia', emphasising the Slavic heritage, especially peasant culture and social values (sense of community). 'Western-isation' is seen as an erroneous attempt by the urban elites to abandon this heritage, and thus 'sell out' to alien western (urban) values.

3   *The geopolitical perspective*: Russia as a bridge between Europe and Asia, with Siberia as the main connecting commonality (stretching across the continent). Asia was seen more as an imperial back-up to Russia's geopolitical aspirations, its 'natural' backyard, rather than an equal part of the state proper. Russians do not consider themselves as having European cum Asian identities.

These different identities have been invoked in various combinations and with various emphases, depending on the target audience and the purpose of projecting that image. Gorbachev, for instance, made use of this multiple identity by projecting one or the other to his different international audiences, all for best political effect. Neo-nationalists want to extend Russia to its maximum imperial extent, using this extended territory to sustain Russia's economic development, thus clearly aspiring to a neo-colonial status. Then there are the neo-Soviets, viewing western values as alien to the Russian nature. Renouncing property ownership and self-centred capitalist (bourgeois) values as 'un-Russian', they advocate a 'natural' socialist Russian identity and way of doing things. Eurasia is seen as the natural, legitimate, and essential Russian homeland.

The other post-Soviet fall-out for Russia's identity and national self-perception is the alienation of some 20 million ethnic Russians within the former borderland republics, who are now considered 'foreign', and legally and rhetorically 'othered' by the host nations, albeit with considerable variations in intensity. It is strongest in the Baltic States, while less so in Belarus or the central Asian republics, for instance. The Russian state sees itself as the legitimate homeland of those ethnic Russians who find themselves transformed from part of an elite within 'their empire' to a barely tolerated (if sizeable) minority with reduced citizen rights (as in the Baltic States). Having not possessed separate representations and offices, in contrast to the other 14 republics, Russia and the Soviet Union have tended to be seen as synonymous, both within the Union and by the outside world. Denying a clearly separate identity and representation to Russia was a deliberate attempt by Lenin to reduce Russia's visibility and evident dominance of the Union, thus reducing potential rifts with, and ill-feelings among, the other nationalities. But the long-term outcome has been that Russia's own perception of, and attitude to, its position within the Soviet Union has become somewhat blurred, and that has raised its desire to re-/gain a visible independence and identity.

Russia sought its own statehood and its own institutions to underpin that, and this very much influenced policies towards the Soviet Union in 1990 and imme-diately thereafter. With the Union looking increasingly weak, the republics, and Russia in particular, sought to affirm their independence, and the issue of sover-eignty became paramount – more so than democracy. As a result, each former republic sought to find its own way of post-communist development, separate from the Union, and with more, or less, democratisation. 'The Union republics of the USSR began to take responsibility for their own affairs' (Smith 1999, p. 17), each following its own path. The centre of political bargaining thus shifted from the Union to the republics, with Russia, not surprisingly, taking a predominant position, thus becoming the main challenger to the Union. Not surprisingly, reflecting the very different cultural and historic legacies in the former republics, in many of them the 'old communist elites managed to convert themselves into nationalists and have continued to rule on the basis of the new ideology. The sovereignty of the republics has thus not been a triumph of democracy and civil society, but the establishment of the borders within which both might later develop' (ibid., p. 19).

The reluctance and inability of the Communist Party of the Soviet Union (CPSU) to reform and become the driving force of the modernisation and transformation of the SU rendered it increasingly out of touch and, ultimately, irrelevant. Changes then took place anyway, without the Party, sweeping it aside as a no longer wanted relic of the past, along with the associated institutions and arrangements; 1991 marked the end to it. The Party was no longer able to retain the vital link between ideology (as legitimation) and organisation (its role in politics and society), and modify and/or re-present it in response to changing circumstances in such a way as to retain a legitimate claim to power.

New administrative-political divisions

The collapse of the Soviet Union reinforced existing, and created new, divisions both of the administrative-governmental and the more statistically based, social-economic kind. Both occurred at the level of the former republics, and also within the newly independent states. The former symbolise the ex-republics' new independence as separate states, with the borders between some particularly 'high', such as between Russia and the Baltic States, where they serve as new 'demarcation lines' between a Soviet and newly emphasised (western) European belonging. This becomes particularly obvious in the new 'Iron Curtain' running through the border town of Narva, marking the now international and European Union border between Russia and Estonia. Here, a new 'high' border separates the previously integrated town straddling the Narva River into two: Ivangorod and Narva. With 93 per cent of the population Russian-speaking, their first response to the separation of their city was to be repatriated with Russia by drawing the boundary around the western edge of the city (Urban, 2003). But this, of course, was resisted by the Estonians. The creation of new boundaries, both physical and mental, has been one critical outcome of Estonia's (and the other two Baltic States') independence – suffering from a lingering anxiety about Russia as the sovereignty-threatening 'outsider' across the border (Vetik, 1998), while also worrying about a loss of sovereignty when joining the European Union as a 'safe haven' (Kuus, 2002). Thus, in Berg and Oras's words, 'at the moment, there are more barriers than gateways on the mental maps' of Estonians, unsure whether they feel as the last bastion of western culture, or as an international, outward looking space linking 'east' and 'west'(Berg and Oras, 2000, p. 623).

Within the new states new divisions emerged too, some initiated through administrative changes and an emphasis on decentralisation, as in Russia. Other divisions resulted from newly emphasised ethnic identities straddling the new state borders, and seeking their own representation and territorial expression – sometimes even with Moscow's blessing, as part of a divide and rule approach to counteract rapidly growing power bases at the regional level (*oblast*) (see e.g. Khakimov, 1996). The ongoing battle between Russia and Chechnya is one such example, fought with particular bitterness and vehemence as it is also seen to be of symbolic value, potentially setting an example to follow, if allowed to secede. But there are also less headline-grabbing processes of division, such as the emergence of greater regional independence within Russia, a process President Putin tries to reverse. The Soviet legacy has been an administrative division into regional entities, drawn up for political and administrative convenience, for the government in Moscow, to permit more effective central control, rather than to serve as an expression of decentralised representation and territorial management.

Since Russia's independence, during the Yeltsin era of the mid-1990s, regions were given greater autonomy to manage their own affairs, especially in terms of economic development, but also politically. This was largely a response to the perceived flaws of the overt centralisation of the Soviet state, widely held responsible for the economic underachievement of the former Union. As a consequence, as Baev (1996) observes, 'the regional elites are converting their former *virtual* [emphasis added] property (administrative rights) into real property (control over natural resources, industrial bases, agricultural lands, etc)' (ibid., p. 372). The need for the central state to woo the regions and their new, increasingly more influential elites, meant further concessions to regional political and economic autonomy, weakening the Russian state's control further, to the point where it was struggling to call in its due share of taxes (Baev, 1996). Because of this new regional control of assets, economic differences are growing, as the uneven distribution of natural resources and the inherited economic structure were exploited, to maximise return for the regional economies and their controlling elites. But any such figures need to be seen against the backdrop of attempts to undervalue regional output figures, so that relative 'poverty' can be claimed and federal aid be attracted from Moscow, while also saving on taxation because of the lower than actual income figures. This emphasises inter-regional differences in economic performance as far as official statistics are concerned, but there is also a political price to pay. 'Quite typically, the richer regions have more political clout in Moscow, and are able to secure the distribution of the meagre resources in their favour; marginalising the poor' (ibid., p. 373).

This decentralisation of economic power does not, however, mean an automatic shift to democratisation and the devolution of power to the people. Rather, it reflects the emergence of new fiefdoms, 'run' by emerging, powerful 'tsars', often in conjunction with equally emerging economic 'oligarchs'. This development challenges Moscow's traditional hegemony. Some Russian commentators have likened this shift from national to regional autocratic power to a form of 'neo-feudal division' (Baev, 1996). Basing their power largely on the rural districts of the regions, the urban districts, through the mayors, are often the only effective political challengers to the patrimonial style of regional governance pursued by these new 'barons'. They have little tolerance for signs of opposition or questioning of their policies, while vehemently defending 'their' regional autonomy *vis-à-vis* Moscow. With electoral procedures and outcomes not always very transparent, usually favouring the incumbent holder of the regional leadership, genuine democratic principles, or even a civil society, are far from obvious. It is not surprising, therefore, that Putin tries to rein in the independence of these new regional centres of power, by making the regional 'tsars' his appointees, rather than elected politicians from within the region. However, this is more an expression of political rivalry, and an attempt by Moscow to repatriate power, than an attempt at tackling the widening regional imbalances in a 'multi-speed and variable-geometry federation' (Baev, 1996, p. 375), and whether the presidential announcement of 2000 to 'transform local government into a lower level of "vertical executive"' (Gel'man, 2003, p. 57) – thus substituting local autonomy (and with it, the potential for the development of civil society) with top-down centralism – will serve that objective, remains to be seen.

Economic divisions as legacies of a colonial-style space economy

The territorial divisions in economic activity across the former Soviet Union meant that after its break-up, the various newly independent national economies would have specific specialisations, but would lack many other essential activities. At least for the time being, some form of maintaining an economic tie-up was considered crucial, even if countries wanted to go it alone politically. The outcome was the Commonwealth of Independent States (CIS) as a loose association of participating countries, unthreatening in its nature to the newly established countries' national autonomies. Inaugurated by Russia, Byelorussia and Ukraine in 1991, it was joined by the other states soon after. Reflecting concern about the sudden disintegration of a some 70-year-old economic and administrative structure, there were, initially, great ambitions for establishing something akin to the European Union in the longer term, but the reality now looks rather different. With new national identities and nationhoods firmly established, and the respective national economies re-connected with the outside world, there is much less sense of sharing common interests and requiring some form of common reference point. The CIS has thus lingered, while silently fading away. The political elites of the new independent states are quite happy with their new-found roles, and would not want to be reduced to the second rank of a mere regional leadership. In addition, a new generation is growing up with a more detached view of the immediate Soviet past. They have fewer 'hang-ups' about that time, its achievements and failures. Instead, they feel more part of a global age and its 'western' symbols (Dawisha and Turner, 1997). Even in Belarus, the most staunchly authoritarian European former Soviet Republic, a Belarusian, as against a Soviet, identity is slowly emerging among the younger generation (Grichtchenko and Gritsanov, 1995). This is the more sur-prising, as the Soviet-style authoritarian leadership seemingly seeks unity with Russia (a customs union exists already), thus effectively undermining any sense of separate identity.

The legacy of the Soviet Union was a tightly interdependent economic space, with clearly allocated economic specialisms in a spatial division of production across the USSR, directed and controlled from Moscow; this got abruptly split into separate parts, with new borders and varying economic policies. Many of the new states had been given particular economic functions as part of the Union-wide organisation of production. In 1990, trade between the republics accounted for some three-quarters of all imports in all the republics, and nearly 85 per cent of all exports, with all trade directed and controlled by Moscow (Schroeder, 1996). This meant a clear economic dependency of the borderland republics, especially those in central Asia, on their connections with the rest of the Soviet Union's space economy. The end of this structural context resulted in considerable structural adjustment problems – the 'post-Soviet ailment' (Primbetov, 1996. p. 164). Others, such as Ukraine, Georgia, Azerbaijan and Moldova, have sought to use their natural resources, and their transport links to western Europe, as a means to gain economic independence from Russia, while also pursuing attempts at securing their independence through new strategic alliances.

The Baltic Republics were the only former republics actively seeking to disengage from the still existing Soviet Union. Russia and the remaining 14 republics were 'catapulted into independence at the end of that year, when the Soviet state dissolved' (Schroeder, 1996, p. 12). Among the Baltic States, there was thus quite a different political background and determination to 'go it alone', giving them a head start in development. But, despite their differences, they all had to tackle the legacies of the Soviet economy – strong militarisation of production, specialised economies as part of the division of production, backward technology, uncompetitive labour skills and practices, massive environmental degradation, and an unawareness of the workings of the 'market' both for goods and labour. In all countries, the massive task of marketisation, liberalisation and stabilisation – the Washington Consensus mantra – has been directed from above, guided by (western) advisers and often copied from elsewhere. Still, each newly independent former Soviet republic developed its own ways of proceeding, based on local power relationships, interest groupings, the political capabilities of the political elite, and available economic resources. Russia's development, as the 'biggest fish in the pond', naturally affected the development of all the other republics', as all economic links and financial interchanges converged on Moscow. The rapid decrease in Russian economic and financial aid to the former republics stimulated their striving for independent, varying economic prospects.

For the newly independent states of the Caucasus and central Asia, economic prospects are particularly challenging (Bartlett, 2001), given their relative geographic peripherality, unfavourable topography, and difficult accessibility (being landlocked). Conventional development strategies, such as resource-based development, import substitution, export promotion, or neo-liberal marketisation *per se*, are unlikely to help in this respect. Bartlett (2001) suggests a conventional regional development strategy, including favourable regional trade agreements and financial aid from international development agencies. But any such strategy requires stable political conditions, especially clear ownership and power structures, and settled ethnic-territorial arrangements, before allowing cooperation between these new countries (Primbetov, 1996) and international finance. Particularly, ethnic rivalries make longer-term economic strategies difficult, and are an obstacle for international engagement in these countries. Another problem is the Soviet economic legacy, with reliance on primary extractive goods (mining) and food production. In effect, these economies resemble those of colonial dependencies. What existed in industrial capacity was largely defence oriented, and offered little in terms of export opportunities (Bartlett, 2001) and required new investment for any conversion to civilian products. Table 5.2 illustrates the limited degree of foreign trade and inward investment. The second main obstacle has been the dependency on the Russian economy, again, suggesting a colonial-style economic structure. This dependency is reinforced by Russian control of much of the energy sources, especially oil and gas.

Against this background, it is not entirely surprising that a ranking by Deutsche Bank, in the early 1990s, of the 'independence potential' of the 15 Soviet Republics showed Ukraine and the Baltic States with the highest, and the four central Asian republics, Kyrgyzstan, Uzbekistan, Turkmenistan and Tajikistan, with the lowest potential for independent, economic viability, and thus political independence (Schroeder, 1996, p. 11). One of the main challenges to independence was, indeed, seen in the economic dependency on one another, and on Russia, as the economic core of the Soviet Union, in particular. In 1990, between 44 and 75 per cent of all

Table 5.2 Economic situation in the Caucasian and central Asian states of the FSU

Country	Population (m)	GNP per capita $US	FDI per capita $US	Export as % GDP	GDP 1988–98 (%)
Armenia	3.8	480	26	19.1	−7.7
Azerbaijan	7.9	490	174	24.5	−11.5
Georgia	5.4	930	40	13.8	−15.4
Kazakhstan	15.7	1310	244	29.2	−7.0
Kyrgyzstan	4.7	250	207	37.0	−7.2
Tajikistan	6.1	350	7	n/a	−11.9
Turkmenistan	4.7	936 (GDP)	102	n/a	n/a
Uzbekistan	24.1	870	16	22.2	−1.8

Source: Bartlett, 2001

imports into the republics came from Russia, and between 37 and 66 per cent of all their exports went to Russia alone (Schroeder, 1996). Creating genuine independence, not just politically but also economically, thus meant massive restructuring of their economies. The main challenges were seen, *inter alia*, in an over-dependency on a narrow range of products, few of which had export potential, technological backwardness, large-scale environmental degradation and a general inexperience with market forces.

Another difficulty is the continued lack of democratic principles and, instead, a continued neo-nationalist autocratic structure with limited accountability. 'Strong-arm presidents [that] dismiss recalcitrant legislatures, intimidate opposition figures, rig their own elections to compile Soviet-type super majorities, and engineer constitutional amendments to extend their terms of office' (Bartlett, 2001, p. 142), are unlikely to attract foreign investment. It is a clearer, and more transparent and predictable, institutional-legal framework, together with an ability to manage political differences between the neighbouring states of the region, that are fundamental prerequisites for necessary economic assistance to become available (on the necessary scale) and translate into effective development strategies and outcomes.

Where successful, the regions, especially those able to exploit their natural oil reserves, are effectively going both global and regional, reflecting the fact that regional elites are becoming more independently minded and more firmly rooted in 'their' regions. Efforts to project a stronger resolve with regard to Moscow are illustrated by the fact that most regions have formed regional alliances to lobby Moscow politically and financially (e.g. the Siberian Lobby), and many regional politicians (governors) seek to use the representation of regional associations as platforms to launch their national political careers.

Based on their political and economic structures, and post-independence 'paths' of development, especially democratisation, the post-Soviet republics may be grouped into five clusters of relative political, structural and geographic-cultural similarities, as described below.

The Baltic States

The Baltic States –'late joiners' of the Soviet Union, with past democratic experience –were clearly determined to seize the opportunity, and leave the Soviet Union as quickly as possible, to regain their aspired independence and tie in with Europe. All successive policies were consistently tailored to that goal, especially EU membership. Their re-orientation towards 'Europe' is reflected in their trade flows, too. While in 1991 their economies depended up to 85 per cent on trade with Russia/the SU, and only 15 per cent with the outside world, the situation was reversed only four to five years later. Their joining of the European Union in 2004 marked the final stage in the reorientation to (western) Europe – legally and politically re-positioned on the 'other side' of the former Soviet Union's border.

The Russian Federation

Russia's transition has largely been shaped by its rather tumultuous move into independence, seen more as the inevitable result of political events than as the primary objective, with a sense of disorientation in its new global context. Its rather protracted and breakneck speed of change immediately post 1991 followed, at first, the paradigm of 'shock therapy', driven by a reform-committed, anti-communist leadership. President Yeltsin's personality, credibility as a 'man of change' and authority at that time were essential elements in the pursuit of this rapid approach. The price was a relatively weakened role of the state, undermining the goal of effective and beneficial marketisation. Only recently, under Putin's leadership, has the position of the state strengthened, *vis-à-vis* the economic elite, albeit in the company of continued centralisation and restriction of democratisation which, in turn, has caused concern in international politics and, especially, among investors.

The Western Republics of the 'near abroad'

Moldova, Ukraine and Belarus sit at the interface between Eastern European and Russian traditions. Considerable differences in their more recent developments have become apparent, with Belarus moving ever more firmly back to an authoritarian, neo-Soviet-style regime, while popular unrest and grassroots demonstrations brought about a shift towards democratisation and a European outlook in Ukraine. In Belarus, there are no credible signs of democratisation and economic reforms, and few signs of new developments. It is not surprising, therefore, that Russia is wary of adopting a struggling economy looking for assistance. Another difficult case, with few signs of democratisation, is Moldova, which is effectively split into two parts – the official territory of the Moldovan state, and the Transdniester region that proclaims itself to be a separate republic. Russia, ignoring the opposed world view, more or less supports the separatists in pursuit of its own strategic interest in the area, and generally seeks to maintain close links and influence.

Similar squabbles about post-Soviet spheres of influence have also affected Ukraine, particularly evident during the people's Orange Revolution of late 2004.

For once, this offered evidence of an (unexpectedly) emerging civil society, and bottom-up revolutionary pressure in the style of the CEE countries of 1989/90. It was surprising, given that in 2002 only about 15 per cent of the population claimed to have participated in local democratic processes (UNDP 2003, p. 121), but it also revealed the division between a Russia- and a Europe- (EU-) oriented half of the country, very much reflecting its position between the Russian and EU cultural and political-economic spheres of influence. Ukraine is, in this way, an interesting case of a post-Soviet definition of identity, and geopolitical and cultural-historic belonging. It also shows the underlying legacies of a post-imperial concern about maintaining established, 'owned' spheres of influence. Internally, Ukraine put in place the principles of more local autonomy, with its statute of 1991, although the subsequent decade was largely driven by centralist policies (Boukhalov and Ivannikov, 1995). As a consequence, local leaders saw little evidence of a genuine shift towards more decentralised decision-making, compared with the pre-*perestroika* days, and there was little immediate evidence of a genuine interest in democratisation (ibid.). In fact, shortly after independence, more than 40 per cent of the local leaders agreed 'that a few strong leaders would do more for the country than all the laws and political speeches', suggesting that there were 'already many signs of nostalgia for the "strong hand"' (ibid, p. 136). This observation supports the outcome of a comparative assessment of the support for transition in Ukraine, Estonia and Uzbekistan, which showed the Ukraine with the lowest level of support for transition (Hopf, 2002). The sudden outburst of grassroots support for democracy and 'Europeanisation' – as against Sovietisation – comes, therefore, as a particular surprise.

But there are also divisions within Ukraine, between the historically 'Russian' eastern part and the more Austro-Hungarian-influenced western part. The Ukraine's urban and industrial east is home to a high share of the 11 million Russians in Ukraine (about one-fifth of the population). Not surprisingly, among these newly 'near abroad' Russians, commitment to Ukraine is less than in the Ukrainian-dominated western part. Still, there is evidence of their identity going beyond the simple dichotomy of being 'Soviet' rather than 'Russian', as there are many smaller-scale, intra-regional variations, reflecting particular local features (see Dawisha and Turner, 1997). With internal variations between sub-regions, issues of identity-building are much more complex than the conventional East–West or Ukrainian–Russian paradigms would suggest.

With virtually no precedent of independent statehood to refer back to, there was no obvious path or model from which to 're-create' a Ukrainian state, such as existed for most CEE countries. This made a simple territorial approach appear politically and economically the most realistic approach. As a consequence, the Soviet-defined Ukrainian Republic was institutionalised as a separate state in its own right, but with different traditions, and different historic and cultural memories and affinities, captured in that territory, building a Ukrainian nation and identity has not been an easy task. The arguments propagated during the Orange Revolution illustrated that quite clearly (Kubicek, 2005). They mirror the country's straddling of the political-geographic (strategic) border between a Europe symbolised by the European Union, and a Russian perspective of Europe (see also Solchanyk, 2001), although the European Union put much less pressure on Ukraine to follow the 'road to democratisation', than it did on other countries (Kubicek, 2005).

Continuing economic, financial and strategic dependencies on Russia, while now demonstrating explicit pro-EU ambitions, clearly demonstrate the difficult position across historic-cultural and political fault lines. Economic decline and severe adjustment problems have shaped much of Ukraine's post-independence development. Most Ukrainian citizens' experience of post-Soviet daily life has been difficult, and anything but the initially hoped for materialisation of western-style conditions (Solchanyk, 2001). But the ongoing economic problems, with widespread socio-economic dislocation and disorientation across all social groups, has created a sense of 'shared grief' among citizens, and thus contributed to the tentative formation of a territory-wide sense of Ukrainian commonality. Not surprisingly, in 2002, more than 40 per cent of Ukrainians claimed that their economic situation had worsened since independence, although many of them were hopeful that things would improve in the future (UNDP, 2003). It is here that the European Union is being challenged in its understanding of 'Europe'.

Transcaucasia (Armenia, Azerbaijan, Georgia)

Post-communist developments in the central Asian republics have shown patterns quite different from those of the former Soviet Republics on the western, European border of the FSU. Here, a number of political, at times violent, crises occurred following armed uprisings against the established communist leadership, with ethnic groupings seeking to resurrect their competing claims for power and influence. Threats, or actual executions, of coups through armed gangs with opposing political interests, at times with ethnic undertones, have continued to shape the climate of post-Soviet developments. Accusations of vote-rigging and violence through gunmen have undermined democratic credentials. In addition, territorial uncertainties have added a sense of insecurity and threat, such as in the breakaway regions of Abkhazia and South Ossetia in Georgia, and the territorial dispute about Nagarno-Karabakh between Armenia and Azerbaijan. Russia claimed the mantle of post-colonial arbiter between conflicting interests, although not without its own political-strategic interests in maintaining influence. But despite struggling with military conflicts, steep economic decline, political uncertainties and slow democratisation, there is still a resolve to progress with economic reform, encouraged by the prospective wealth resulting from exploiting their natural resources for international markets. Reforms have been pushed through, although at different paces, facilitated by external (financial) support and the revenues from oil exploration and export. The role of the respective political leaders (presidents) has been instrumental in shaping the nature and progress of development. Thus, for instance, Azerbaijan has embarked on more visible economic change towards marketisation to appease and attract international finance.

Central Asia (Kazakhstan, Uzbekistan, Kyrgyzstan, Turkmenistan, Tajikistan)

Unprepared for the effectively 'forced eviction' from the Soviet Union's political and economic context, these comparatively poorest and economically most dependent republics faced a particularly difficult transition period. This is largely marked by generally slow progress towards market reform under continued control through authoritarian leaderships. Turkmenistan and Tajikistan are being run, effectively, as dictatorships, while Uzbekistan, although sharing an authoritarian leadership, has embarked on economic change to secure future development. There seems to be a widespread belief that strong leadership (or a 'strongman') equals a strong state, but that need not be the case (Fish, 2001). Often, the state is then reduced to effectively little more than an instrument of power for the 'strong man' leader.

Central Asia was among the first places to witness outbreaks of ethnic unrest and violence during the times of *perestroika*, reflecting a surfacing of underlying, suppressed tensions, and disaffection with conditions established under Soviet control, especially the lack of congruence between state territories and ethnic geographies. Here, the multi-ethnic nature of the Soviet Union became a difficult legacy for transition away from communist rule. This does not necessarily mean a shift towards western-style democracy, although there are examples of competitive politics and signs of democratisation in some countries, with Inner Mongolia being one, albeit a seemingly unlikely, candidate. Situated in a virtual geopolitical no-man's land between the Russian and Chinese spheres of influence, it was able to defy the regional trend of authoritarian leadership (ibid.). As also demonstrated in the former Yugoslavia, the end of communism allowed ethnic tensions and gripes to re-emerge, after the communist regime's 'lid' on such pressures had been removed through *perestroika* and, later, the collapse of the Soviet state altogether. With little or no experience of independent statehood, and with a sense of 'having been made independent', rather than having actively gained independence, the post-Soviet central Asian states have now embraced their independence from Moscow, and begun to exploit their growing financial muscle through the now possible export of petroleum under their own auspices. Financial independence has raised their confidence and brought a more assertive position towards Russia.

But the ethnically diverse states have faced internal difficulties with ethnic regions seeking to reclaim their own ethnic identities, and thus territorial-administrative separateness. The result has been competing claims to state control and national leadership. Independence from (European) Moscow's control had become the primary goal in the early 1990s, while democratisation was considered an optional (western-borne) 'luxury', and thus was awarded lesser immediate concern. Indeed, many of the incumbent regional communist leaders reinvented themselves as nationalistic leaders, and held their grip on power. Evoking their achievement of national independence and elaborating their fight against Soviet central control, served as their legitimation to retain office. As a result, democratisation effectively fell off the political radar screen, allowing the old communist guard to carry on, albeit 'relaunched' under the mantle of nationalism (e.g. White *et al.*, 1993).

Apart from Tajikistan, the other four remaining central Asian republics have largely succeeded in moving towards post-Soviet statehood without violence, although the outcome seems to be a very clear drift towards formalised and institutionalised authoritarianism, especially in Turkmenistan, Uzbekistan and, increasingly, Kazakhstan. Presidential election outcomes of well in excess of 80 per cent approval ratings, point to 'controlled' democracy familiar to other formal democracies in Asia or Africa, for instance. There is also a shift in geographical-political-cultural affinity, away from Moscow and its European outlook, towards the Middle East and Asia on the back of a re-emerging Muslim identity, although this is a nationally varied process.

New divisions and diversities at the local (urban) scale

As can be generally observed throughout the other post-communist states, whether in Central and Eastern Europe, the former Soviet Union, or China, the cities have been the main arenas of change, showing the newly emerging socio-economic differences in detail, and acting as the main foci of economic transaction between the countries and the outside world. This applies, in particular, to the capital city as the main switchboard of power and thus, at least initially, the most attractive place for new inward investors and foreign visitors, not least because of its role as the focus of public administration. The capitals benefit from inherited relatively better living conditions from the communist period, when the main cities attracted most new investment and allocation of resources through central planning.

Similar to other industrialising countries with quite considerable differences in quality of life between city and region, the largest and economically most active urban areas have attracted large numbers of the domestic population from the less well-to-do rural and peripheral parts, in the search for better conditions. Thus, the urban population increased from 17 to 70 per cent (Shaw, 1999) during the communist industrialisation period, distributed among 168 cities with populations exceeding 100,000 each, located mainly in the European part of Russia (ibid.). In the immediate aftermath of the Soviet regime's collapse, people left the cities as a survival strategy, having lost their jobs, by going back to their roots in the rural communities, at least for the time being.

As with most capital cities, their concentration of administrative and economic opportunities and decision-making capacities has always been their main attraction, and this is especially so in the case of Moscow. Here, reflecting the typical features of 'transition', the old spatial social-economic and political-administrative arrangements have given way to new forms, which emerged concurrently with the decline of the old ones. Physically, these changes became evident in a proliferation of low-level locations in the form of kiosks and ad-hoc-built warehouse-style outlets – a 'capitalism without capital'. It was an immediate response to the collapse of the existing state-managed distribution systems, and very much represented an immediate self-help solution. Increasingly, however, these have been replaced by more permanent, often internationally funded developments in strategically selected locations, and built with considerable capital investment. Since the late 1990s, these also include more and more high-profile office developments, although their success depends on the 'right' location. Indeed, there is evidence of 'trendy' office blocks being abandoned unfinished as investment ruins (author's own observation in southern Moscow). This may be interpreted as the gradual displacement of the 'economics of transition' with 'universal economic mechanisms and strategies with global effects' (Rudolph and Brade, 2005). The main arenas of these developments and changes are the central business district and the pre-revolution residential districts of central Moscow, as well as the newly developed locations on the periphery along the orbital motorway – especially the junctions with the main arterial roads into the city centre. Moscow's attraction as the main and most diverse labour market in the Soviet Union, as well as the source of many products and services unavailable anywhere else in the Soviet Union during the communist period, contributed to its steady growth and development pressure.

Not unlike their western European counterparts during the 1960s and 1970s, city and regional planners sought to contain that growth and redistribute it to other parts of the country. The restrictions include controlled residents' permits – a system also used in China, for instance (see Chapter 6) – making it difficult for migrants from elsewhere within the country to settle within Moscow's administrative boundaries. Again as in China, this has set in train a rapid growth of other settlements just outside Moscow's administrative area, yet within easy commuting distance (Rudolph and Brade, 2005). While this process has, effectively, been carried over from the communist days, distinct processes of social polarisation are a typical feature of post-Soviet times. As in many other post-communist capital cities and, indeed, other large cities generally, considerable income inequalities emerged, exacerbated by the much higher income levels achievable within the main cities, and capital cities, in particular. In Riga, for instance, a western-owned up-market department store, with western European price levels, sits within a few minutes' walking distance of an open-air market, whose stalls offer counterfeit designer ware, and home-produced goods at rock-bottom prices. The atmosphere there, as experienced by the author, is decidedly rural and 'Russian', rather than cosmopolitan. Such contrast corresponds with growing social segregation, resembled in property prices and image, with property being the main vehicle of reproducing social inequality in clearly visible physical differences across the former Soviet Union. In the central Asian republics, such as Kazakhstan, ethnic factors overlay this growing social segregation, spelling potential problems with rising tensions in the future (Gentile, 2002).

In Moscow, an 'elite living' category has emerged in the property sector, referring to pre-revolutionary period property in the prestigious central districts, but also peripherally located 'gated communities' (Rudolph and Brade, 2005). The latter are not entirely new phenomena, in principle, as the Soviet-era elite (*nomenklatura*) also preferred physical separation from the rest (the 'lowlier' parts) of society. Under post-communism, merely the residents have changed. In contrast to its post-communist counterparts, especially in the CEE countries, there has been no large-scale suburbanisation. Too valuable is the status and privilege of being a Muscovite, erecting a strong psychological barrier to moving 'out' beyond the city limits. Suburban development is driven by those who cannot obtain Moscow residency, and need to reside outside the city limits as commuters, to work within the city (ibid.).

Post-communist transition under post-Soviet conditions: some concluding comments

Developments in the former Soviet Republics after their independence have shown considerable variations in the ways in which they translated the propagated 'Washington Consensus', 'one-size-fits-all' path of post-communist transition into reality. The three credos of liberalisation, marketisation and democratisation were by no means the common outcome. Differences reached across the former Soviet Union, varying from continued Soviet-style authoritarianism in all but name (Belarus) to delayed grassroots pressure for democratisation (Ukraine). In addition, there are strong issues of ethnic territoriality and competitiveness for power, especially in the central Asian states. But what they all have in common, contrasting the post-Soviet states with those of Central and Eastern Europe, is the top-down initiated, elite-driven process of transformation, with little, if any active involvement of 'the people'. The end of the Soviet Union was, in effect, an inadvertent result of 'tinkering' with the main pillars of the Marxist–Leninist system, especially absolute authority and control, a strong, closely interconnected bureaucracy (*nomenklatura*) which identifies with the state, and adherence to dogma and ideology, even if to the detriment of responsiveness to changing circumstances and policy challenges.

It is also interesting to note that, despite widespread disillusionment with the everyday reality of Marxism–Leninism, there was no evidence of tentative challenges to the system. The length of ideological 'conditioning' of the public, some 30 years more than anywhere else, is seen as one of the main contributing factors to the particular Soviet way of ending communism, and embarking on a rather variable, winding road of change with no clear destination in sight. The transition process itself has thus been rather protracted, often contradictory in direction and destination, and challenged by strong elite interests, while the general public was largely reduced to bystanders to these developments.

'In Russia, the consequences of market transition have been especially destruc-tive', states Burawoy (2001, p. 288), when comparing developments in Russia with those of the CEE countries. There are several reasons for these differences. For one, some of the CEE countries had made tentative moves towards a lesser degree of centralisation, and some minor injections of market elements into an increasingly stagnant system, as early as the 1970s. Examples include Poland and Hungary. Nevertheless, others, especially East Germany, sought to be a stalwart of Soviet-style centralism and state control. In fact, several CEE countries had focused their economic planning increasingly on fiscal aspects, and less on physical planning, while allowing some limited marketisation in agriculture, trade and retail (e.g. through farmers' open-air markets, or permitted private business, if not employing paid labour). In the FSU, no such modifications had been attempted until 1986,

under *perestroika*, when it was effectively too late to rescue the system, and *perestroika* turned into a *de facto* 'policy of self-destruction' for the regime, as no new system was put in place to take over from the dismantled old one (Lane, 1996, p. 126).

Thus, there were quite different legacies in the FSU and the CEE countries when the communist states disintegrated. While there were some with more recent experiences of market elements in several of the CEE countries, most had at least a recollection of pre-communist experiences of marketisation. No such experiences, either recent or as living memory, existed in Russia or most other former Soviet states. As a result, 'when the party-state disintegrated in Hungary and Poland, it revealed a flourishing entrepreneurial economy, whereas in Russia it augmented the power of the large monopolistic conglomerates that continue to dominate the economy', personified through the rise of the oligarchs (Burawoy, 1994, p. 289). These differences in 'helpful' legacies for post-communist marketisation and democratisation influenced the choice of transformation strategies. Russia's relative backwardness in moderating the state-planned system meant that it had the biggest step to make to establish a liberal market economy as the pronounced goal, while others, like the CEE countries, could 'get by' with more modest steps. They had at least some limited elements of entrepreneurial thinking to refer back to. And where a more (too) radical approach was taken, such as in Poland, the next elections brought back more socially oriented, 'soft transition' minded governments.

Given the developmental gap, it was not entirely surprising, therefore, that the Russian leaders were gripped by something of a panic not to 'miss the boat' and fall hopelessly behind. The plan of Shatalin, the Russian economist and minister, to reach a market economy in a mere 500 days illustrates this somewhat panicky approach. The outcome was a form of administrative anarchy, as the changes overwhelmed the existing capacity of the administration. This undermined the credibility of the state's policies and, especially, the legal framework, causing foreign direct investment to take a step back and 'wait and see'. Free markets alone do not seem to be sufficient for attracting FDI (Daniel and Reid, 1998). The contrast with China's approach of maintaining the political-administrative structure while introducing market reforms could not be stronger. There, the state drives the economic development process, whereas in Russia, in a rather ill-conceived love affair with liberalism, the state largely withdrew from the economic arena, leaving it to the vagaries of market forces. However, rather than encouraging the market to swing into action, no longer constrained by state intervention, it caused a sense of insecurity and power vacuum, which constrained effective market operation. Burawoy (1994, p. 289) thus contrasts 'Russia's involuntary transition without transformation' with China's 'developmental transformation (but) without a transition to a market society' (see also Chapter 6). It is the notion of *society* that matters here in particular – society as a *shared*, communitarian construct, embedding the individual into a wider social context of shared values, ambitions and aspirations.

It is here, as Brudny (1997) points out, that Russia's adoption of a full-blown neoliberal market approach differs most significantly from the way marketisation was adopted in the CEE states. There, it went alongside the construction and new imagination of national identity, autonomy and (generally) re-empowered nation-hood after the withdrawal of Soviet hegemony. Marketisation, and its uneven distribution of financial rewards, were thus presented, and perceived, as part of the price for gaining independence. Thus, despite the new inequalities, newly reinforced national consciousness and identity maintained a sense of 'us', including state and society, embarking on a joint journey into a self-determined future.

And it is here that the experiences of Russia and the other post-Soviet states differ. In Russia, in particular, the collapse of the Soviet Union resulted in at least mixed feelings – a new, if somewhat distant, sense of strengthened Russian autonomy and 'self', mixed with a sense of relief from the burden of dependent republics, but also with a sense of loss – loss of an empire, and the associated political stature and authority in the world. There was no clearly developed sense of nationhood and identity as 'glue' to keep an increasingly more differentiated and unequal society together. Without such common underpinnings, neo-liberal markets could reduce society to an unconnected pool of individuals – the end of society in the conventional sense, which Margaret Thatcher, a most ardent force of neo-liberalism, publicly envisaged as the way forward.

In Russia, in contrast to the CEE states, the notion of democratisation has been associated with much more mixed sentiments and experiences. Not having been out in the streets to claim it in the first place, had given the essentially politically driven 'arrival' of democracy somewhat different values already. Combined with a sense of the loss of an empire, and an unclear notion of a new (lesser?) Russian statehood and identity, the process of marketisation, especially privatisation, became a more separate process in its own right. The massive inequalities resulting from the particular process of elite-centred privatisation missed the 'cushion' of national resolve and strengthening observed in the CEE countries. Instead, a growing sense of 'them' and 'us' – those that, often ostentatiously, benefit from the changes, and those that do not – has emerged, undermining the development of a broad support base for democratic principles and a willingness to engage.

'Adoption of a democratic conception of membership, identity and boundaries of the nation is, therefore, as crucial to democratic consolidation as the formation of a market economy and the creation of a multi-party system ... This issue, however, was the Achilles heel of the Russian democratic movement' (Brudny, 1997, p. 312). Much of it was dominated by political and personal power struggles, such as between Yeltsin and Gorbachev, or the different political groupings seeking to attract the Russian president's following. It is the nature and quality of the political leadership that matters too, therefore, to take 'transition' beyond '*economic* transition' (Dabrowski and Antczak, 1997). But the predominant focus has clearly been economic, leaving other issues like ethnic minorities, the position of Russians in the former republics, and the impact of the lost superpower status largely unaddressed, although they clearly matter to the individual.

The result has been a (continued) sense of separation between people's needs, concerns, and interests, and those propagated and pursued by the political elite 'up there'. It is not surprising, therefore, that public interest in participating in political processes and decision-making is rather limited. Democracy, and with it, the experience of transformation, has not generated the galvanising and engaging impact of a reinforced national identity, and the experience of a common resolve and destiny, despite all the obvious differences in transition outcomes for the individual. The recent, quite unexpected outburst of a grassroots democratic movement in Ukraine, and its impact on national politics, has highlighted the importance of a sense of nationhood and nation-based commonality for 'genuine' democratisation as a transition outcome, beyond the predominance of liberalisation as the paramount goal.

6

'DUAL TRACK' TRANSFORMATION IN CHINA

Gradual marketisation and political status quo

Introduction: Maoism and market

China's spectacular changes in its global economic presence over the last ten years have attracted much attention to the fundamental policy changes behind this phenomenon. These have transformed China from a closed country, separated from the global economic (and political) community, with images of Mao's army of uniformed peasants in communes dominating the public images in the West. Since the 1980s, however, China has increasingly opened its borders, albeit geographically selectively, and embraced western culture and, especially, capital. The repatriation of the symbol of successful free trade, Hong Kong, in 1997, and its continued existence as a clearly separate entity and political-economic system, highlights China's attempt at maintaining a balance between the communist 'old' and embracing, albeit carefully, the capitalist 'new'. It is this strategy of riding two horses simultaneously that has, over the last 20 or so years, characterised China's post-Maoist (economic) transformation. The underlying new policy has sought to carefully introduce marketisation, while maintaining strict political control of a one-party state, including the rhetoric of communist values of Mao Zedong Thought and its political morals.

China is thus a very particular case of 'post-communist' transition, as this chapter sets out to demonstrate. Its experience is certainly different from that of the CEE countries and the former Soviet Union (Pei, 1994), and questions may be raised as to what extent China *is* actually 'post-communist'. The recent reaffirmation by China's still quite new president, Hu Jiantao, of the continued validity of communist values, and the need for party cadres to be re-educated in Mao Zedong Thought and Marxism–Leninism (Eimer, 2005), rather than engaging in all-out capitalism, highlights the somewhat contradictory reform agenda. On the one hand, China shares features with other 'managed transitions', such as most former Soviet republics and, indeed, Cuba (which, very tentatively, seeks to follow in China's footsteps) in its attempts to move the economy away from communist doctrine and 'the plan', while on the other it retains the principles and rhetoric of a one-party state under the rule of the Communist Party. In this way, it is essentially a divided transition, seeking to separate economic and political-governmental spheres, encouraging market forces and entrepreneurial ambition at both individual and corporate (public and private) level. At the same time, the leadership seeks to muffle any signs of independent political ambitions and developments and thus evidence of an emerging politically articulate civil society. This 'dual track' approach is so fascinating because of its inherent contradictions, as well as challenges to the widely held paradigm of post-communist transition following automatically the rules of the Washington Consensus (see Chapter 3). China seems to demonstrate that democratisation and marketisation are not necessarily two sides of the same coin.

China's approach to reforming a centrally planned economy has so far been very successful in terms of overall national output and revenue, being able to avoid the economic contraction associated with such swift and fundamental systemic restructuring as attempted elsewhere. Yet, despite the headline-raising economic success story, there have been considerable social costs, especially strong geographic variations between the 'winners' and the 'losers' of this transition, and new societal stratifications. It is this ever bigger inequality that concerns the Chinese leadership for fear of possible uprisings. Being seen to be in control is thus an important part of maintaining power for the Party (Eimer, 2005). The psychological pressures associated with the need to be seen to be successful in the competitive marketplace has in China also had clear social costs: some 3.5 million people a year attempt suicide in China for that reason, with 250,000 succeeding. The casualty rate is highest among men in the 20–35-year age group, that is those who grew up under the new dual system and now find themselves literally in a divided, almost schizophrenic world (ibid.). Inevitably, such signs of desperation and anxiety raise questions about the sustainability of the changes with all their inherent inequality. It is the increasingly evident exclusion of a large part of the population from gaining a 'fair' share of their country's economic success that causes concern among the Chinese leadership who fear possible Ukraine-style popular challenges to their political hegemony.

China's 'post-communist' economy: combining 'conventional' development policy and Mao Zedong Thought

China and the FSU embarked on very different routes to tackling the increasingly evident economic problems of their centralised planned economies. Essentially, China has sought to graft the 'market' on to the existing system as *economic* driver, while retaining most other aspects of the communist one-party state. By contrast, the former Soviet Union abandoned the old system altogether and sought to build up both a market economy and a democratically organised state. Attempted continuity thus contrasts with deep discontinuity, pitching economic growth against immediate economic collapse with subsequently only slow recovery. Against this backdrop, several key differences become apparent in the two largest (ex-) communist countries that circumscribe today's scope for development (see also Pei, 1994; Sachs and Woo, 2001).

First, the two countries entered their economic reform programmes with very different legacies of economic development. China's level of industrialisation was much lower than that of the former SU, reflecting the structural characteristics of a developing rather than an industrialised country. The legacy of Stalin's near obsession with industrialising the Soviet Union as much as possible was largely responsible for this difference. This also translated into the greater availability of surplus rural labour in China. Both countries had in common a deliberate, ideologically driven absence of foreign investment and thus a reliance on the agricultural sector financing industrialisation. The net effect of China's lower degree of 'old industrialisation' was a blessing in disguise, as the inevitable structural adjustments following liberalisation meant that there was less of 'old industry', usually in the form of large state-owned industrial plants, to become redundant. This meant less 'baggage' to carry over into the new market-led economic system.

Second, demographically, China has a much greater ethnic homogeneity than the former Soviet Union, resulting in less tension and competitiveness between central and regional powers. The lower degree of urbanisation, but higher literacy rates, potentially favours a more dispersed response to economic opportunities than the much more urbanised situation of the Soviet Union – again, a result of deliberate development policies under Stalinist communism. In Russia, some three-quarters of the population live in areas defined as 'urban', while in China it is less than one-fifth (Sachs and Woo, 2001). From a marketing point of view, this is a disadvantage, as supplying a more dispersed population is more resource-intensive. Also, in terms of finding a broader, more diverse labour pool, a higher share of concentrated urban population is generally considered an advantage, in particular for more specialised economic activities. For more basic, low-skill activities, however, as associated with initial phases of development, that may not be so problematic. Chinese entrepreneurship has reflected this dispersed structure with smaller-scale production facilities and equally dispersed entrepreneurialism.

Third, the sequence of 'transition' is the inverse of that in the former Soviet Union, with reforming economic conditions first, while maintaining Marxist–Leninist principles for the political arena until further notice. In the FSU and, especially, Russia, political change came first, introducing new post-communist political mechanisms, while economic transition towards market principles followed later, driven by the new political leadership (see Chapter 5).

The fourth important difference in circumstances was the immediacy of the 'revolutionary ideal': China was ruled by first-generation revolutionaries (Mao Zedong, Deng Xiaopeng) until very recently, while in the FSU it was the fourth generation (Mikhail Gorbachev) under whose auspices communism ran its course.

Yet, despite these differences, there were important, fundamental commonalities of the two countries' communist era, in particular the institutional-political structures of a single-party state, centrally planned economies and a rhetorically and practically dominant Marxist–Leninist ideology. But central planning in China was never as fully developed and all-embracing as under the Soviet regime. The Chinese system controlled a mere 1,200 commodities, while the Soviet central plan had more than 25 million commodities (Sachs and Woo, 2001). In a way, China benefited here from the 'advantage of backwardness' in turning ideology into reality.

Another important factor in carving out China's way of reform has been contacts with the many expatriate Chinese, as well as the proximity to Hong Kong and the booming Asian markets. The Special Economic Zones in South China were deliberately planned as competition to Hong Kong, and thus they attracted large amounts of low-end manufacturing activity, helped by a common language among China, Taiwan and Hong Kong. In its essence, the Chinese development model follows that of the newly industrialising countries of Southeast Asia, such as South Korea, Singapore and Malaysia. Their evident success as the 'Asian Tigers', together with a culturally greater proximity and familiarity than exists with the Soviet Union, encouraged the Chinese leadership to seek inspiration from there, rather than from Moscow. A similar situation applies in Vietnam, which, under its reform process of renovation, or *Doi Moi* (Thayer, 1992) during the later 1980s, focused entirely on the economic part of the equation, while ignoring the political context, as if it 'developed in a vacuum' (Kolko, 1995, p. 29). This has been the case especially since the reform policies in the former Soviet Union under Gorbachev's leadership in the late 1980s, seen with displeasure in Beijing. Believing that Moscow was on the wrong track with those reforms, the authoritarian developmental models in the East Asian states offered a much more familiar and acceptable alternative. They gave a more positive, hopeful role model for China's leadership and its ambition to retain political control while pushing the economy forward.

China's relatively more successful transformation is seen as casting some doubt over the wisdom of the Washington Consensus. Does perhaps the Chinese model offer a superior approach, thus challenging the initial capitalist triumphalism displayed early in the 1990s (Wiles, 1995)? The CEE countries and the former Soviet Union experienced a recessionary economic transformation, with an L-curve shaped development of rapid and deep decline first and then only slow recovery. China's transition is thus seen primarily in its economic dimension, particularly now, with its growing pressure on western producers bringing its economic model to widespread publicity. Much of the literature and discussion on transition in China, and its performance in relation to other post-communist economies, has concentrated on economic indicators, without much consideration of other, 'social' factors and costs (see also Sachs and Woo, 2001).

Comparisons between the Chinese and FSU's paths of post-communist transition offer a particularly interesting dimension, as they sit (or did at least in the mid-1990s) more or less at the opposite ends of the different modes of post-communist change. Pei (1994), for instance, points to the different degrees of economic development and the underlying structures, especially the degree of centralisation and thus inherent scope for entrepreneurial decision-making, but also the degree of social protectionism. The Chinese system, especially for the large rural population, provided only minimal social security, much less than their Soviet counterparts received. They thus had to be inherently more innovative and self-reliant. Both factors mattered in shaping mindsets and attitudes towards the loss of state provisions and exposure to competitive market forces (see e.g. Chen, *et al.*, 1992; Gelb *et al.*, 1993). The growth of China's economy, despite ongoing changes to its governance, is a unique factor among the countries embarking on post-communist economic reform. Pei (1994, p. 3) points out the strict adherence to the principle of 'the dual process of democratization and marketization' In China, while these processes have been advocated and pursued as two sides of the same coin, they are treated as strictly separate, allowing a market economy to sit alongside Marxist–Leninist ideology. Outside the immediate Chinese geopolitical sphere of influence, which comprises Laos, Vietnam and Cambodia, Cuba is the only other country where the communist regime seeks to maintain control while, albeit gradually, allowing marketisation, if not democratisation. Economic necessity, and evidence of popular dissatisfaction, not an ideological sea change, account for the 'softening' of the Marxist–Leninist stance in both China (Zhang, 2000) and Cuba (see Chapter 8).

The particular communist legacy of Maoism: a rurally based communist development model

Rooted in a rural, agricultural tradition, the key features of Mao's regime were, as in Stalin's Soviet Union, the claim to total control of state and society and thus the repression of civil society and any political debate other than reciting officially sanctioned statements. Several means of coercion were used, including the work units (rural collectives or urban factory compounds), political campaigns and public events (including show trials), and, as elsewhere under Soviet-inspired communism, surveillance and informants within society. But the main concern of Mao's ideology was to bring people to aspire to a better society and, in the process, become 'better' people themselves. The emphasis was thus on an active contribution to building a new society through one's own life. In this, Maoism differs significantly from Stalinism, where the population was reduced to the passive role of being objects of total control by the state (party). Instead, they were to take a more proactive attitude. As a result, a key element of a functioning market system, personal ambition, was readily available when economic reforms began with the end of Mao's rule in the mid 1970s. The Chinese population was thus a more fertile seedbed for the installation and operation of market forces than its Soviet counterpart.

The Chinese system became less obsessed than Stalinism with total planning and central control of *all* aspects of public life and allowed elements of decentralised economic decision-making and local governance. Some small and local businesses were permitted, many based in villages and part of the rural economy (Dacosta and Carroll, 2001). This relative decentralisation of economic planning and decision-making generated regional variations and economic development by reinforcing variations in institutional ability and resources. It had also become apparent by the late 1950s that the Soviet economic model was delivering only mixed results.

At the end of the Maoist period, China was a mosaic of more and less developed areas, with a largely disillusioned and pragmatic population, especially after the difficult experience of the Cultural Revolution (1966–75) and its ill-fated attempt at reinvigorating revolutionary, that is 'anti-bourgeois', zeal and political societal awareness of the 'correct path' of development. It was against this experience that the reform course emerged through the new, pragmatic political leadership of

Deng Xiaoping. His focus was less on ideology and 'the promised land', than on *de facto* socialism on the ground as experienced by the people.

The immediately implemented changes introduced greater economic freedoms and autonomy for producers in all sectors, including the permission of a parallel economy with market principles for products 'surplus' to the planned (prescribed) output. Furthermore, the doctrine of self-financing modernisation was abandoned and foreign capital invited, if initially only in clearly defined 'special economic zones' as test cases. Overall, there was a shift away from a planned economy towards a state-controlled (managed) market economy. Pragmatism prevailed (Gray and White, 1982).

The result of the reforms of 1978 was the reinvigorated economic and, by extension, political life of a stagnant regime. An integral part of the new strategy was a dual approach of facilitating the development of large enterprises while also encouraging small unit production. This is known as 'walking on two legs' (Saich, 2001, p. 37), and provided an important skill resource during marketisation of the rural economy during the 1980s.

The Cultural Revolution's legacy includes a general distrust of the political machinery and a low political interest of the people in state affairs. This, to some extent, let the leadership 'get away with' its dual track approach of reform, allowing consumption and market forces, while retaining the socialist rhetoric. In a way, the experimentation with reforms and *marketisation* may be argued to be the logical extension of Mao's variable understanding of socialism as a transitional mode between capitalism and communism, experimental and flexible by nature, requiring contradictions in policies and, at times, also *voltes-faces* (see also ibid., p. 47). Viewed in this context, the reform process of post 1978 does not at all mark a seismic shift in policies, but rather a logical extension, another phase in the evolutionary process of seeking to develop a socialist state and society.

Transition Chinese-style: marketisation without democratisation – the strategy of 'market authoritarianism'

It was the continued balancing between 'market' (i.e. the reformers) and 'plan', favoured by traditionalists seeking to maintain maximum control for the Party, that embodied the unique feature of the Chinese model of post-communist transformation. The zigzag course in reform policies reflects the negotiated relationship between two roughly evenly influential elites, where neither is in a position to see through its policies without challenges (see Chapter 3). The oscillating policies of reform and reaffirmation of Maoist ideology reflect an inherent nervousness, especially since the events in the Soviet Union of 1991, about losing control in a self-propelling reform process. But overall, the reformists could push their agenda through.

Reformers pursued partial or dual track reforms largely out of political necessity, and endeavoured to constantly expand the market segment of the dual-track system. Without such a market-oriented commitment at the highest level of Chinese decision-making, it is scarcely conceivable that the market force could have expanded so fast and have been sustained for so long.

(Zhang, 2000, p. 34)

Market orientation had become the main paradigm, but it was 'orientation', not ad hoc introduction *in replacement of* the previous system. The ideological battle was between 'more of the same' (planning), albeit more effective, and 'something new', albeit carefully edged in.

The leadership had realised, supported by academic research, that there was no 'best practice' or blueprint answer, waiting to be copied and delivering positive results immediately. The difficulties with modernising planned economies through the limited, somewhat half-hearted introduction of market elements had become evident from experiences in Yugoslavia, exploiting its arms-length relationship with Moscow under Tito's leadership, and Hungary's attempts at opening up its economy in the 1980s. Both countries' leaders sought to maintain the principles of state planning, but with a dose of 'market' added to overcome bureaucratic inertia. In the end, however, this proved too little too late, and the lack of time for slow reform was recognised by Chinese leaders (Zhang, 2000, p. 37) who thus adopted a more determinedly market-driven approach.

The Chinese approach to transition has thus been gradual, experimental and, most importantly, localised. This small-scale and 'messy' approach contrasts fundamentally with the (externally prepared) 'designer approach of Russia and eastern Europe' (Chen, 1993, p. 139).

A key element in the marketisation strategy was the devolution of responsibility to the local level of administration and even further than that, to individual families or production units. Utilising elements of a latent local sense of economic responsibility, mainly at commune level, which had been permitted by limited decentralisation for small economic projects, provided a fertile ground for more locally administered economic initiatives. From central government's point of view, this devolving of responsibilities meant also a safeguard against having to shoulder responsibility for 'things going wrong'. The devolution of responsibility, and blame, has changed the role of the local administration and encouraged a more entrepreneurial spirit. Yet, it also reinforced underlying differences in opportunities, personal capabilities and the institutional capacities of the different localities and their administrations. Differences have been further magnified by the second main policy plank of allowing some to get rich first, be they individuals, localities or regions. This encouraged entrepreneurialism, enterprise and ingenuity but, not surprisingly, the outcome has been a reinforcement of underlying inequalities – intra-urban, inter-urban, intra-rural and between regions. Urban–rural inequalities, in particular, continued to grow, albeit steadily, leaving enclaves of poverty between the successful areas (Khan and Riskin, 2000). The reduction of the inter-regional equalisation scheme between 1978 and 1994 exacerbated this unevenness, so that the Shanghai region, China's most prosperous, contributed about 8 per cent of its revenue to the central budget in 1993, whereas it was about a quarter in the preceding years, and the net contribution by the newly rich Guangdong province was only 0.4 per cent of GDP (ibid). Ultimately, however, the central state holds all the leads, especially financial (ibid.), although a link between economic success and financial reward has been established for the relevant territory.

The new emphasis on autonomous economic management and entrepreneurialism by sub-national governments sits somewhat uneasily with the legacy of the communist-era state structure and administrative practices (Blecher, 2003), especially of the hierarchical, top-down form of implementing policy reform. These arrangements are capable of offering no more than a rather weak institutional framework for effective, supportive governance, which is needed for successful market-driven economic development. Without administrative-governmental reform, actually existing structures 'on the ground' may, if perceived to be very different from promised realities, undermine entrepreneurialism.

Special feature of China's transition

So what, then, makes China's transition so special? A particular feature of the Chinese (economic) reform programme is the tentative, trial-and-error approach and the avoidance of adopting any 'good practice' from western institutions. And a key feature of this approach is, as Zhang (2000) points out, a sequence of 'soft' and 'hard' reform initiatives. The former may be seen as preparing the ground for the latter, attempting merely some lesser medications to existing structures and ways of doing things. If they proved useful, leading to expected results (economic improvements), then larger-scale initiatives with more fundamental changes would follow. A good example of this 'kite-flying' approach is the establishment of the Special Economic Zones at the end of the 1970s as strictly guarded test beds for the reform programme towards marketisation. Only when this proved beneficial was the reform rolled out to more areas, and the basic principles of market-driven economics extended across China.

'Soft' reforms, that is gradual reforms of familiar features, included administrative and fiscal decentralisation, as there had been earlier attempts under Maoism to move in that direction. According to Zhang (2000, p. 43), 'soft reforms are essentially growth-oriented, but they involve elements of transition. While hard reforms tend to cut deeper (see 'shock therapy' discussed in Chapter 3, for instance; also Kolodko, 2000a) into the old system and are therefore more transition-oriented, they may also bring about desired development.' In many instances, soft reforms preceded and prepared the way for hard reforms. Essentially, soft reforms seek to not rock the boat too much and maintain the political order, with the state retaining ultimate control of the economy, albeit at arm's length.

Inevitably, as these reforms are a compromise solution, they cannot offer the highest efficiency in pure market terms, but a rushed, immediate adoption of market principles may just as easily lead to sub-optimal allocation of resources. By the same token, the political 'wobbles' among the leadership during the 1980s and

1990s created an atmosphere of uncertainty and were counterproductive to the efforts of stimulating inward investment and thus economic growth. As a consequence, they found themselves under increasing pressure to implement further-reaching, fundamental 'hard' reforms. But policy objectives also changed over time and, therefore, keeping options open was important to allow changes in response to 'learning on the job'. Thus, at the beginning of the reform process, there was no questioning the feasibility, or desirability, of the command economic system *per se*, but merely its implementation 'on the ground'. This meant especially the lack of incentives to increase productivity and to 'try harder'. If that was improved, so the rationale went, the system overall would improve and thus remain stable. The economic difficulties were thus not seen as system-inherent. Instead, attention was focused on modifying the operation of the system at the micro level, rather than changing its basic principles – but response to the initiatives was, not entirely unexpectedly, varied.

The resulting gradual development, carefully edging forward and constantly seeking feed-back from the results of earlier decisions, was possible because there was no public pressure to change the political *and* economic system together, unlike in eastern Europe and the FSU. Instead, as also now tried in Cuba, the basic ideological principles of authoritarianism are to stay in place, albeit in modified form, to accommodate the changes towards a market economy as the essential pillar for the further maintenance of the system. The question, therefore, is how far can/will these changes to the governmental administrative system go, before there is a threat to the very principle of the one-party state and the hegemony of the Communist Party?

It is here that the biggest and most important difference lies, compared with eastern Europe and the former SU states. 'Radical transitions in Eastern Europe and Russia are in part driven by their explicit political objectives to eliminate the legacy of Communism. In fact, shock therapy and mass privatization have actually been built into some political programmes' (Zhang, 2000, p. 52). The political system *per se* needed to be seen to be subject to change, and this as quickly as possible. 'Fiddling on the edges' was no longer a feasible option as soon as people realised the scope of getting rid of the system for good. Adopting neo-liberalism and neo-classical (New Right) economics had gained support so quickly, because these principles represented the antidote to the Soviet-imposed communist system and thus a maximum shift away from it. In China, such systemic change is not wanted by the political elite, and economic policies seek to maintain the political-administrative status quo, while offering some modification to the economic arena as a 'safety valve'.

In China, as in Cuba and Africa, the communist movement was home-grown (if Soviet inspired), and had been instrumental in overcoming colonial occupation. The system thus had much greater inherent legitimacy and acceptability as part of an indigenous revolutionary liberation movement than in eastern Europe and the former Soviet states outside Russia. There, communist rule was associated with the opposite experience: occupation and external control through a foreign force. It is here that Cuba's and China's systems share common features, helping to explain their continued grip on power, if only on the political and no longer economic side. Effectively, in China, continued political control has been bought off with marketisation and much-enhanced consumption for the average Chinese who thus feels improvements, albeit to varying degree, in living conditions. 'In short, what China has experienced is a dynamic gradual *reform* of the existing institutions, not a *revolution* as in Eastern Europe and Russia' (ibid., p. 53).

The continuing balancing act between market and state control has brought an inherent dynamism to the gradualist approach, with variations over time in response to (unexpected) developments and a political struggle between modernisers and those seeking to maintain the status quo. The subsequent repeated changes, at times U-turns, in policy are an indication of the uncertainty of how best to go about 'moving towards' a market economy. The result has been a China-specific combination, re-interpretation and amalgamation of many different economic development measures adopted from elsewhere and then adapted to their own agenda.

Following China's long isolation from global markets and economic processes, there was a general lack of experience with market forces. But:

> the experimental approach, with the more successful sectors or regions setting an example for others [and thus serving as support for arguments in favour of more marketisation], helped reformers to gradually develop their experience and expertise in reform and avoid the risk [political] of large-scale paralysing failures. It has gradually increased the reform programmes' acceptability to a larger population.
>
> (ibid., p. 52)

However, the absence of a 'shock' to the political-economic system of governance allowed inefficiencies through mismatches between new and old ways of doing things (Lin 2004). Despite these essential differences in the smoothness of change, all post-communist countries share the considerable challenges of newly emerging inequalities in economic opportunities at all spatial levels, sub-local to international. In the Chinese case, such inequalities were initially wanted and accepted as part of the so-called ladder-step approach. This encourages the coastal regions to forge ahead with their development, hoping that growth will eventually extend to the remoter, western regions (Lin *et al.*, 2003). Increasingly, however, the leadership is becoming concerned about the much wider than anticipated cleavage between the 'haves' and 'have-nots' and the resulting potential dangers to the stability of the political system *per se*.

The outcomes of reforms: curses and blessings of a two-speed economic miracle

In the mid-1990s, the income differential between coastal and inland provinces was 1:15, and more than 90 per cent of FDI was concentrated in the southern and eastern coastal provinces (Ho and Lin, 2003). Social divisions are illustrated by the fact that some 3 per cent of the population hold savings equal to the entire savings of 800 million peasants (ibid.). Much of this inequality is a direct result of government policy, especially the designation of areas with special tax status and other support. The special conditions for cities include the four SEZs (Shenzhen, Zhuhai, Shantou, Xiamen) of 1980, and the 14 'open coastal cities' of 1984. In addition, there are 'free trade zones' in coastal cities (1993), 'open border cities' and 'open free trade zones'. All of these policies included tax incentives and reduced bureaucracy to attract international capital. Because of the uneven (and sparse) distribution of urban centres across the country, and therefore their disproportionate economic impact on the regions, any change in the urban–rural disparity translates immediately into regional inequality (Wu, 1987). Initially, the government's expectation was that a trickle-down effect would disperse economic growth to the less advantageous areas, but this has happened only to a very limited extent. Instead, a growing divide between the 'winners' and the 'losers' of this reform process has emerged (Wei, 1999). But even within the generally less advantaged rural areas there are local variations between individual communities, that is those with successful community-owned enterprises, serving as 'extended workbenches' for outside investors, and those with less such activity (Shen, 1998).

For some regions, such as the northeastern part of China and the Sichuan region, the changes have even meant a reversal in fortunes. Favoured under Maoist industrialisation policies, these heavy-industry areas have become the rust belt of today's economic development. By contrast, regions largely peripheral under the communist planned economic development strategy, such as southern Guangdong region, have become the wealthiest within China (Saich, 2001). The result is a very different experience with the reform process, pitching insecurity about job losses and future prospects against evident westernisation and seemingly continuous growth. 'For many, reforms have meant bewildering choices, loss of security, rising crime and declining personal safety' (ibid., p. 16), and may require them to migrate to new, unknown, urban places in the coastal provinces.

Inequality *per se* is nothing new, reflecting considerable geographic variations in resources and comparative advantages, and having been quite pronounced under communism, too. What makes the changes more fundamental and disruptive is the inverse allocation of relative economic advantages. This meant a falling from grace of the interior and northeastern provinces as the old centres of centrally planned and strategically located (away from potentially invaded coasts) heavy industry. Under market conditions, the underlying lack of genuine comparative advantages has been ruthlessly revealed, evident from factory closures, neglect and uncertain future prospects. In contrast, the previously neglected urban trading centres along the coast, vilified by communist rhetoric as 'decadent' and 'imperial', have become (again) the main centres of new growth and internationalisation. But even under the redistributive planned economy, the inherent advantages of the coastal cities prevailed, continuing to widen the gap with the other provinces' economic conditions, as cities were turned into industrial rather than administrative and consumptive, places. Also, engrained statist administrative practices and corruption have added to the reinforcement of underlying inequalities (Gong and Li, 2003; Lewis and Litai, 2003).

Table 6.1 Uneven economic development across China's economic (development) regions (GDP per capita in yuan/head)

Year	China (all)		Western		Central		Coastal	
	Urban	Rural	Urban	Rural	Urban	Rural	Urban	Rural
1978	811	285	738	259	768	261	876	321
1984	1155	646	1086	500	995	620	1311	759
1995	2158	909	1978	585	1817	777	2552	1232
1978–95								
annual change	5.9%	6.6%	5.7%	5.0%	12.7%	6.7%	12.7%	6.6%

Source: Yao *et al.*, 2005

Social change: social stratification – the emergence of a new entrepreneurial middle class

The changing policies not only defined a new geography of opportunities, but also considerable social changes. A growing shift from cradle-to-grave state provisions and the guarantee of the Iron Rice Bowl as proverbial symbol of guaranteed nourishment, to limited-term contracts and lesser job security changed the social outlook for many Chinese workers, and certainly reinforced differences in opportunities. Particularly for those working in the old industrial state firms, conditions deteriorated markedly, contrasting with the new riches found by entrepreneurs and employees in new 'trendy', western-style businesses in the big metropolitan areas. The result was a clear stratification of society into 'winners' and 'losers' from reform (on new urban poverty, see Wu, 2004). The new inequalities are reflected in the fact that in 1995, the top 10 per cent of the urban population earned 27 per cent of all income earned, and the bottom 10 per cent only 3 per cent (Démurger *et al.*, 2002).

This social stratification, however, is not merely an inevitable, incidental outcome of marketisation and liberalisation; societal diversity is now explicitly permitted, even welcome, as a stimulus to personal competitiveness and raised ambitions. In a way, it connects with the Maoist believe in personal motivation and striving for betterment, albeit now for economic, rather than idealistic gain. This ambitional aspect of Maoism marked a key difference from the communist system of the Soviet kind, a difference that carried over to the post-communist economic system, and laid the basis for the acceptance of competitiveness and 'striving' among the population, and in business and administration. But social differentiation is not entirely an exclusive responsibility of the 'market'. Thus, for instance, back in the 1970s, unemployed urban dwellers were encouraged by the state to become self-employed, leading to an explosion in the number of small businesses, so-called 'getibu' (also in rural areas). By 1986, there were some 5 million such small businesses in urban and rural areas (Chen, 1993, p. 146), and in 2000 this number had more than quadrupled for the urban areas alone (Démurger *et al.*, 2002).

These small businesses form an important bedrock of a newly developing entrepreneurial middle class. Some of them managed to 'upscale' by successfully engaging with the privatisation process, resulting in a number of new business tycoons with considerable financial means. Most new entrepreneurs, however, operate at more modest scales. Their successful engagement with market conditions has widened the gap in economic performance and wage levels to the conventional state-owned (usually large) businesses. This compares with the situation in other, non-socialist developmentalist-authoritarian Asian countries with their mixed state-market economies, such as Taiwan or South Korea (Herr and Priewe, 1999). These new businesses thus add capacity, rather than replace that of state companies, as was the norm under the Washington Consensus-inspired economic liberalisation.

The observed growing problem of urban poverty at a time of rapid economic growth, reflects the changing nature of economic growth. It is increasingly 'jobless growth', as international competition pushes for least cost and thus minimal use of labour. This problem also applies to other ex-communist economies, such as Poland, for instance, which, despite its annual GDP growth rate of over 5 per cent, retains a 20 per cent unemployment rate (*The Independent*, 26/09/05). In most transitional (post-communist) economies, poverty has been viewed as first and foremost a direct result of economic restructuring and disinvestment, but economic growth with continuous or even growing levels of high unemployment has certainly baffled politicians. The former German Chancellor, Helmut Kohl, expressed his surprise at the steadfastly high unemployment in eastern Germany, despite massive new investment to replace defunct old structures. In addition, as Wu (2004) points out, there are particular systemic legacies that exacerbate the fallout of structural changes for the labour market, with the 'losers' falling between the end of the old state support system and the beginning of the new, market-driven provisions of employment.

The new poverty is thus related to the 'disjuncture between the old welfare system and the new labour market' (ibid., p. 404). But there is still no official definition of 'urban poverty', suggesting hesitation in acknowledging this phenomenon, but also ignoring the potential problems in the future for urban society. This 'poverty of transition', as Wu (ibid., p. 418) explains, is a result of the specific characteristics of 'transition', in particular the 'institutional root of socialism' and the void left between old and new mechanisms of social support. More significantly, this new urban poverty is not just a temporary feature of 'adjustment', and there are distinct geographic locations of these urban poor in the city structure: the inner-urban dilapidated areas with high residential densities, often in immediate proximity to new, upmarket residential and commercial developments, in old industrial areas with their workers' housing compounds, and in the enclaves of the rural migrants, often on the outskirts of town (Yan *et al.*, 2002).

This inequality, as in other transforming societies, is most likely to affect people's attitude to the new conditions and those representing them. Among the less favoured, the old system appears in a rather more positive light, while the winners of the change, not surprisingly, favour the new system and strive for signs in the direction of democratisation. So far, little progress has been made towards democratic principles and, indeed, at the moment the signs are of reinforcing the Party's grip on power (Eimer, 2005). One reason for this is 'that most Chinese were not convinced it [the reform] was the answer for China', a view reflecting a general lack in democratic tradition (Nathan, 1997, p. 66). So far, 'the democratic experiments were few in number, short in duration, and limited in their democratic characteristics' (ibid., p. 65), but as the Tiananmen incident of 1989 has shown, at least some parts of the population seem to value pluralism and civil society. These attempts are thus not a representation of the whole of Chinese society, if one can still speak of *one* society at all. But there is evidence that elements of civil society and political culture are tentatively emerging, a process also evident in neighbouring Vietnam (Thayer, 1992), which has sought to follow closely its bigger neighbour's policies.

China's post-communist reform: for the cities only?

Post-reform developments in China have largely benefited the metropolitan areas and encouraged their further growth, thus widening the developmental gap to the rural areas. The restrictive national policies of controlled residential choices encouraged suburbanisation as the main mechanism of growth and gaining access to the urban labour markets. However, businesses, too, have increasingly chosen suburban locations. On the other hand, the city centres have increasingly seen redevelopment through new office and retail space, as well as residential gentrification (see e.g. Yan *et al.*, 2002). Economic preferences and 'attractiveness' set their own locational priorities, and with their new policy-making possibilities and associated financial rewards, cities are competing vigorously for the attention of international capital. In so doing, they seek to outdo each other by offering incentives, usually conventional locational advantages based around cost savings, such as tax breaks, cheap land and labour. Such ultimately ruinous competition effectively benefits the investor at the expense of the local tax payer and deprives the locality of revenue for future investment. Chinese cities are still largely competing at the cost level, and incentives are thus uniformly of the tax-break type, but in the dash to lure inward investment, 'soft' factors, crucial in the higher-end investors' market, such as local culture and history as beacons of local identity and recognisability, are thus in danger of being ignored or permanently lost (ibid., 2002).

Not surprisingly, given the dash for the same kind of investment, the development and marketing strategies adopted by the cities are very similar, ignoring any locality-specific factors. 'For example, both Shenzhen and Guanzhou proposed that they will create a "Silicon Valley" for Guangdong Province. Evidently, the construction of two Silicon Valleys 100 miles apart is unnecessary and redundant' (ibid., 2002, p. 48). This very much reflects the localist, individualist perspective taken by the cities' policy makers, with few signs of cooperation and communication between competitors. This situation thus essentially reflects an urban economic policy approach found widespread in western European countries in the early 1980s.

The absence of any competitive pressures under the communist regime meant that no attempts were made at differentiating one city from the others. There was no need to be 'attractive'. The seemingly uniformly available potential investment essentially continued the notion that local particularities (and identities) did not matter. Administration and bureaucrats (planners) could therefore continue with their established principles of technocratically focused planning without the need for major strategic (longer-term) considerations. As in the other former communist countries, changing this mindset and demonstrating the need for constructing a distinct, local cultural environment and urban profile, including citizens' participation, is therefore a major challenge of the reform process (Yan *et al.*, 2002).

In contrast to other developing countries, China has historically had a low degree of urbanisation – even though it possesses 12 large metropolitan areas with populations of between 2.8 and 10 million (Logan, 2002a). The state's response to urbanisation has been varied and at times contradictory. The choice is between new growth poles and attractions to inward investors, or to simply provide continued locations of concentrated industrial activity at the behest of the state. Since the beginning of reforms, suburbanisation increased through moves from the city centres and incoming rural migrants. Thus, for instance, in the early phase of the reforms, between 1982 and 1990, Beijing and Shanghai grew overall, but their centres declined by 3 per cent , while the suburbs grew by 40–60 per cent (ibid.). These rapid expansionist developments have also borne their costs: inadequacy of public transport, estranged human relationships, and poor adaptation of migrants to city life.

Both the decentralisation efforts and opening up to international markets (capital) since 1978 have shifted power and responsibility among the three main types of actors (Fu, 2002) – foreign, national and local capital as drivers of economic development – from the national level up and down the spatial hierarchy, substantially weakening the state's role in economic development (see also Lin 1999, 2000): inevitably, there is an inherent danger in these losses in direct (planning) control, because they may translate into a general loss in political standing and control. With new actors entering the stage with a new set of bargaining chips (promise of investment), new growth coalitions emerge between local players and global capital, joined by their shared interest in economic activity in a city (locality) (Fu, 2002). Using that new-found power enhanced by new coalitions with international capital, local government no longer feels an obedient executor of centrally imposed decisions and plans. There is now a direct connection between local decisions and economic rewards, triggering more enterprising local policies. The skill now rests in bringing together the various group-specific interests and matching them up for particular projects as part of a new coalition-building process.

This negotiation-based, informal approach to policy-making contrasts fundamentally with the old centralist system, where the five-year plan was the sole base of investment decisions and actor engagement. Now it is down to personalities to identify scope for development alliances and bring them together for agreed, individual projects. Given the nature of market-led development, and the fiscal weakness of many cities, new developers have an influential bargaining position in such coalitions, and property-led developments, such as in Shanghai (Zhang, 2002), are thus the most visible outcomes of these 'coalitions for growth'. As in western circumstances, such network-based, investor-capital-oriented and often informal arrangements raise questions about the legitimacy of policies and the role of the public (democratisation). In many ways, public participation has not increased much with the shift from central planning to the new corporatist bargaining and decision-making. Projects are decided outside the public realm, and there is little in the way of 'ownership' of place. Local planners and technocrats are thus the main guardians of local interests beyond the immediate concerns of the property development (Fu, 2002), but there are questions about the ultimate sustainability of such disempowerment of the general public, with few, if any signs of a willingness by the government to move towards at least some degree of democratisation.

The coastal city regions as designated 'capitalist playgrounds' and test beds for reform

At the end of the 1970s, with the beginning of the new reforms towards marketisation, established communist-ideology-inspired development priorities and policy responses lost their relevance and appropriateness. Thus, the nature of 'planning' had to be redefined. It could no longer be understood as a directive, top-down applied instrument of determining development in all aspects. There needed to be more dynamism and flexibility, to be able to respond more effectively to the needs of a rapidly expanding and changing economy. Decentralisation of responsibilities meant new pressures on local bureaucrats to develop and formulate their own policies and agendas, and accept responsibility, rather than simply relying on orders from above.

This shift from a national perspective of a uniform economic space to a regional scale of evident inequality acknowledges the inherent inter-regional differences in economic development potential. This inequality is not an entirely new phenomenon, but goes back to pre-reform days (Friedmann, 2005). Together with the new emphasis on city-focused economic development, urban-regional development agendas were the inevitable outcome. Planning was often at least as much about satisfying political and career ambitions of local *nomenklatura*, as it was about local needs. This dilemma has, in its essence, continued until today, as the absence of local democracy and democratic accountability of the local administration has, in effect, continued central control over local development. For instance, while the move of manufacturing/industrial uses from the central areas of cities to the peripheries is a local responsibility, the economically so important greenfield sites are a central control issue. Effectively, therefore, the limits of urban growth remain centrally determined, with the centre retaining 'the last word' in central–local relations.

Examples of a newly emerging city-regional agenda, based around city networks, include the 'Master Plan of Cities and Towns in the Shanghai Economic Region' and the 'Plan of the Urban System in the Pearl River Delta'. These structure plans serve as guidance for the individual localities' plans. New infrastructure works (roads) also link the coastal cities with their hinterlands, thus encouraging city-regional development. There are still elements of the old centrally managed regionalism about, aimed primarily at tackling the legacy of abandoned old industrial districts. This inherently redistributive approach may gain more acceptability in future with the rising concern about the growing inequality between the booming and lagging regions (Tian, 1999), but the showcases of the post-communist economic transformation processes are in the (coastal) city regions.

Towards the end of the Maoist period, Chinese cities contained three main structural components, which reflected past development policies: workers' residential compounds around factories, the old city areas and the shanty areas on the outskirts of the cities. But economic change and migration broke down that socialist-era arrangement. Luxury residential developments, and commercial and recreational uses now compete with high-density old neighbourhoods. *In extremis*, this pitched the newly affluent urban entrepreneurs immediately against those having lost out from the economic transition. Many traditional urban areas (inner areas) have attracted a higher concentration of a marginal, poor (and new) population, which sits immediately next to the newly developed enclaves of western-style 'trendy' lifestyle. The latter involves a highly concentrated and spatially confined 'playground' for the *nouveaux riches*, found in many of the larger post-communist cities. 'It is the absence of positive social objectives in urban redevelopment that induces the great conflicts of displacement and relocation' (Wu and He, 2005).

The radically redirected flows of investment have fundamentally altered the dynamics within China's space economy, and differences have become more dramatic. As a result of the post-Mao 'Step Ladder Doctrine', allowing the coastal regions to resume their historic economic pre-eminence, China has effectively been divided up into three regional clusters: the coastal regions with their role of being leaders in development; the central regions with an emphasis on conventional 'old' industrial production and activities such as coal and steel and energy generation; and the western cluster with a more medium-term development target. The new reliance on growth impulses from designated (urban) 'growth points' in the coastal areas was effectively an extension of the tentative, strictly localised test cases of marketisation in the early 1980s. Any perceived 'misdevelopment', so the rationale went, could then be apprehended by pulling the emergency brakes on the reform process of opening up to global capital. The Pearl River (Shenzen, Hong Kong) and Yangtze (Shanghai) River deltas have become the showcases of China's breathtaking speed and extent of economic transformation.

The Pearl River Delta as 'virtual region' of urban-centred economic transformation

The Pearl River Delta is a 'virtual region' (Herrschel, 2005) in the sense that it is an economic (and cultural) region without a corresponding administrative-governmental structure. There are also no geographic commonalities for the whole region. It covers eight prefectures of the Guangdong province, including the Special Administrative Areas of Hong Kong and Macau, whose status means their boundaries have a quite considerable barrier effect. The eastern half of this region includes the highly developed areas of Shenzhen (SEZ) and Guangzhou, whereas the western half is much less developed. Two of the four original SEZs (Shenzhen and Zhuhai) are in the Pearl River Delta region, giving it a certain status as a pioneering region of the economic reform policies.

Boundaries do matter in this region, delimiting the economic 'experimental cases' of the Special Economic Zones, including Shenzhen as the pioneer of that policy. The Zones are almost exterritorial areas, certainly for Chinese migrant workers. Requiring a special permit for entering the zones was meant to control contact between the new market economies and the rest of China, so that an unwanted spread of a 'western market bacillus' could be prevented. Shenzhen was established in 1980 to try out the new policy agenda of a 'socialist market economy' as a 'controlled experiment'. With no clear blueprint to follow, 'learning as you go along' was the only realistic option (Bruton *et al.*,, 2005), and this trial and error was felt 'safe' in a limited scale experiment only.

The new city of Shenzhen sits within the Pearl River Delta Open Economic Zone in Guangdong Province, right on the border with the Hong Kong Special Administrative Zone. It was established there so that it could benefit from Hong Kong's proximity and its Chinese population's historic links to its northern neighbouring province. The result was the unique situation of a communist-controlled developing area bordering an affluent capitalist system looking for new economic opportunities during the booming 1980s and early 1990s. This complementarity of interests at the time has given the region a distinct head start in its economic transformation towards 'the market' in relation to the other parts of China. The learning effect of the Hong Kong model on Shenzhen city brought an economic integration between the two systems and immediate economic legacies on both sides of the Chinese–Hong Kong border. Although this border has ceased to be international since 1997, it nevertheless marks a distinct administrative and systemic division between the now Special Administrative Zone and mainland China.

Rationale and operation of the Shenzhen Economic Zone challenged and changed the scope and meaning of 'planning' in the context of Chinese cities by removing the certainties of state planning and subjecting it to the uncertainties of capital markets and business decisions. Under the Maoist industrialisation strategy, the emphasis was on turning a consumptive city into a socialist productive city, and providing the land for this development to happen was the expected role of planning. But this no longer holds true, as new aims and means of planning had to come in its place to match the new purpose of the Shenzhen urban area (see e.g. Xu and Yeh, 2003). The new challenges to urban planning also came from the previously unknown scale of migration from rural areas to Shenzhen, bringing the population to 5 million. They located primarily just outside the SEZ, as access to the Zone is tightly controlled, but still within the urban area. This shift was underpinned by a massive economic growth rate of some 40 per cent annually between 1985 and 1992 (Gar-on Yeh and Li, 1999; Nanto and Sinha, 2001), adding to the challenges to somehow manage the ensuing physical expansion of the city.

The first development plan was very much in the mould of a socialist-era technocratic, centrally defined holistic approach to manifest future development, both social and physical. Developments on the ground, however, overtook the neatly laid-down socio-economic plan. As a result, the city adopted a less formalistic and prescriptive comprehensive plan in the mid-1980s, focusing primarily on the built environment, but this still proved too inflexible and unimaginative to accommodate the unabated increase in population and continued building activity on the border of the SEZ within the growing Shenzhen conurbation. For effective development control, cooperation with the surrounding municipalities has become increasingly urgent, and Shenzhen came to recognise the importance of a regional approach to boost international competitiveness. This includes seeking to attract and retain higher-end business users, especially in the service industries, and developing Shenzhen into an international city with a clear identity and image reflecting that ambition. Realising the temporary nature of the competitive advantage based primarily on offering cheap labour, the city now recognises the importance of environmental quality as an important selling point. Maintaining that requires effective, but not necessarily technocratic, planning. The city thus continues its original role as a test case of China's new policy and developmental directions, here a shift from a welfare state to a development state in the mould of the Asian developing countries.

Generally, as a result of the changing national policies, Chinese cities are experiencing a transformation from places of production to places of consumption, thus effectively reversing Maoist development goals. As elsewhere under the impact of structural economic changes, the outcome is functional and structural decline, adding to the task of 'promoting urban development and local economic growth, restructuring urban spaces and transforming urban functions' (He and Wu, 2005, p. 4). At the same time, local governments, having been given the task of local development promotion, have engaged in increasingly proactive, acquisitive economic policies. 'The concept of entrepreneurial government has thus been introduced to urban governance, and marketised operation and competition have been encouraged' (ibid., p. 5).

But established administrative structures, practices and 'cultures' are inevitably slow to follow the changes initiated by the reforms. Despite formal decentralisation, the state continues to hold important controlling powers through the hierarchical organisation of the civil service and there, especially, the Party. Promotion of local officials depends on a display of loyalty and satisfactory 'performance' with regard to set political goals, and is assessed by higher-placed officials. The result is that local officials tend to be more concerned with centrally defined targets than with local needs, as there is no democratic legitimation required locally. In the pursuit of these goals, local actors seek to ensure meeting their targets, and thus key growth industries (e.g. high-tech, automotive) have a stronger voice than the declining textile sector when it comes to pushing their respective interests with a city government (Zhang, 2002). There, it is basically just a consideration of the 'most profitable' use of a piece of land, rather than an evaluation of any wider implications – socially, environmentally, or otherwise. The main impact of the reforms has been a change to the nature of local growth coalitions so as to include not only the whole government hierarchy, but also non-public-sector actors, especially business representatives.

Decentralisation, marketisation and political legitimation have transformed China's governmental structure and workings. This concerns in particular local governments, which changed from being agents of centrally defined policies to independent 'local states' (Zhu, 1999, p. 424) that pursue developmental policies. In this, they follow the East Asian capitalist model of development with the strong, at times interventionist, role of the state in driving economic development as the dominant objective, together with increased productivity and national competitiveness. China's local governments are now responsible for local prosperity as a result of reforms and decentralisation of responsibilities. In response, 'China's local governments have become economic interest groups with their own political agenda, and thus a local developmental state' (ibid.). The task for the state is to find its new role between being omnipotent and thus inhibiting the market, and being too weak to make an impact and provide predictability and reliability as the foundation for an effective functioning of the market. Overall, 'therefore, decentralization, pro-growth legitimization, competition between localities and the uncertain tenure of local chiefs combine to make China's local government the local developmental state' (ibid., p. 429). Against this backdrop, can the claim to an unchallengeable socialist core to state and society be maintained, and thus the continued division between economic and political-societal spheres?

China as a two-track developmental state between market realities and authoritarianism in socialist guise: some concluding remarks

China's development under the reform process shows distinct, uniquely Chinese features, but, of course, also processes and phenomena shared with other parts of the world, whether post-socialist or not. The particularly Chinese characteristic is the attempt at separating an economic and political-societal sphere, and pursuing very different, ideologically opposed development strategies for each of them. Effectively, the Chinese government is trying to ride two horses, at varying speeds, and with a fence between them. Obviously, this is a difficult, risky task with unpredictable outcomes, with most observers likely to conclude that 'this cannot work out'. Evidence so far suggests, however, that the Chinese government has been able to keep the two horses on track and on course, and many critics are now having second thoughts. There are no signs so far that the separation of the two tracks is going to be removed. If anything, recent statements by the Chinese government seem to reaffirm a resistance to political reform and, instead, affirmation of adherence to Mao Zedong Thought in a thinly veiled 'crackdown' on liberal tendencies within Chinese society.

Economic liberalism is not permitted to 'jump the barrier' and undermine the established political order by encouraging the development of an emergent civil society. This marks a major difference from the transformation model applied elsewhere, such as in eastern Europe, where post-communist development has clearly been interpreted as 'westernisation', that is marketisation and democratisation as inseparable parts of the same story. In China's case, the perceived economic opportunities of the gigantic potential market and future political capacity has made it much more difficult and unrealistic for western institutions and politicians trying to pressurise the country's leadership into following western ideas of 'right' forms of post-communist transformation. For the other former communist countries in the region under obvious Chinese geo-political influence, as well as for Cuba, the 'Chinese model' offers the only realistic alternative to the western model of a liberal market democracy, as propagated under the Washington Consensus. But questions arise about the future of the authoritarian political system governing a society pursuing a capitalist economy, especially when faced with ever greater spatial and social inequalities. Will it be possible to continue suppressing the emergence of a civil society and its claim to political influence? Will western influence, imported with western capital and exercised through international links, push for democratisation?

Nathan (1997, p. 12) summarises the contradictions of the Chinese transition process as part of a regime that has 'internationalized its economy while fostering nationalism; expanded economic freedoms while violating political rights; and decentralized bureaucratic power while rolling back a nascent civil society'. As a result of the strong emphasis on economic development and decentralised responsibility as the main engine of change, 'power has gravitated into networks of elites, giving rise to what some scholars call "local corporatism", a form of rule operating by personal influence and corruption' (ibid.), rather than democratically accountable processes. But the transformation process is continuously modified and adjusted to circumstances, reflecting the 'learning on the job' approach followed by the regime. Thus, 'China remains in transit, but its political direction no longer seems as clear as it did in 1990' (ibid.). The evident emergence of a new middle class does not necessarily point towards an emergence of democracy. 'China's new bourgeoisie is not only dependent on the state [and the created so favourable conditions] but also lacks a grounding in legally secure private property from which to grow toward an independent status in the foreseeable future. If this continues to be the case, China's modernization may turn out to be … "conservative modernization"' as a late twentieth-century form of a nineteenth-century model (ibid.). The issue of a clearly institutionalised framework regulating property ownership is a recurrent theme with post-communist developments elsewhere, especially in the former Soviet Union.

As a result of the changes, considerable inter-regional and intra-local disparities have emerged, reflecting varying participation in the changing economic conditions. Looking ahead, one of the main challenges, and not only for China, seems to be finding a path of development that can be sustained politically, economically *and* socially. This may include establishing stimuli of development other than external capital alone. As Friedmann (2005, p. 118) observes with regard to the future uncertainties of development:

> Although the stasis of the former regime has been broken, given the diversity of China, the dynamism of its people, the declining faith in the Communist Party as the moral center of society, the unstable mix of the traditional and the new, the emergence of new forms of social stratification, and the shattering of old verities, anarchy remains a possibility.

With power and the range of actors becoming more pluralised, the Communist Party may no longer be able to claim the central role as moral and political beacon in the development of China. Continuities thus seem to be as important as new beginnings. Important is the connection and complementarity between the two. This includes the need for state officials within the hierarchy not just to look up, waiting for instructions on how to proceed and thus avoiding responsibilities, but also to respond to the bottom-up pressures of an increasingly more articulate and impatient public.

The requirement is for more horizontally working relationships between the various actors, very much in the sense of 'governance', organised around territory, rather than adhering to a strictly hierarchical, vertical arrangement centrally controlled, with rather limited territorially based input. Increasingly, cities, especially those along the coast, search for their own place in the changing economic environment, seeking to step out of national dependencies in a rather 'un-Chinese' way. Not surprisingly, some are more ingenious and determined than others (see Friedmann, 2005, ref. to Wuhan and Kunming). The more enterprising cities are increasingly beginning to act as entrepreneurial cities and growth promoters, thus challenging established practices and political-institutional dependencies on the national government.

With a growing number of actors and places involved, a new civil society is slowly emerging, 'defending their own interests and making demands regarding what they expect their government to do for them' (Friedmann, 2005, p. 123). This, of course, contradicts and challenges the national government's continued claim to unrivalled political (and moral) leadership, and so it is not surprising to see the current Chinese leadership seeking to reaffirm its control by emphasising Mao Zedong Thought and communism as the only permissible public doctrine underpinning the nature and operation of society. In addition, the nature of China's economic growth looks increasingly likely to change, and with it the potential rate of participation of the Chinese public in the financial benefits of market-driven development. As Friedmann (2005, p. 127) observes, 'with what is now effectively jobless growth, China's entire modernization project is put in jeopardy. People, and not only in China, expect growth with equity. Failing to get this brings the threat of instability.' This also applies to the development of democratic structures, as the case of Cambodia demonstrates. Promising betterment is the key to political office, especially in the areas outside of towns (Roberts, 2001), and this encourages short-term, quick-fix solutions to please the electorate immediately, rather than engagement in longer-term development projects. In a society with a historically strong patron–client relationship (Curtis, 1998), reward for loyalty is crucial for retaining positions of power, and this mechanism becomes the stronger, shaping any attempts at democratisation, the more dependent a client becomes on patronage for pure economic survival. Western understandings of democracy and democratisation do not then apply. This, together with the Chinese experience of marketisation without democratisation, has fundamentally challenged the simplistic and self-centred perspective of western governments and institutions about the nature and progress of 'democratisation' in post-communist 'transition' countries.

Overall, the emerging Chinese model of economic transformation from state socialist centrally planned economy to a marketised socialist model has started a unique process of 'stop and go' changes, reflecting a form of trial-and-error approach to economic marketisation. The new model abandoned Maoist doctrine and radicalism about communitarian ownership and egalitarianism, Stalinist economic control and mobilisation, but also the neo-liberal doctrine of privatisation, marketisation and democratisation as an integral model. Effectively, Chinese market socialism marries two seemingly contradictory and mutually exclusive concepts of economic management. It combines economic liberalisation with continued political centralisation and totalitarian state control. Underlying these contradictions are considerable variations in their form and extent, but also a negotiated outcome between urban and rural places, as well as across the provinces, with the main urban centres of the southern, coastal provinces going the furthest in economic development.

7

AFRICANISM, MARXISM, POST-COLONIALISM

Pragmatic use of communism and post-communism as a label

Despite the liveliness of debates on 'post-socialism' over the last decade or so, Africa has very rarely featured in the discussions. Yet African states developed quite interesting modifications to the European model of a communist state, based on different experiences with socialist ideology and communist regimes, that go beyond evidence of a 'developing country factor'. This chapter will explore these African particularities, and their translation into 'African socialism' and 'African communism' respectively and, consequently, 'African post-communism'. This will include discussions revolving around the two main rationales behind the adoption of Marxism–Leninism: post-colonial 'Africanisation', and the strategic exploitation of the Cold War competition between the superpowers for 'spheres of influence' in Africa.

A good illustration of the understanding of *African* 'post-socialist/communist' development is provided by the special issue, of the journal *Communist Studies* of June 1992, 'Marxism's Retreat from Africa'. At that time, the claim of the adherence to Marxism as the legitimating rationale for autocratic states in Africa was quickly disappearing with the end of the Cold War divisions, but this did not mean an automatic shift towards more democracy, as expected by the 'West' as the 'natural' follow-up to the end of the socialist/communist regimes. Instead, changes were relatively slow and erratic, and often yielded little effective change to the reality of governance. As a consequence, 'transition to pluralist democracy and to viable market economies remains less certain' (Hughes, 1992, p. 2). This uncertainty, however, primarily affects the political, state-institutional side, and reflects the predominantly pragmatic, technical-managerial perspective taken by those African countries claiming adherence to socialist or Marxist–Leninist principles. Their adoption of one or the other ideology, be it socialism or market democracy, needs to be seen primarily as in essence a pragmatic choice in a divided Cold War world. 'African socialism' and, subsequently, post-socialism, are very much part of the colonial and post-colonial legacy of seeking visible state autonomy, while still being dependent economically and strategically. But, in 1990, the rules of the game changed with the end of the bipolar world, replaced by the political and economic hegemony of the US (see also Ottaway, 1999). Against the reality of a divided Cold

War world, the notions, and rationales, given for the adoption of 'socialism' and Marxism varied, along with the very concept of 'socialism' *per se*.

Key features of African socialism, according to Mohiddin (1981), are (with reference to Kenya):

- absence of political-ideological dogma, and thus adaptability of the system to changing requirements, and developing pragmatic responses to *actual*, rather than idealistic-hypothetical conditions;
- non-alignment to the two main political blocs, and thus avoidance of a dependent satellite relationship;
- prevention of class formation, and thus societal divisions, through more egalitarian policies;
- search for a synthesis of Marxist socialism and free market economy as the 'African way', drawing on the particular qualities of African tradition and culture.

Given the economic and political instabilities in post-colonial Africa, especially the challenges to territoriality and boundedness of statehood, and its relation to the question of nationhood, quite formidable challenges exist for the further development of Africa in a 'post-communist', single-superpower world, where the market-led, international capital-dominated economic paradigm is paramount. There is little alternative to that, it seems, especially given the continued strong dependence of African countries on external help to pursue developmental policies and overcome the considerable structural economic problems of 'underdevelopment', underinvestment, and inequality. This, of course, makes these countries particularly susceptible to external financial and political influences.

Less clear is the political question of democratisation as a central plank of western development support. A whole range of options sits between 'full democracy' and 'authoritarianism', whether personality- or institution-led, albeit now without the suffix 'socialist' or 'Marxist'. More recently, however, as discussed below, there have been signs of growing public pressures towards democratisation, and a clearer definition of 'statehood' and its power *vis-à-vis* the population. 'Post-communism' and 'transition' have thus quite particular meanings in an African context, differing from those in Europe. Yet at the same time, as a colonial legacy, the European concept of 'socialism' has contributed to the post-colonial political landscape and the emergence and operationalisation of Africanness, against the background of European cultural and idealistic legacies. Many of the post-independence African leaders were educated in their former colonial mother countries, and those studying in Paris experienced French communist ideology at first hand.

The ideological dimension of Marxism–Leninism as a post-colonial development strategy of convenience

The ideological construction of post-independence Africanness began immediately after the first wave of independence of the 1960s, and was propagated as African socialism. Despite its obvious reference to the Marxist ideology, the actual concept behind this label had little overlap with the Soviet-style Marxist doctrine. The main influences were a re-awakening nationalism and pan-Africanism. It was very much a response to the success of the struggle to end colonialism. There was thus a close link between these freedom fights and the first Africa-based discussion on statehood and its role *vis-à-vis* society. In most of these socialist countries, the anti-colonial revolutionary leaders had gained office and thus established an important psychological link between the anti-colonialist struggle and socialism (a phenomenon also observed in Latin America, especially in Cuba; see Chapter 8). 'Socialism at this juncture was hence both Afro-centric and non-aligned. It shunned the unselective transfer of socialist terminology (such as class struggle) to Africa and at the same time laid claim to a universality of political ideals' (Ottaway, 1999, p. 161), while 'the early socialists stirring in Africa were [therefore], in many respects, the intellectual counterpart to the quest for political autonomy' (ibid.). Effectively, there was a division between ideology and idealism in the aftermath of independence, and pragmatic responses to the new circumstances. This idealism sought to address and reconcile a broad and diverse range of concerns: developmental-programmatic, nationally focused, popularly appealing, and politically uniform objectives. Inevitably, there was an inherent contradiction between the projected, developmentalist notion of African socialism, as forward-looking and reformative, and the Leninist belief that party officials and functionaries are by far the best suited to drive and implement such an agenda from the top.

By its nature, African socialism became very much entangled with its leading proponents, all of them personalities of the anti-colonial movement. While this gave them instant authority and credibility among the public, it was also its shortcoming. There was little in terms of a general ideology that could be carried over to another leader, as each of them created their own hotchpotch of selected political and historic mythologies, to create a new, Africa-centred approach. Not surprisingly, these personally designed imaginations barely survived the demise of their initiating leaders into the mid-1970s. They had not propagated a standardised and generally accessible version of socialism *per se*, but rather national identities and African pride, however defined. The impact of these first post-colonial approaches can be felt in the pan-African communist ideology of later years, which resulted from a fusion of both external political and economic pressures, and internal societal political and ethnic interests.

External influences in the shaping of African communism

'The "Cold War" saw a return to Soviet and Western European communist parties championing anti-imperialist struggles led by the national bourgeoisie' (Hughes, 1992, p. 6). African politics effectively sat between the 'hesitancy and manipulativeness on the part of the Western communists and the blatant mixture of cynical self-interest and "proletarian internationalism" on the part of the USSR ... ' (ibid.). In its essence, communist ideology was not a locally grown concept, but was associated with external, even imperial, influences and legacies, and this alien association made it appear 'un-African', tainted by imperial connotations. Only if associated with positive experiences, such as the 'freedom struggle', was socialism (or Marxism–Leninism) acceptable. Prior to independence, for instance, in French colonies, radical African critics of French colonialism were portrayed as nothing else than Moscow's puppets, so as to weaken their credibility and question their 'Africanness'. But in reality, links with Moscow were usually frayed, even if receiving economic or military aid, albeit with many strings attached.

Against this backdrop, claims to 'communism' were largely opportunistic moves by those holding or seeking power, rather than an expression of a political programme and ideology. 'African radicals, during both the colonial and independence periods, tended to adopt a cautious as well as calculating attitude towards the communist powers. Essentially, they were out to win power for themselves rather than to place their movements and countries under external communist domination' (Ottaway and Ottaway, 1986, p. 6). Consequently, then, *post*-communism means much less of a paradigmatic political-cultural shift than in Central and Eastern Europe. In Africa, the association with outside powers was, in effect, often seen as continued imperial domination, and this shaped public attitudes to the communist ideology originating in (colonising) Europe. The difference between the 'capitalist' West and an 'anti-capitalist' Soviet communism had thus to be elaborated to make the concept acceptable. Nevertheless, the leading role of the often European-educated political elite as the main driving force of any socialist movement meant that any such shift was essentially top-down and urban-focused and elite-driven, rather than the outcome of popular demands. There was no mass movement or bottom-up revolutionary pressure for adopting Marxism–Leninism. Thus, even in 'good' socialist examples, such as Mozambique, 'the acceptance of Marxism as a national ideology and a framework for political and economic management was less than complete and hedged with contradictions, eclecticism, and compromise' (ibid., p. 9).

Pragmatism and the socialist/communist label for national politics

In the 1960s, the only other model of post-colonial government and policy was pragmatism, which realised existing dependencies and refrained from developing a grand vision of Africanness. Here, the new leaders sought to continue using the established administrative and economic expertise of the colonial period, and simply projected these into the future, if under their own control. This included continuing external trade links, very much in the mould of the colonial days. These pragmatic understandings were no less state centred than the populist idealist socialist model, with both using administrative centralisation to run state and society. Given its more compromising, accommodating nature, African pragmatism has survived much better than socialism. The independence-inspired, milder, more idealistic views were giving way to a new generation with more rational, functionalist views of state policy and development. And this included a more realistic assessment of the dependency on capitalist markets. Inspired by study visits to France, and thus contact with French communism at the universities, many of the younger post-independence (urban) political elite revisited socialism in a much more scientific, pragmatically technocratic way and, as part of that, borrowed more ideas from Soviet-style Marxism–Leninism. The Soviet model of development seemed to promise an alternative to succumbing to western capitalism, and thus, in the eyes of many leaders, offered an effective break with a continued colonial-style dependency. In addition, the military gained in influence as a result of their role in the battles for national independence and, subsequently, the territorial integrity of the newly independent states. There was an active drive to expand Soviet influence as part of the USSR's outwardly directed policy of empire building in different parts of the world, in direct competition with the US (see Chapter 5).

The five former Portuguese colonies – Mozambique, Angola, Guinea-Bissau, Sao Tomé and Principe, and Cape Verde – illustrate the Soviet rush for Africa. This new political-ideological wave of expanding and consolidating a Soviet sphere of interest was less eclectic and idealistic than African socialism, and was rather pragmatically driven by geo-strategic considerations (Albright, 1980a). This reinterpreted understanding of socialism was now propagated as Afro-communism, with explicit reference to Marxism–Leninism and the societal focus of the ideology. Within the declared socialist or communist countries, attention had shifted from independence and the immediately following euphoria about national self-determination, to re-shaping society and overcoming the still prevalent colonial structures. Class struggle thus gained in currency, albeit adjusted to the fact that there was no industrial proletariat. Instead, the peasants and the bureaucratic class (civil service) were pitched against each other as competing class elements of society, with the former being portrayed as the suppressed victims of the exploitative and privileged bureaucrats with their alleged inherently colonial attitude. Propagated in the right way, as a move towards more equality, such an agenda was obviously going to be popular. At times, this populism outshone any reference to socialism or Marxism–Leninism. Instead, moral values, tradition, nationalism and anti-imperialism were the main ideological cornerstones of this strand of 'consolidative' African politics.

African socialism or Africanised Marxism–Leninism?

The inherent conceptual competition between 'Africanness' and 'socialism' is summarised in the title of Babu's book 'African Socialism or Socialist Africa?' (1981). This refers to the different emphases between the two concepts – the indigenous, post-colonial sense of Africanness and the 'imported' political-economic regime of socialism. In particular, Babu (1981) explores evidence of a more generic, post-colonial experience with socialism, where geography matters less than common historic experiences. For this, he compares the Asian with the African post-colonial experience, and points to the key difference between the two continents – the degree of economic independence and capacity. China and (North) Korea, despite their association with, and receipt of aid from, the Soviet Union and Eastern bloc, maintained a distinct distance. They did so in the attempt to steer their own paths between the hegemonic worlds of the two superpowers. In Asia, as in Africa, discussions arose around the suitability of the theoretical principles of an industrially based socialism for the particular conditions of the pre-industrial, post-colonial (feudal) countries/cultures. It was from there that Mao Zedong developed his version of communist ideology, tailored to the conditions of an agricultural society (see Chapter 6). The parallels to 'African socialism', especially Tanzania's Ujamaa, are obvious. Indeed, there were different African 'versions' of socialism, reflecting the different colonial legacies, for example between Tanzania and Kenya. Such legacies include the artificial nature of most boundaries (and thus potential disputes about territories), a multiplicity of ethnic and cultural-political divisions in societies (raising the question of nationhood), economic weakness and a dependency on outside forces, and the legacy of a colonial memory as common reference point (Chazan *et al.*, 1999).

In Tanzania, there was little in terms of a European-style bourgeois class, with peasant farming remaining the dominant feature. In Kenya, by contrast, there was a strong new 'colonial class', consisting of white yeoman farmers and civil servants. The African peasant farmer in Tanzania was an integral part of the country's economy, and not merely an adjunct to European-owned plantations (Mohiddin, 1981, p. 43). As a result, Tanzania achieved its independence with a weak middle class, whereas Kenya had a very prominent, powerful such class. Thus, the Tanzanian idea of Ujamaa as a form of traditional African communitarianism seemed more relevant to its own situation than that of Kenya, from where the concept originated. The Tanzanian president, Julius Nyerere, the author of the Ujamaa concept, expected the international capitalist system to facilitate the country's transition from colonialism to socialism through extending its continued business interests in Kenya. This, however, did not happen, given the aversion of 'western' capitalism to the notion of 'socialism', however defined.

> Thus, in Tanzania, as in Kenya, socialism, its content and practical implications, became a political issue. It could no longer be left to allusive official guidelines, sporadic conflicting and confusing statements by politicians. The people demanded a clear-cut definition of socialism and its practical meaning for them.
>
> (ibid., p. 58)

Nyerere's 1967 Arusha Declaration was the response. But generally, the dependence on international capital and trade meant that Tanzania gradually slipped back into its old colonial role as a producer of primary products, dependent on external capital and demand.

Despite the overt differences in the emphasis, implementation and rationalisation of socialism in Africa, Chazan *et al.* (1999) identify four key themes reaching through the four decades after the end of colonialism, albeit with considerable variations between countries. These include, in particular, the lack of political-ideological coherence, with some being more eclectic and pragmatic than others, anti-colonial sentiments, the predominance of more parochial world views, and, in the 1990s, a return to liberal ideas and a conscious effort to translate them into African terms (Mohiddin, 1981, p. 167). Reaching beyond these immediate conceptual concerns, the main commonalities between African countries were nationalism and an emphasis on Africanism. This included anti-colonialism, and critiques of economic dependence and underdevelopment.

Given this rather 'laid back' nature of 'African socialism', both in the 'West' and the 'East' there was scepticism about the rationale, ideological depth, and nature and destination of what, in the 1960s, became known as African socialism. Disillusioned with the lack of ideological vigour, Moscow coined the more ambiguous phrase 'non-capitalist road of development' (Ottaway and Ottaway, 1986, p. 6). Nevertheless, by the mid-1970s, the idealist 'African socialism' had given way to more Leninist input, reflected in the label 'African communism (Marxism)'. This reflected a shift in political-ideological emphasis and leadership. The earlier, post-independence North–South perspective, had given way to an East–West contrast, defined by the two superpowers' spheres of interests. This also meant an end to the doctrine of non-alignment propagated by the Afro-centric, inward-looking 'African socialism'.

The Marxist–Leninist leaders of the 1970s and 1980s abandoned the idea of an independent 'African way' of socialism and sought to utilise the Cold War rivalry between the two superpowers to their advantage. They thus 'considered the Soviet Union and other communist countries to be their "natural allies" in the ongoing struggle against colonialism, neo-colonialism and imperialism' (ibid., p. 9). Yet, nationalism was far from dying out as a major political force. 'We do not intend to become another Bulgaria', stated a top Mozambican politician in the late 1970s (quoted ibid.). This highlights the somewhat ambivalent relationship with external forces, viewing them as a useful source of support at the international level, yet also fearing new dependencies resulting from such engagement. The dual approach to domestic and external (international) issues under 'Afro-communism' (Ottaway, 1986) stands for an attempt to engage with, but also keep its distance from, the Soviet Union and its political hegemonial aspirations. The parallels to 'Euro-communism', with its similar emphasis on individuality and different national paths to a Soviet-led Marxist–Leninist society (state), are obvious (Mandel, 1978).

African socialism as programmatic 'Africanist' development strategy

In the 1960s, at the time of African socialism's greatest popularity, the then Tanzanian president, Julius Nyerere, stated that African socialism was essentially about the extended family, that is 'familyhood' or 'Ujamaa'. This was the focus of his propagated idealising concept of the Ujamaa model village as an expression of traditional African, pre-colonial communitarian society. This notion became the centrepiece of his and other post-independence African socialist governments, drawing on both Africanism and socialism. The concept of class and the class struggle, the central plank of Marxism, was ignored, thus removing one of the key tenets of 'Marxism'. 'The task of an African socialist leader, then, was to restore African society to its pristine classless self', meaning, largely, a return to African pre-colonial times, with communal ownership and an absence of individual control of the means of (agricultural) production (Okoko, 1987, p. 16). '"African socialism" – was born of the post-colonial crises of economic development and national identity. In this sense, like the classical varieties of socialism, African socialism also underscores the primacy of the economic element in socialism' (ibid., p. 12).

Economic development and socialism had thus become widespread synonyms in African politics, as part of decolonisation. Indeed, economic development was seen as the path to national self-determination and true independence, and socialism offered a credible alternative developmental strategy to that, offering a path around the seemingly universal leadership of the capitalist system. Being non-capitalist, socialism could therefore be portrayed as inherently associated with liberation from western domination and independence. These values added a specifically African dimension to 'socialism', and are opposed to those of Central and Eastern Europe. So, it is not too surprising that many African leaders were using 'socialism' and 'economic development' interchangeably (Okoko, 1987). Socialism promised a new start, independent of old structures and dependencies. Given such far-reaching associations, it does not come as a surprise that 'socialism' meant many things to many people. With its promoted rootedness in African communitarian tradition, socialism, independence, national identity building, self-assertiveness, and historic-ally derived idealism, were all rolled into one. The result was a specifically *African* 'brand' of socialist rationale. But, as Okoko (1987) points out, African socialism is utopian in its refusal to accept that African society has not been as homogenous and communitarian as portrayed in the 1960s post-colonial political discourse. The aim was to reinvent socialism as a post-colonial, pre-industrial and, most importantly, in spirit African development strategy. But this idealistic, perhaps somewhat naive view also meant the ultimate failure of this approach.

Crucially, there was no single thinker or implementer of socialist principles under African conditions, and no hegemonic state able to set an example and offer effective patronage. African socialism is rather a more or less coherent fusion of many different views, developed independently by different politicians in their varying national contexts. The absence of a clear ideological yardstick against which to measure claims to 'socialism' allowed each leader to use the term 'socialist' as a label, yet still adopt a political-economic regime as they saw fit at a particular time. But there have been attempts at linking theory and practice, and providing a theoretical, strategic underpinning to the claims' of 'socialism', notably in Tanzania, under Julius Nyerere, culminating in the Arusha Declaration of 1967, which officially adopted a socialist path of post-independence development. At that time, many Tanzanians felt a sense of initiative and resolve after the end of colonialism, and 'saw themselves in the revolutionary vanguard in Africa' (Cliffe, 1972). But this was the exception, rather than the rule.

The 'Afro-Marxists' responded to the conceptual weaknesses of 'African social-ism' by adding a dose of pragmatic geopolitical realism, as well as the 'socialist classics' of class difference, class struggle, and proletarian dictate. This 'transitional socialism' (Okoko, 1987) was the link between African socialism and the much more Soviet-oriented and orthodox Marxist–Leninist Afro-communism. The latter bore all the hallmarks of Leninist communism, as practised in the former USSR, and thus was a political-ideological import. Political-economic realities, especially the lack of promised and expected economic development, and the difficulty in establishing the idealised communitarian societies, had made such direct policy transfer more acceptable than during the early years of more hope-filled post-colonialism. Greater emphasis on coercion was considered the necessary instrument to turn theory into reality. The other reason for this emphasis on ideology was the realisation that economic and political dependency on the former colonial masters, or international capital in general, effectively continued, and that bolder steps needed to be taken to break that dependency. The Soviet model appeared to offer that possibility, through greater reliance on a developed national economy and mutual economic assistance among the communist states. Ultimately, this resulted in a shift from a post-colonial 'North–South' perspective, with pan-African ambitions, to a superpower-led 'East–West' perspective, embedded in Cold War geopolitics. What the Afro-communists wanted was an outright alliance with the Eastern bloc, as a bulwark against the capitalist world and its perceived imperialist ambitions. But this entailed subscribing to another political hegemon.

Reviewing the situation in three Marxist states – Angola, Mozambique and Ethiopia – in the mid-1980s, after roughly a decade of these states' claim of adherence to Marxist principles, Ottaway and Ottaway (1986) examined whether Afro-communism was 'good' for them. Persistent shortages of food and other essential goods, as well as continued power struggles between ethnic groups, were suggesting otherwise, but then the situation was not that different from those countries subscribing to western capitalism.

The effects of external pressures

By the late 1970s and early 1980s, the economic situation in Mozambique caused the government to realise the need to engage with global, western capitalism, to gain investment and bring the necessary drive for economic development, and thus 'mend its tattered economy' (Saul, 1985a, p. 130). This confirmed a strong dependency on external links – about 80 per cent of the country's trade was with the US and western Europe (ibid.) – highlighting the continued 'neo-colonial' dependencies, and challenging the political drive to socialism and independence. As a sop to the socialist rhetoric and public imagineering of the country's destiny, it was propagated that any foreign investment would have to further Mozambique's economic plan, guarantee the training of Mozambican workers, and generally be socially conscious, 'friendly' and allow nationalisation at the end of a contract-defined time (ibid.). Not surprisingly, foreign investment remained rather unimpressed. To improve things, in 1983 (before the time of Gorbachev in the Soviet Union), the Mozambican president toured western Europe to invite business investment and seek military cooperation. Claiming a socialist political identity did, therefore, not at all preclude collaboration with the class enemy, if it promised advantages for Mozambique. Such pragmatism, with obvious contradictions between theory and practice, is fundamentally different from the much more dogma-driven approach of the eastern European states. But this apparent contradiction also embodied one of the main difficulties of this version of socialism – the lack of a clear and consistent conceptual underpinning of its main claims and the state organisation. But the country, like most other African 'socialist' countries, felt that it had made a choice, rather than having had socialism imposed on it, as in eastern Europe. This is why a Mozambican socialist party (FRELIMO) official could claim, 'It is our experience which led us towards Marxism–Leninism ... We have, on the basis of our practice, drawn theoretical lessons' (ibid., p. 136). Theoretical arguments and reasoning of Marxism–Leninism thus tended to be understood as an *a posteriori* organisation and rationalisation of practical policies and requirements, often appearing more 'muddling through' than following strategy.

But there was a second reason for the growing prominence of Marxist–Leninist rhetoric, if not ideology, among African socialist states. The inherent centralism and state-centred, bureaucracy-driven and administration-controlled form of government suited many African governments seeking rationales to justify and legitimise their usurpation of power. The Soviet rationale of Marxism–Leninism offered that opportunity. Such countries were 'rebranded' as 'Afro-communist' by analysts (Ottaway and Ottaway, 1986), or 'Afro-Marxist'. Examples include the 'People's Republic of Congo' (established by disaffected radical soldiers in 1968), Somalia in 1970 (proclaiming itself as a 'socialist state'), Dahomey in 1974 (renamed the People's Republic of Benin), and Ethiopia and Madagascar in 1976. Upper Volta, renamed Burkina Faso, followed as recently as 1984.

The drawback of focusing on the maintenance of power and structures meant, however, that Marxism–Leninism had lost most of its responsive, dynamic and innovative nature, and instead, was essentially a 'frozen Marxism' (Saul, 1985a, p. 139). Not surprisingly, terms like 'democratic centralism', 'productive forces' and 'vanguard party' point to the first and foremost *technocratic* understanding of communism. The notion of a proletariat and 'class struggle' had little public resonance. Nevertheless, imported sound bites and phrases were used, whether they fitted the country's circumstances or not. When the Soviet Union collapsed and eastern Europe vehemently rejected 'textbook Marxism–Leninism', in Africa, the ideological rug, albeit imported and an uncomfortable fit, was pulled from underneath Afro-communism of whatever mould.

Overall, the ideological and implementational eclecticism makes it difficult to speak of one version of African socialism, or communism, as there are so many interpretations. Nevertheless, two main waves of post-colonial claims to some form of 'socialism' can be identified: 'African socialism', or 'populist socialism' (Young, 1980), in the 1960s and early 1970s, followed by a more 'mainstream' approach to socialism in the Marxist–Leninist mould in the 1970s–80s, also referred to as 'scientific socialism' (see Table 1.1 overview). Nevertheless, African leaders favoured more 'home-grown' strategies and justifications for maintaining power, especially references to anti-colonial struggles. This reflected a 'high degree of opportunism among leadership groups, the low level of popular attachment to socialist ideas, the weakness of state institutions, and the continued need for western economic assistance' (ibid, p. 10). As a result, many of the first-wave African socialist countries were swept away by military coups in the 1960s.

Often, the claim to Marxism–Leninism by the revolutionary party occurred only slowly over the years, usually in response to growing economic problems and few signs of developmental progress. But, although these countries adopted the official 'socialist rhetoric', in none of these instances did the USSR actually impose its form of government upon them, contributing to the so different perception of the nature of socialism/communism. And this led to very different outlooks on the value and nature of post-communism.

Overall, Ethiopia was perhaps the most committed, moving closest to the Soviet model of Marxism–Leninism, while Angola and Mozambique kept a more visible distance, and maintaining that closeness while also, driven by need, accepting US food aid, albeit without further political 'repayment'. By the mid-1980s, therefore, one could conclude that 'Afro-communism is still very much alive as an idea' and as an effort to establish Marxist–Leninist institutions (Ottaway and Ottaway, 1986), albeit with distinct variations between countries in response to their particular histories: societal-political factionalism in Angola, revolution-inspired popular participation in political decisions in Mozambique, and a more state-centred Marxist–Leninist approach in Ethiopia. Societally, circumstances varied, too. While Ethiopia, with its large landless peasant group, came closest to the model of a 'mass proletariat', in Mozambique and Angola peasants were much more likely to own their own land. Their enthusiasm for communist-style cooperatives or collectivisation was therefore rather muted, while meeting the end of communism much more positively than Ethiopia. There, the regime had little immediate impact on the economic opportunities of large parts of the landless population. Land scarcity thus acted as an important facilitator in making Marxism–Leninism more acceptable to the population. But, irrespective of the differences in implementation and popular support, experiences since the mid-1970s had shown that Marxism–Leninism did not offer any miracle cures for Africa's political-economic weaknesses and difficulties, and neither did capitalism. What it did seem to offer was a belief that an alternative route to development was possible, even an African solution, although, ultimately, this turned out to be more of an idealistic illusion than an economic reality. The end of socialism as a credible alternative to capitalism, however, has removed that theoretical scope for an alternative route, whether embarked on for genuinely idealistic reasons, or for pragmatic political expedience.

The legacy of African socialism and Afro-communism for 'African post-communist development'

The general conclusion from reviewing Afro-communism is that it is a rather more superficially adopted form of communist state-societal organisation, compared with the commitment and rigour applied to the ideology's implementation in eastern Europe, the Soviet Union, or China, for instance. The official end of communism in the main stalwarts of African Marxism–Leninism, in 1989–92, may therefore be expected to have been much less dramatic in its political-economic and governmental implications. This will be investigated in the following sections.

The flexible interpretation and implementation of doctrine and policy by the leaders of the Marxist countries, in response to economic pressures, may well have helped maintain these regimes until the end of Soviet communism. While maintaining the official public rhetoric of Marxism–Leninism, underneath repeated 'modifications' usually meant a 'watering down' of Marxist principles, except in the cases of the staunchest defenders of the Marxist–Leninist doctrine – Angola, Ethiopia and Mozambique. In fact, Ethiopia seemed to become even more dogmatic, *vis-à-vis* its economic problems, in the mid-1980s. Pressure for greater orthodoxy, by leftist government factions, became a somewhat petulant response to the obvious pressures for change, not least those emanating from the Soviet Union itself. Political institution-building was a key agenda in the Marxist states, but with varying results. While Ethiopia and Mozambique made progress, Angola did not, resorting instead to greater dictatorial control, and the dogged implementation of (yet more) orthodox Leninist economic principles, such as forced collectivisation and extended state control. The difficulty in building effective political institutional structures, however, is not a particular problem of Marxist regimes, although it appeared to be by far the most common shared experience among all African countries, 'communist' or not, and reflects a wider difficulty with statehood and state-building in Africa. Often, they were subjected to an array of ad hoc, impromptu modifications and responses, increasingly obscuring the original nature of 'socialism'. The starting points for post-socialist/communist developments are thus equally diverse and unpredictable.

Another key particularity of African communism/socialism was its relative political distance from the Soviet Union, and lack of cooperation within a 'communist bloc' within Africa. Distrust between more 'hard-line' communist states and the more pragmatic ones, willing to deal with the 'West', if necessary, made such cooperation unlikely. Thus, Ethiopia sought ever greater approximation to the Soviet Union in its fierce anti-western sentiment. Angola and Mozambique were much more integrated with the western economies, while maintaining a clear distance from Moscow to allow enough scope for developing their own versions of communism, rather than simply transplanting a ready-made model from Europe (Ottaway, 1980). Their dependence on 'western' famine aid, which exceeded that available from the Soviet Union, also created a sense of frustration and embarrassment about this dependency on the 'imperialist West' among the socialist leaders. Thus, Mozambique, for instance, joined the World Bank and International Monetary Fund in 1984, in the realisation that economic development support was more likely to come from there, rather than from the Eastern bloc. Ethiopia, by contrast, stood steadfastly by the Soviet Union, probably out of the realisation that existing Soviet influence in the central African region (Horn of Africa) was offering greater potential political-economic benefits than a collaboration with the US. But all of the states, despite their ideological differences, sought superpower back-up militarily, to defend and support contested state identities and territories, and spheres of regional influence.

Generally, by the mid-1980s, 'Afro-communist' (Ottaway and Ottaway, 1986) regimes had become pragmatic, juggling internal economic and security problems, and external dependencies, while often struggling to stay intact as states *per se*. By that time, it had become obvious that 'Marxism–Leninism was no more of a miracle cure for Africa's ills than other ideologies or remedies tried by African regimes or prescribed by foreign agencies' (ibid, p. 243). As a consequence, the collapse of communism in Central and Eastern Europe and the Soviet Union was much less of a symbolic act of liberation in Africa, than it was in Europe. African communism had been more of a long-distance, non-ideological affair, rooted in post-colonial world views, resentments and ambitions, and not class struggle and Sovietisation. Both were never really imposed with the same rigour and ruthlessness as in eastern Europe or the Soviet Union, and had a very different political and social-economic context. And this, of course, meant that, consequently, post-communism has been much less of a dramatic, epoch-marking development, than remembered in Europe. 'Muddling through' and overcoming economic and political crises in an ad hoc way has been the effective political agenda among African states, whether 'communist', 'socialist' or 'western'.

The end of African communism – and choices in development?

Developments in Africa since the end of the Cold War have been shaped by two main factors: the end of politically motivated subsidies to friendly governments as proxies for the two superpowers' strife for geopolitical influence, and the growing insistence on the adherence to New Right policies propagated by the main sources of aid, the international development and financial institutions (i.e. the IMF and World Bank). Politically motivated support thus gave way to economic consider-ations, often through rather technocratic procedures. Among these institutions, the neo-liberal doctrine of development ruled supreme in the late 1980s, with the Washington Consensus (see Chapter 3) of the mid-1980s offering the yardstick by which national economies and related governmental policies were assessed. This is similar to the political-economic pressures and appraisals applied to the former communist states in Central and Eastern Europe and the former Soviet Union.

The end of the Cold War brought about a fundamental shift in the perceived importance of Africa on the political agenda of the USA and post-Soviet Russia. Strategic interests had changed, and the need to obtain a regime's political-strategic friendliness was no longer considered necessary. The Cold War competition for Africa between the two superpowers had run its course, and the two superpowers' respective client states had lost their bargaining power. The end to the bi-polar Cold War geopolitics meant that they were no longer needed in the ideological and geopolitical competition between two superpowers. As a result, Africa fell off the geopolitical radar screen. Thus, existing assistance, mainly economic and military, was steadily reduced, leaving those regimes without the revenue that had been crucial for their survival for some time. For instance, Zaire, Morocco, Ethiopia, Somalia and Kenya had received favourable trading terms from the US as client states (see Thomson, 2000). In terms of overall volume, the immediate financial impact of US aid to Africa, for the continent as a whole, has been rather limited. It was near the amount spent on Brazil alone (ibid.), but that money was concentrated on a limited group of countries, those amenable to US policy. Reducing, or even terminating, those support measures inevitably weakened the international and domestic political standing of the relevant regimes, and their policy-making scope. But 'with the collapse of mentor regimes even the most obdurate of Afro-Marxist governments could no longer hold out' (Hughes, 1992, p. 18).

Not surprisingly, a number of former US clients lost power in the 1990s. Growing global economic competitiveness made it increasingly difficult to offset these lost incomes, and assistance came now with political conditions attached (Thomson, 2000). There was no longer the need to tiptoe around the political sensitivities of Africa's leaders, and thus their likely allegiances. Authoritarianism was no longer quietly tolerated, if the respective rulers declared their 'right' allegiances. Pressure mounted, instead, to push more ethical concerns, especially democratisation. The general enthusiasm for regime change towards democratisation and liberalisation, demonstrated with regard to the CEE and FSU countries, was effectively extended to politics towards Africa. The new credo afforded by western governments was that of good governance instead of good relationship, and this underpinned the availability of assistance for African governments. Continued economic and financial dependencies, despite post-independence rhetoric and ambitions, granted external political and economic interests an important foothold in African politics and governance. The new political climate resulted in some fundamental shifts in Africa's political landscape, challenging authoritarian rule and pushing for more democratic principles. For instance, in Zambia, Kenneth Kaunda had to surrender his grip on power after nearly 30 years, when pressures for democratisation made Zambia the first Anglophone country to topple its government through democratic defeat.

Thus, 1991 also meant the end to an autocratic regime going under the banner of socialism (Bates et al., 1998). The effect was a wave of democratisation, enabled by weakened authoritarian governments in the aftermath of reduced international financial and military support and thus their scope to dispense patronage. Instead they imposed unpopular policies of economic restructuring and financial austerity. These were the main pillars of the Structural Adjustment Programmes imposed on African countries by the World Bank and the IMF, in return for alleviating a mounting pile of debt. But despite the new, economy-driven rationality, political considerations remained never far away. Thus, for instance, the French government continued to extend selective financial aid, and western governments chose to ignore the abandonment of democratic processes in Algeria in 1991, because this meant containing the rise of Islamists to power (see Thomson, 2000).

This continued external interference with the politics of supposedly autonomous states raises questions about their genuine degree of autonomy. Furthermore, it reinforces the ruling elites' hold on power, because they can use these connections to their advantage. The states, thus, were inherently undemocratically elitist, with a weak civil society and weak autonomy. This questions their actual statehood, as seen both externally and internally. Internally, the regimes coming, from the 1980s sought legitimacy within their own societies, either through bribing an electorate with politically motivated hand-outs, or by ideological justifications. Tighter financial conditions meant that the bribing approach was increasingly difficult to apply, and with it the attempt to claim and sustain political control over a territory.

As a result of the relative, and visible, political weakness of many African states, some observers see these states as devoid of the crucial ingredients of statehood, reducing them to quasi states (Jackson, 1990). These states depended on external approval and subsequent support to be able to maintain control, and Cold War politics meant that the interests of African regimes, and those of the superpowers and former colonial rulers, converged, be it under the banner of socialism or of democratisation. But the end of the Cold War removed the basis for shared interests, and with it the system of political patronage. New political priorities afforded by the western-controlled donors exercised pressures on the existing regimes, ultimately leading to their succumbing to internal challenges and political reform in the 1990s. In effect, with the New World Order, civil society joined state elites in becoming a beneficiary of the attentions of external interests (Thomson, 2000). Increasingly, donors bypassed existing state elites (who were no longer useful) and addressed civil society directly, while claiming the moral high ground and following missionary ambitions of promoting western-style democracy. This includes funding programmes for voter education on the principles of democracy in practice. The outcome has been a shift in the power and control of state governments, from regimes depending on external support for their hold on power, to governments drawing on domestic support. But economic realities undermined the democratisation ideal, as the main global financial institutions intervened with national economic and monetary policy to have their neo-liberalist views implemented.

It is at this point that interesting parallels emerge with the post-1990 transition process in Central and Eastern Europe. Similarly to most of communist eastern Europe, Africa was suffering from a crippling debt mountain, undermining the provision of public services and, incidentally, the principles of good governance that western governments have been claiming to support (ibid.). Structural Adjustment, similar to the post-socialist doctrine in eastern Europe, was the 'magic word', and liberalisation, democratisation and marketisation the equally repeated mantra. This involved free markets, a minimal state, and market-driven provision of public services, as well as a purely microeconomically assessed value and efficiency (and thus viability) of state enterprises. These changes, as in the CEE countries, were deemed to be the solution to the continent's economic (and subsequently social and political) problems, generating economic development and growing affluence, which would filter down to benefit everyone. The main difficulty with this new, externally defined political agenda was that it undermined the political status quo in Africa, and the well-established practices of governance. The state's historically central, all-important institutional role in providing services, jobs, opportunities, help and patronage, as a legacy of colonial practices of administration and control, was being challenged. And there was little difference between declared socialist/communist and capitalist states.

With such an overbearing state, client–patron interdependencies provided the social 'glue' holding together the state-societal arrangements. Structural adjustment policies, however, sought to remove just this very arrangement, by weakening the state's authority and ability to offer patronage in return for allegiance. Instead, established certainties in public life and social interdependencies were removed, leading to political instability with all its various implications, such as public strikes, ethnic tensions, and even riots. Weak multi-ethnic states, just as in Central and Eastern Europe, permitted new social, ethnic and economic divisions to emerge, often threatening to challenge a state's territoriality. Specific African problems included serious food shortages, a legacy of ineffective policies, and particular social mechanisms. But with tight fiscal control imposed on states, their scope for devising nationally responsive policies was seriously reduced. Effectively, decisions of public policy now had to be made in consultation with IFI (international financial institution) officials back in Washington, DC (ibid.), and this included, in particular, the rolling back of the state and, by presumed, idealised implication, the rolling out of civil society. Reality, however, proved to be much more complex and unpredictable.

The outcome has been an emergence of different types of regimes, such as administrative-hegemonic, pluralist, party-centralist, personal-coercive, or populist (Chazan *et al.*, 1999). Socialist countries adopted regimes with administrative hegemony (African socialism period) and party centralism (African communist period), or scientific socialism. Administrative hegemonic regimes were established in the early 1960s, and were thus shaped by the immediate legacy of colonial administrative practice, and the then idealistic enthusiasm for moving towards a brave new world of an autonomous Africa and communitarian societies. Although centralised, key groups in society are consulted to maintain the integrity of the state. The upshot of this is a propensity for bargaining, both internally and externally, and this ability has contributed to these states' relative longevity (ibid.).

Party-centred regimes are quite distinct arrangements. There is extensive central control, with the unitary (communist) party apparatus above the administrative structures, often with an influential role for the military. Although the executive remained important, the unitary party's importance was paramount, encompassing all public life, and following a small elite agenda (see also Chapter 2). This system was more susceptible to personality-based changes in power structures and arrangements, and was thus inherently more volatile, than the more bureaucratic, civil-service-based model with its lesser single-party dominance. Indeed, following the collapse of the Soviet Union, the communist party state has disappeared from the African landscape (ibid.), giving way to a variety of modifications of the state–government–people relationship.

The collapse of socialism/communism in the Eastern bloc brought with it a rapid loss of the values and political currency attached to this perceived alternative path of development to western (market) domination, which resembled colonial history too closely. The response to this varies from initiating reform programmes towards a western-style market economy (encouraged now by western donors), as in Benin, Angola and Mozambique, to military coups and the establishment of an authoritarian regime, albeit without the label 'socialist/communist', as in Burkina Faso or Ethiopia. This diversity makes it difficult to draw comparisons, as Allen (1992) points out, not only because of their varying circumstances, but also, and especially, because of the fundamentally unequal interpretations and implementations of 'socialism'. It is therefore difficult to compare like with like, and resorting to comparisons of structure, rather of than policy or ideology, may be the only realistic option

Many of the changes in Africa were obviously connected to the demise of the Soviet Union and its sphere of influence, but the disengagement and re-definition of its relationship with Africa, and its interpretation and application of 'socialism', had begun several years earlier. During the 1960s and early 1970s, Cold War politics meant that strategic military and competitive considerations were paramount, with economic concerns of lesser importance, leaving little scope for *choosing* client states (Light, 1992). But changes had begun under Gorbachev's reform policies, realising that the attempted global reach by the Soviet Union was economically unsustainable, as too many subsidies and other economic support measures had to be offered to the socialist/communist client states. Instead, cost-effectiveness became part of policy evaluation. The end of the Cold War and the dispute with China about the 'correct' implementation of Marxism–Leninism, meant that Africa seemed much less 'useful', effectively accelerating the withdrawal.

Examples of post-socialist development: authoritarianism, socialism and communism

Ghana is an interesting example of the waning scope for *dirigiste* governments under a *socialist* banner. It also reflects the evident failure of the rather simplistic developmental models of the 1950s and early 1960s, with their 'big push' and import substitution policies (Young, 1980). In Ghana, a centrally led, controlled economic development was propagated as the way forward to overcome structural dependencies in a western economic system. Successful economic development was considered the key to overcoming the problems of underdevelopment (poverty, ill-health, etc.). As discussed above, in the 1960s, most newly independent countries emphasised their 'Africanness' in policies to mark the end of colonial dependency. 'Together, Pan-Africanism and socialism would coalesce into a progressive ideology for building a new Africa. The new political system had to be "scientifically" formulated and vigorously propagated' (Haynes, 1992, pp. 41–62). After just a few years, in 1966, the initial African socialist government, with its unclear socialist policies, was overthrown by a military coup. Although the military government continued to pay lip service to socialism, market forces, and personal ambition and prosperity, were the main political driving forces. Another military coup, in 1981, brought about the slow, gradual abandonment of socialism, even as a rhetorical reference point in politics. In effect, the new government consisted of two competing groups, advocating socialist and liberal market policies respectively. Nevertheless, western-style democratisation was considered unsuitable for Ghanaian conditions by the political elite, and they thus adopted the Libyan approach of a 'Third Theory' of a 'state of the masses', without much in the way of ideological or philosophical underpinnings. This 'Third Theory' was meant to offer a middle way between socialist and capitalist approaches.

As part of these changes, the late 1980s brought decentralisation policies, emphasising sub-national government by establishing 110 directly elected District Assemblies. Still, there was no genuine competitive party politics, with those in power claiming leadership as their natural right. The end of the Soviet Union, however, weakened their claim to power as part of historic determinism, because they depended on Eastern bloc economic and political support. Instead, pressure mounted for the introduction of competitive politics and democratisation. Overall, during the 1980s, Ghana saw fundamental political changes. Beginning the decade as a socialist-oriented country, the state evolved over time into a 'developmentalist' system, with a focus on the local level and non-party development. This included attempts at rolling back the system of patronage. Evident economic improvement supported the rationale and public acceptance of the new system.

Burkina Faso (or Upper Volta until 1980) is another example of a stop-go process of post-colonial development, with phases of military dictatorships alternating with more democratically oriented periods. Until 1980, the then Upper Volta had managed to avoid authoritarianism, which had otherwise become an integral part of post-colonial development across Africa. But in 1980 it, too, succumbed to the almost standard process of an authoritarian coup by the military, and subsequent disbandment of any democratic principles as an 'act of emergency' to protect the state (Otayek, 1992). Until 1987, they controlled the state as a self-proclaimed 'revolutionary' regime with a Marxist label, when it was violently displaced by another military coup for a more democratically minded movement, the Front Populaire (ibid., p. 83). This process began the shift towards a multi-party democratic regime, by first dropping any reference towards Marxism–Leninism, and then establishing a liberal constitution and presidential elections, in 1991. There was thus a steady shift from authoritarianism to a democratic regime, a process driven from 'within' rather than imposed from the outside as, for example, a condition of a financial aid package.

External events, especially the collapse of the Eastern bloc, of course, helped the democratisation process along, but they did not kick-start it. 'Democratic opening' was a central theme in the Popular Front's claim to power in 1987, largely on the back of worsening economic conditions, as in many other African countries. The evidently growing loss in credibility of Marxist propaganda contributed to the weakening of the incumbent government, and helped a military coup with a more democratic agenda, a rare combination in any case. Subsequently, 'opening' became the new buzzword summarising the new political agenda and climate. This included combining 'democratic' and 'revolutionary' as political programmatic paradigms. Popular measures, such as widening and supporting the public sector, and cooperating with the trade unions, brought about political support for the new

179

regime, together with a general enthusiasm for change after the end of the nominally Marxist regime.

The sense of change and participation in the country's affairs, albeit initially merely at a more limited level, was one important driver in bringing about support for the new regime. The other was the political context, with the revolutionary regime of the 1980s in deep organisational trouble, undermined by internal divisions about the 'right' political and ideological path. Initially, the post-1987 regime continued the authoritarian approach, albeit less confrontational, co-opting the various parts of civil society. 'Democratic opening' thus was more about bringing together different parts of society (and their respective interests), especially the urban elite and the civil service, rather than establishing a democratic process *per se*. This only gradually evolved, driven along by the events in eastern Europe.

Improving the lethargic economy was the main focus of policies, not least because success here would offer a broad legitimation for the government's claim to power, but the difficulty was the ambiguity of policies, torn between the Marxist tradition of state-driven *dirigisme*, and the need to open up to global trade and adopt a more 'market-led' approach. The gradual shift was also possible because the economy was in far less difficulty than in many other African countries (see Mohiddin, 1981, p. 89). The negotiations with the IMF, in 1989, were thus against a background not of chaos, but of managed change. Not surprisingly, the IMF ordered its usual medicine of neo-liberal policies: stabilising public expenditure, liberalisation, and negotiated investment priorities. The required changes, however, set in train a process of democratisation and 'opening-up' that went far beyond the originally more limited objective of combining 'market' and 'plan'. The IMF's imperatives and a general weakening of the Marxists' case, *vis-à-vis* the events in eastern Europe, ultimately resulted in a democratic government, in 1991.

The two key factors underpinning this process were, first, as an external factor, the collapse of the Marxist model of state governance in the Soviet Union, and, secondly, the domestic process of political challengers being able to infiltrate the state machinery and direct it in its favour. But rather than idealistic arguments for the virtues of democracies, it was economic improvement that proved to be the main 'selling point' for any political changes, including democratisation. 'In Burkina Faso, as elsewhere in Africa, the population is waiting for democracy to prove itself superior to authoritarianism, especially bringing improved economic and social well-being. In other words, there can be no democracy without development' (Ottaway and Ottaway, 1986, p. 100). And this link is not unlike that observed elsewhere. Anything else would be 'subsistence democracy' (ibid.). This, however, is a formidable challenge, because liberalisation, as evident from other post-authoritarian (socialist) states, can trigger substantial economic costs through restructuring and 'adjustment', and lead to considerable divisions in society between those that benefit from the changes and those that do not.

Ethiopia is one of the few states in Africa that followed a more genuine Marxist–Leninist approach, resembling, in some ways, the 'actually existing' socialism of eastern Europe. Established in 1974, at the end of the revolution ending colonialism, it lasted until 1991, when it gave way to new democratic movements and market principles in the economy. Until then, 'the seriousness with which this [Marxism–Leninism] was pursued by government and opposition alike, indicates that we are dealing not merely with a pragmatic response to the needs of a Soviet alliance, or cosmetic "Marxism–Leninism", but with an ideology perceived as having a real application to local conditions' (Clapham, 1992, p. 106). Ethiopia was more genuinely committed to the idea of Marxism, because the doctrine resonated with the particular conditions in the country. It 'offered – or at any rate appeared to offer – an extremely attractive set of integrated solutions to the problems that Ethiopia faced in the early 1970s. In many ways, indeed, both the problems and the apparent solutions mirrored those of early twentieth-century Russia' (ibid.). At the same time, it offered a 'doctrine of revolution' (ibid, p. 101), a notion of a progressive ideal that challenges the status quo, that is post-colonial status, and offers an alternative (better) future. The precise details and philosophical and economic reasoning were of lesser importance. It was the projected image of socialism (and Marxism) as the challenger to capitalism and, by extension, colonialism, that, as in other African socialist countries, resonated with the population.

Marxism also offered a 'doctrine of development', which involved the masses, however defined, and allowed the people to see themselves as participating in the country's development. The third factor, following the Soviet Union's example and ideological paradigm, was the emphasis on multi-ethnic nation building, something of particular importance and concern in Africa. Creating a national or regional identity out of a group of different ethnic cultures, seemed a very positive quality of Marxism. The fourth attraction of Marxism–Leninism was, emphasising on the Leninist tradition, the strong role attributed to the state. A strong state was attractive to those seeking assertiveness and control, especially the military groups who had been important players in de-colonisation. Centralisation, hierarchies and lines of command are to their liking. Marxism–Leninism thus seemed to offer a powerful state with an even distribution of resources among the population.

Not surprisingly, therefore, Ethiopia made a serious attempt at installing a truly Marxist–Leninist state, modelled on the Soviet Union. The enthusiasm for this approach included ideology rather than pragmatism driving economic development decisions, such as the villagisation programme of 1985 onwards. This programme set out to forcefully relocate a dispersed (living) rural population into new cooperative, centralised villages as agricultural production centres. There, among other things, people could more easily be monitored and controlled. Parallels to the Soviet *kolkhoz* system become evident. The fifth perceived advantage of Marxism–Leninism was its international connectivity to one of the two superpowers, and economic and military support from there. Many Ethiopians resented the USA's support for emperor Haile Selassie prior to the revolution. External mentorship also provided important back-up and credibility in the continuous internal struggle in Africa to connect territory to ethnic identities.

The socialist project in Ethiopia eventually was overtaken by events in the later 1980s. While some of the goals, like creating a strong state and providing a revolutionary agenda, had been achieved, the other goals, especially economic development, had not. It was this fact, particularly the economic difficulties, that, in the end, had shown the weaknesses of the Marxist–Leninist model of development. This disappointment was exacerbated by the unpopular villagisation programme, which was forced on people for ideological, rather than economic, reasons, and alienated most. The outcome was production losses, which, in turn, in good socialist tradition, were blamed on inadequate levels of socialisation and commitment, leading to just more of the same medicine.

The loss in popularity of the socialist model became apparent towards the end. When, in 1990, Ethiopia's president, Mengistu, announced the abandonment of socialism, the laboriously constructed, but vehemently resented, cooperative farms disappeared virtually over night, as peasants helped themselves to anything they could carry away, and set up their individual farms again. The collapse of the system in Ethiopia, at that particular time, was mainly coincidental to the collapse of communism elsewhere, as Clapham (1992, p. 116) points out. The main reasons for this timing were the loss of arms and state control to separatist factions (Eritrea), seeking to wrestle control of the territory from the government. Coincidentally, the challengers also subscribed to the Marxist ideology, and their political ambitions succeeded. But most of these leaders had no conceptual understanding of Marxism, and used it more as a convenient strategy or label simply to gain power. Crucially, however, the public had become tired of *dirigiste* socialism, especially after the villagisation programme, which had seriously undermined the government's claim to power. A shift towards promoting private enterprise almost certainly contradicted the principle of state socialist centralism.

The most challenging change concerns the status of minorities. 'Whereas under the socialist system the formal recognition of nationalities was in principle balanced, and in practice totally subordinated, by a centralized hierarchy of political organization, economic management and military control, in the post-socialist era the nationalities are left on their own' (ibid., p. 124). The implications of that have become all too apparent in the break-up of the former Yugoslavia, for instance, and continue to affect the relationships between the countries of the FSU. It is in this respect that Ethiopia shows some interesting, if rare, parallels with developments in Europe. Most other conditions and factors are very different between the two continents, but Ethiopia is also quite unique in an African context, because of its quite strict adherence to a purer form of Marxism–Leninism when compared with other African post-colonial states. 'In each case, socialism served most basically as an ideology of state consolidation, which could draw on an existing tradition of statehood ... , but it proved lamentably ineffective, either at developing the economy, or at creating a common sense of identity among the peoples within its borders' (ibid.).

Communism, urban development and after ...

Urban areas, like elsewhere (see e.g. Chapters 5 and 6), have developed their own dynamics within the national framework of communist and post-communist development. This includes their relative economic advantages, which attract both migrants from the rural parts, and any outside investors who prefer the wider range of opportunities in urban areas, preferably the national capital. With the threat of a potentially much increased developmental gap between 'urban' and 'rural', governments introduced control systems to maintain the necessary levels of rural population, the main instrument of control being access to land for building and housing. Such regulation, of course, only works through adherence to official permits and restrictions, but fails to become effective if bypassed through illegal building activity, a self-help phenomenon widespread in developing countries (Jenkins, 2001). Ethiopia is one of the world's least urbanised countries, with a mere 15 per cent estimated to live in urban areas (Woube and Sjöberg, 1999), although there have been higher increases in growth rates than in most other African states between the 1960s and the communist revolution of 1974 (Kloos and Adugna, 1989). Afterwards, urbanisation had slowed down, not just because of state control, but also because of deteriorating economic conditions, making cities appear less attractive for rural migrants, who instead preferred to stay within the network of kinship support within their home villages. This situation changed in the late 1980s/early 1990s, when communist rule weakened, and, subsequently, a large influx of a migrant population pushed the urban population figures up considerably. This development effectively reversed attempts under the socialist system, since 1974, to limit rural–urban migration – although their effectiveness is unclear (Woube and Sjöberg, 1999).

The main mechanism of control was urban–rural land reform, to make staying in the rural areas more attractive, while imposing stricter controls on the urban housing market, although the latter's effectiveness is also unclear (ibid.). Rent controls were another means of raising the barrier for potential migrants. Yet the availability of subsidised urban housing (if limited in numbers), price-controlled food and generally deteriorating rural conditions all combined to maintain cities as potentially attractive destinations for rural migrants. It thus took more regulative measures to 'stem the tide': identity cards, household registration books, and associated mechanisms of control were put in place, such as the establishment of *kebele*.

The *kebele* are local communities of about 300–500 households at the sub-local level, and sat at the bottom rung of the strictly hierarchically organised state administration under Marxism–Leninism. The line of command went from the top, right down to these local groups, which also served as the 'eyes and ears' of central government, and thus represented a typical form of control employed by communist regimes. This included keeping detailed registers of inhabitants and their houses. Society had thus effectively been centralised, although the communes may have been seen as a sign of devolution, but records were not always kept up to date. One of the main reasons for the limited availability of land for development had been the outlawing of any transfer of property rights through sale, rent, inheritance, and so forth. As a result, the property market collapsed and neither residents nor (former) owners had an interest in maintaining their property. As elsewhere with communist systems, state provision faced a number of difficulties, especially limited capacity and, with continuously growing demand, increasing shortages. Demand increased, despite attempts at limiting migration from rural to urban areas through formal regulation. Housing shortages were thus pre-programmed. In any case, obtaining access to land and housing required proper registration with the designated locality. The aim to limit urbanisation reflects an inherent hostility of communist ideology to cities, because of their close association with the bourgeois class. This rejection contributed to a lower degree of urbanisation, also referred to as 'under-urbanisation' (Murray and Szelenyi, 1984), and it had become a hallmark of communist regimes, largely irrespective of a country's stage in development.

But there are also particular national features in the shaping of communist urban policy, as Murray and Szelenyi (1984) argue. Control mechanisms put in place by the communist state, and carried on thereafter, do not seem to have been very effective. One of the main reasons for this rests in the particular nature of socialism/communism (see also ibid., p. 41). With bureaucrats trying to regulate and plan all aspects of life, a tired and disillusioned population tries, in imaginative ways, to bypass and overcome this *dirigisme*, pursuing their own personal goals and overcoming endemic shortages. Effectively, therefore, socialism (or communism) encourages individual strategies. Outcomes of state managerialism have been inflated migration figures, to obtain higher block grant allocations (based on head counts), and, second, an expansion of the informal land and property market – in open defiance of the state. This was possible because representatives of the local level of administration (*kebele* level) tolerated such manipulation of the truth, as it boosted the locality's and thus their standing.

This shows some parallels to the measures employed by the Chinese government, but there, as in Ethiopia, migrants responded by seeking to circumvent such obstacles put in their way, and resorted to informal ways of obtaining land and homes on the edge of the cities. The political challenge for the government has been to provide these illegally residing urban poor with formally approved land plots within the city, but limited administrative capacity meant a continued gap between demand and supply. For instance, Maputo's land cadastre had not been updated between 1985 and 2000, thus effectively failing to provide legally secured land parcels for existing and new dwellers, leaving land development without permit and legal status the only *de facto* option (Jenkins, 2001). Consequently, there was no consistent provision of land for low-income groups, making their move from informal 'shanty town' settler to legalised urban resident very difficult. Jenkins (2001) blames this situation, which has remained unchanged throughout the 1990s, on a combination of lack of adequate legislation, weak institutional capacity, inadequate policy instruments, and a generally limited interest in taking initiatives to alleviate the problems. Instead, the limited availability of officially 'approved' land opens the gates for speculation and bribery. Nevertheless, an embryonic land market is emerging, with officials in a strong supplier position (ibid.) and, not surprisingly, land developers, who are among the main beneficiaries of this supply-led market, in a politically influential position. But their interest is in the middle and upper end of the market, not the needs of the urban poor on their informally acquired, and illegal, plots of land. Other agencies are necessary here, outside the traditional state machinery, as part of a 'horizontal' civil society (ibid., p. 645), that is essentially an emerging broader urban governance structure. But the details of such developments are still unclear, as a local state is still in the process of finding its feet and role within the state hierarchy. How far 'inherited' communitarian arrangements, such as the *kebele*, may serve as 'cells' for an emerging local civil society, is not yet quite clear.

Summary and conclusions: African communism, pragmatism and after

Africa's experience with communism and subsequent post-communism has had quite unique characteristics based on the combination of overlapping ideologies, idealisms, legacies and realities. The experiences with colonialism and its out-comes, especially state administrative structures, territorial arrangements and national ambitions of autonomy, have had a major impact on the response to, and adoption of, communist ideology and its practical implementation. Particularly, in the early years after the end of colonial rule, strongly idealised perceptions of African identity and history prevailed in connection with the move towards inde-pendence, including the related struggle itself. African socialism was portrayed as a natural fusion of the Europe-derived ideology of Marxist socialism, and the pre-colonial African tradition of communitarian village life. Drawing this link served to legitimise the adoption of socialist principles for the new states as part of a process of Africanisation, despite the obvious origin of the concept in (colonialist) Europe. Failure to achieve promised successes, especially in quality of life, 'on the ground', undermined the credibility of this idealistic concept of African socialism, giving way to more radical, second-generation leaders and their widespread autocratic ambitions. The strongly centralised, highly autocratic system of Leninist-style communism appealed to them as a more pragmatic, realistic means of exercising power and controlling national development. This shift to African communism had the added advantage of visibly inviting Soviet support. As part of the competi-tion between the two superpowers for influence in Africa, the Soviet Union was interested in encouraging and supporting fledgling communist states, if for nothing else than propagandist reasons to demonstrate the spread of communist ideology. The pragmatic use of the label 'communist' aimed at attracting Soviet development aid, while also serving as a convenient justification for running an authoritarian one-party state. The spectacular collapse of the European communist regimes, and the termination of the Soviet Union, removed both pragmatic reasons for main-taining the label 'communist'; in fact, they turned into a detriment.

With only one superpower left, and liberal marketisation and democratisation the only available options for obtaining development assistance at the international institutional level, abandoning references to communism were opportune. In addition, the obvious failure of the system in the Soviet Union itself had severely undermined its credibility as a likely road to economic improvments among the African population (as elsewhere). Post-communism in Africa has thus been at a much lower key than in Europe, as it had not developed a deeper ideological resonance with the population – it had been in place for merely a decade or two.

There were thus only a limited amount of structural legacies, especially in the economic sector. So much of their fundamental structural characteristics did those countries share with their 'non-communist' counterparts as to make them almost indistinguishable. Only Angola, Mozambique and Ethiopia maintain more visible post-communist features, as communism there was implemented more rigorously and consistently according to the European model. Consequently, Africa is much less evidently shaped by the effects of post-communist transition, which are easily overshadowed by the general challenges of development and competitiveness in a globalised world.

8

LATIN TRADITION, EUROPEAN MARXIST DOCTRINE AND MARKET FORCES

Communism and after in Cuba

Introduction: towards a 'revolutionary socialism'

Socialism as a state regime in Latin America, has, as in Africa, been shaped by its colonial legacy, albeit of an earlier period, but much more by the immediate hegemonic effect of the United States' political and economic strategies aimed at its perceived 'backyard', Latin America. This made it particularly difficult for 'revolutionary' socialist regimes to establish a functioning state, as they had to overcome not only the inherent contradictions and flaws of the principles of Marxism–Leninism as regime-operating principles, but also active countermeasures by the US to prevent 'Moscow' from taking a foothold in the Americas. Ideology and political antagonisms during the Cold War were such that any political movement 'to the left' was considered a potential Trojan Horse for Sovietism's global ambitions (McCaughan, 1997), and if just outside the United State's territory, this was seen as political-ideological teasing, as illustrated by US responses to the Nicaraguan Sandinista revolution, for instance (Weber, 1981; Smith, 1993).

The internationalisation of the Nicaraguan experiment with socialism stood against the background of the Cold War, and thus local political events became part of the global competition between the two superpowers, even if more imagined than real. In the case of Nicaragua, the Soviet bloc had very limited interest in getting politically involved, and since the end of European communism, Latin America has been handed back by the western Europeans to the US as their 'natural' sphere of influence (Smith, 1993). The Sandinista government itself sought to keep its distance from the Soviet bloc to not unduly alarm Washington policy makers. For instance, they sought to buy their military equipment from western rather than communist bloc sources (Walker, 2003). Some commentators, like Ruccio (1987), even question the degree to which Nicaragua actually was 'socialist', given the relatively weak position of the state, especially as a centre of accumulation. The strong, continued external economic influences undermined the state's capacity to control national economic and political developments, quite unlike Cuba.

The Cuban Missile Crisis at the beginning of the 1960s was the pinnacle of this competitive posturing. Cuba, just off the United States' territory, has since its 1959 revolution been stuck between the ideological fronts of 'East' and 'West'. At the same time, Cuba also sought to shift towards a more independent post-colonial existence, while rooted in the Latin American tradition of the strong role of the military and 'strongmen'. The country thus sat on the intersection between North–South and East–West political-economic paradigms and realities. This has shaped its identity, politics and development over the last nearly 50 years, leading to its particular status as 'icon' or 'obstacle', depending on one's political view. A few other Latin American countries sought to follow a revolutionary 'socialist' path of development in the 1960s and 1970s, too, with Chile, Nicaragua and El Salvador gaining particular international attention (Midlarsky and Roberts, 1985; Corr, 1995; Skidmore and Smith 2001). Apart from Cuba, none of the 'revolutionary socialist' systems has survived both the inherent system-specific difficulties and substantial external interferences with domestic regime building.

All this has had implications for politics and the nature of the regimes following the end of these attempts at creating a particular Latin American form of socialism, which could, perhaps, be described as 'revolutionary socialism' or 'revolutionary communism' respectively, i.e. a combination of an ideology rooted in European industrial society and Latin American post-colonial political traditions and societal structures. It is against this background that this chapter will explore the nature of these regimes and the effects of the end of the Cold War and the Soviet Union as pivot of the communist world on their development. Particular attention will be placed on Cuba as the most prominent example of a communist regime in Latin America, which continues to operate, albeit with distinct signs of adjustment to the changed global political and economic conditions after 1991. Its 'special status' revolves around its attempts to follow an alternative path of post-Soviet development to that prescribed by the 'westernisation' of almost all countries whose regimes had subscribed to Marxist–Leninist doctrine. While the huge size of China may give it the scope to pursue its own political-economic strategy, Cuba is a much less likely candidate for pursuing an alternative strategy. Are there specific Latin American, or Cuban, features that may help explain a 'Cuban path of post-communist transformation' (Centeno, 2004)?

Hoffmann (2001) refers to Cuba's 'double identity' as Latin American and socialist in nature, thus sitting at the interface between 'transition' research and 'transformation' research. While the former, in his eyes, focuses on the 'two waves' of democratisation processes in Latin America and southern Europe in the late 1970s and early 1980s, the latter engaged with the post-communist changes in Eastern Europe and its 'dilemma of simultaneity' (Offe, 1994). Cuba's 'double identity' means that it 'differs from the transitions in Latin America in that the political and national questions are posed simultaneously and are tightly connected to each other' (Hoffmann, 2001, p. 2), and it also differs from the Eurocentric paradigm of post-communist transformation. Centeno (2004) points to several key features of Latin American statehood and development that, as distinct legacies, shape Cuba's further development. These legacies include not only poverty and political instability, caused by weak states and frustrated attempts at democracy, but also military intervention and, especially, US interference. It has been the exception from these that has marked Cuba's development under the banner of

Fidel Castro's version of 'socialism', or 'Castroism'. Under this, Cuba has retained its exceptional position among its Latin American peers since the collapse of Soviet-sponsored communism in 1990/1. But the strong dependence of Cuba's status on the personality of Fidel Castro raises questions about Cuba's continued ability to steer its own path of transformation after his demise, rather than joining the generally observed Latin American 'standard' of statehood. This includes, in particular, a re-emergence of strong social and economic inequalities, symbolised by a 'dollar apartheid' (ibid.) as the most visible sign of the introduction of a dual economy – dividing those with and those without access to the US dollar. For some years, there have been clear signs that Cuba's communist regime is shifting from Castro's version of Marxism–Leninism towards something else – something not yet very clear (Radu, 1995). What is clear, however, is a recently significant reduction in references to 'socialism' or 'communism' and, instead, more emphasis on national perspectives, including a 'Cuban way', however vague in its concept.

Dual legacy: Latin post-colonialism and European Sovietism

Latin America's and, especially, Cuba's, adoption of communism, and subsequent developments, draw on two main traditions – those of Latin American post-colonial societal-cultural legacies and economic dependencies, and those of Eurocentric Marxism–Leninism. In principle, this is not dissimilar to the background to African socialism, where, too, the relative influence of the two strands varied between countries, and over time. Given the direct linkages to the Soviet Union, Cuba became part of Eurocentric post-communist discourse and analysis. As a result, Cuba's development of communism happened against a very different background to that in other Latin American states, leading to a particularly close and tight inter-connection between the questions of national identity and the dominant political discourse (Hoffmann, 2001). But it is also different from the Eurocentric paradigm of post-communist transformation and, especially, its main causes.

There has, more recently, been a stronger influx of Latin American culture and history, including the authoritarian (military) component. In addition, external factors encourage a sense of national togetherness and resolve, especially *vis-à-vis* a hostile US foreign policy towards the island state. The Cuban government has utilised this national resolve as a trump card, by cloaking itself in the rhetoric and sense of nationalism. Thus, very much in contrast to true Marxism, both socialism and nationalism were merged into a particular image of statehood, and the communist government projects itself as the vanguard of defending national independence. Effectively, the particular mix of ideologies straddles two historic trajectories – that of a post-colonial Third World country, and that of communist Second World (European) countries. 'This double identity still marks the political, economic and social structure of the country' (ibid., p. 3).

Cuba brings together key features of the different qualities of 'communism' and 'post-communism' in Europe and Africa, and, going by its more recent policies, in China too. But Cuba is not just an incidental amalgamation of the different traditions of actually existing communism. It actively developed its own interpretations and applications, fusing Latin American post-colonial traditions of governance with a European developmental ideal of the industrial age. The outcome of this fusion is the Cuba-typical combination of nationalism with socialism (Pollis, 1981). While the re-interpretation of Marxism from a post-colonial perspective *per se* has not been a uniquely Cuban invention, as evidenced by several such attempts in Africa (Chapter 7), it is the particular combination of nationalism and socialist ideology that has become a hallmark of Cuban politics. In its Latin American tradition, 'the [communist] revolution represents the embodiment of the national struggle for independence' (Hoffmann, 2001, p. 3) – a notion that continues under 'post-communism'. In contrast to African socialism, however, nationhood and the nation-state are not interpreted as an idealised continuation of pre-colonial communitarian (tribal) conditions, but rather a defensive concept *vis-à-vis* a hegemonic external threat to national autonomy. But both interpretations, regardless of their differences about the role and nature of the nation-state, differ considerably from the one developed in Karl Marx's theory of communism. They thus reflect a re-interpretation and adjustment of that theory to suit the particular post-colonial, pre-industrial conditions shared by all developing countries. In so doing, they added a completely new, conceptually alien, component to Marxism, because 'Marx and Marxists in general have not dealt with the phenomenon of nationalism, viewing the nation-state essentially as part of the superstructure created by the bourgeoisie to serve their interests' (Pollis, 1981, p. 1009).

Making 'Castroism'

After the US-backed authoritarian Batista government had given up its defence against Fidel Castro's revolutionary forces in 1959, Castro sought to gain the support of the wider population with the usual 'bribes' of raising incomes, halving rents and freezing prices, while socialising most of the (largely American-owned) land and industry. The main challenge was to keep the bourgeois middle classes in Cuba, and thus retain their skills and entrepreneurial spirit. Although the younger sections of society (students) had been the main recruiting ground for the Castro army, the establishment obviously felt threatened by the changes and loss of property (Fitzgerald, 1994). After 1970, Cuban socialism went through a period of ideological affirmation ('rectification'), seeking to re-emphasise the revolution's popular appeal. At the same time, there was a degree of Sovietisation (McCaughan, 1997), as Cuba sought to fit into the COMECON system and, especially, to comply with the Soviet Union's terms of trading as a condition of increased support. This meant, *inter alia*, a stronger centralisation of the state, albeit with some elements of bottom-up participation for the less important issues. This 'democratic centralism' included a growing 'technocratisation' of government through an expanded role for professionals within the administrative machinery. The increased Soviet profile in Cuban communism mirrored the growing dependency on Soviet economic assistance to fulfil the five-year plans (Brundenius, 1981), as well as strategic support. It thus acknowledged Soviet hegemony, something initially resisted in the post-independence euphoric attempt at running a more autonomous course for Cuban politics and development.

An important feature of Cuba's political landscape, shared with other Latin American countries, is a strong leadership with clear idealism, but also pragmatic skills to make ideas 'turn to reality'. The propagation of the 'revolution' and its ideals, especially national independence and self-determination, has become the key agent in maintaining the political system under Fidel Castro, going far beyond the superficial claims to socialism seen in Africa. Although in both cases personalities were an important force in shaping 'socialism' in the respective countries, Castro's sheer length of time in office has provided a sense of continuity and robustness. This, together with a sense of national defensiveness, helped to overcome acute economic and political crises, such as the severe economic decline immediately after the collapse of the Soviet Union at the beginning of the 1990s, triggered by an abrupt and drastic reduction in economic support; Cuba's GDP fell by up to 15 per cent annually between 1991 and 1993 (Hernández-Catá, 2000). Thus, the regime was able to keep the population on its side without having to resort to excessive forms of suppression. This suggests an underlying, inherent acceptance of the system in principle, if not necessarily the detailed conditions of everyday life.

In the late 1980s, the relationship with Moscow had begun to become more difficult, following the reform agenda under Gorbachev, of which Fidel Castro disapproved. The Soviet Union was abandoning its outwardly directed drive for influence and representation of superpower status (see Chapter 5) and, instead, was concentrating on internal restructuring and the establishing of a new regime. Cuba was thus no longer so important as an outpost on the doorstep of the US, and as a bridgehead for expanding influence into Latin America. Moscow wanted to reduce the cost of its international political engagement, and this included cutting back on subsidising the economies of its struggling strategic allies. Ideological affinity was no longer the key to Soviet financial help. By that time, Cuba's government had begun to tentatively introduce some market elements into the economic system to counteract the ever more evident stagnation and shortages, which promised to get worse with the reduction in Soviet support. But then, by the late 1980s, as if scared by its own courage, the changes were rolled back to a more orthodox approach, seeking to reaffirm the Party's control of all matters Cuban, as if fearing changes were getting out of control. This mirrors the principal features of the reform process in China, with a similar, trial-and-error zigzag course. Both felt alarmed and vindicated in their hesitancy by the events in the Soviet Union, where reform attempts developed their own dynamism and, inadvertently, swept away Marxism–Leninism.

It was not surprising, therefore, to see a toughening of rhetoric and ideological control, propagated as the 'rectification of errors', such as in 1986 (Fitzgerald, 1994). With an eye on China's reform process, Castro was concerned that consumerism and the people's strive for material gain could encourage individualism, and push the revolutionary consciousness of communality and national achievement to the margins (see Azicri, 2001). Castro thus now publicly wondered about the salience of copying a system that has obviously different cultural roots and societal-economic realities from Cuba's.

In a way, this was a defiant reaction to the disappointment about the fact that the Soviet Union had walked way from Marxism–Leninism. But it was also a reflection of concerns about losing control. The current Chinese government's new emphasis on the teachings of Mao Zedong Thought, next to a market-driven economic strategy, stems from the same unease about developments slipping out of hand (see Chapter 6). The schism between Cuba and the Soviet Union went as far as Castro banning Soviet publications in the late 1980s for being too critical of communism (Azicri, 2001). The ensuing policies in the early 1990s thus need to be seen as an attempt by Castro to re-emphasise Cuba's independence, expressed in its ability to choose its own way, irrespective of the main external hegemonic forces. This notion of slipping out of a hegemonic dependency, albeit triggered involuntarily, and thus regaining 'Cuban-ness', seems to have been an important factor in galvanising people to the state, despite the ongoing serious economic problems. But independence from the Soviet Union has also strengthened Cuba's position within Latin America, as it is now seen less as the Soviet 'poodle' than as David against the North American Goliath (McCaughan, 1997, p. 154).

Insisting on continuing with the established path outside hegemonic dependency thus has become a sign of stronger independence and self-determination, and goes back, albeit in a rather tentative way, to the Cold War days. A more distinct profile of Cuban communism has emerged, drawing on its Caribbean legacies and culture, and projecting a more nationalist resolve. Coercing was certainly an important factor too, but judging by events in East Germany, applying pressure indiscriminately would have seriously undermined state authority and, ultimately, triggered its downfall. Instead, the Cuban government managed to project its idealistic view of Cuban independence, through the continuation of Marxism–Leninism, onto the general public.

Maintaining the communist system with all its paraphernalia has thus been transformed into a general national course. Despite frequent dissatisfaction with many aspects of their daily lives, Cubans seem to have accepted the propagated need for national unity for maintaining independence, especially *vis-à-vis* the overtly strong neighbour, the US. The Cuban leadership may see consolation in the seemingly successful Chinese model of targeted and controlled economic liberalisation *alongside* continued one-party control, and there are signs that some members of the Cuban government are interested in the Chinese model (McCaughan, 1997). But although China is supportive of the Cuban cause in principle, relationships remain at arm's length because the Chinese government is concerned about having to provide costly economic assistance. Yet the Cuban leadership, ageing veterans of the revolution, are hesitant to shift away from their ideals and dreams. This contrasts with Vietnam, for instance, where a new, younger generation of leaders, who feel less attachment to the past ideologies of the communist revolution (Agarwal, 2004), push for export-oriented development.

Factors underpinning Cuba's communist regime and defining its difference from Eastern European communism, include (Radu, 1995): (1) the Cuban regime is younger than the eastern European and, especially, Soviet system, allowing the revolutionary generation to still be involved with decision making. (2) Cuban society is less isolated, having access to media information from Latin America. (3) The Cuban communist revolution had genuine support from the population, whereas eastern European communism was imposed by force through an occupying force. (4) The institutionalisation of Cuba's communist regime rested on Fidel Castro's personality, rather than the Communist Party. The Party existed through Castro, rather than the other way round, as in CEE and the FSU. Castro developed his own legitimacy to power, independent of the Party, drawing on his revolutionary credentials. (5) Cuba sought to maintain its own identity and scope for policy-making within the communist bloc, styling itself as a springboard for communism to expand into Latin America.

Since the collapse of communism in Europe, Fidel Castro has sought to empha-sise these cultural and ideological differences from Europe-based communism, and thus developmental pathways, as a way to rationalise why Cuba's socialist project need not necessarily follow European events and thus be doomed. 'Havana is not Warsaw. Communism in Poland was imposed by Soviet troops, which had committed atrocities that were resented by the Polish people', Castro stated in reference to the Pope's imminent visit to Cuba, given his attributed impact on the downfall of communism in Poland (Azicri, 2001, p. 272). The focus is thus on making the communist endeavour at least appear Cuba's own, as a choice, rather than imposition. Castro thus seeks to portray a very different meaning of, and associations with, communism and its associated objectives, contrasting almost diametrically with those of Eastern Europe. Nevertheless, modifications had to be made to reflect the new post-Soviet circumstances. And this meant a less dogmatic and hard-line interpretation and implementation of communist principles, especially in the economic sphere. Azicri (2001) sees the biggest challenges now for the Cuban regime in this contrast between the difficult economic reality and the resolve to maintain Cuba's political independence. This tension sits in front of an increasingly more influential Latin American background. In response, socialism or, for that case, communism, has been recast and re-imagined as 'late socialism', sold to the Cuban population as an advanced phase on the trajectory of national post-revolutionary development, rather than an abandonment of doctrine for the sake of the regime's survival.

Responses to the changing geopolitical environment: inequality, selective marketisation and 'dollar apartheid'

For 30 years, from 1960 to 1990, Cuba followed the doctrine of Marxism–Leninism in economic management. This meant a strict central planning of all resources, 'intercepted with periods of anti-market radicalism and experimentation with moral stimulation' (Mesa-Largo, 1998, p. 857). As a result, 'in 1989, Cuba had the most collectivized, egalitarian, externally dependent and Soviet-subsidized econ-omy within the socialist world; the anti-market movement and the recession made even more difficult a way out of that situation' (ibid.). With the disintegration of the communist trading bloc, COMECON, and the subsequent collapse of Cuba's main supply lines and trading links, finding new ideological soulmates elsewhere was of paramount importance for the development of alternative sources of supply and revenue. By the late 1980s, 85 per cent of trade was through COMECON, and 66 per cent of its imports came from there, thus reflecting the firm embeddedness of Cuba's economy in the Eastern bloc's spatial division of production.

The 'Special Period', initiated by the Cuban government in 1990 as an emergency measure in response to the serious supply shortages through missing imports from the Soviet Union, sought to ration the meagre supplies available, while also stimulating all available resources. Introducing market reforms and reorganising foreign trade was the first line of defence, with foreign investment legalised in 1992 (Klinghoffer, 1998). This concentrated, initially, on developing newly designated tourism compounds for western visitors. These compounds are effectively gated communities, accepting the US dollar, while keeping out the local population. The new tourism industry is instrumental for Cuba's economic recovery, with its revenue-earning capacity between 1990 and 1996 increased by a staggering 500 per cent, from $250 million to $1.3 billion (ibid.). This marked an important shift away from the established doctrine of rejecting western capital and reflects the moves towards liberalisation and 'westernisation' of the other former communist countries elsewhere.

Cuba's 'Special Period' and its underlying rationale of economic policy, mark a 'fourth path' of post-communist transformation, differing from the post-Soviet and eastern European model, the Chinese model, and the African post-colonial model. Thus, Cuba has not moved much in the direction of democratisation, as has been the case in many of the CEE countries, but, instead sought to emulate China's policies. China and Cuba share the determination to continue with authoritarian one-party rule, and apply any reforms only gradually and carefully. But there are also distinct differences beyond sheer scale. Thus, China is much more a rural country, with a mere 20 per cent of its population classed as 'urban', contrasting with Cuba's 85 per cent (ibid., p. 178). Considering the generally different attitudes to politics between rural and urban populations, this difference is significant for likely political activism and the potential emergence of signs of civil society. Urban communities tend to be more fertile in this respect. Furthermore, Cuba did not go down the route of mass privatisation, although it is now actively engaging in attracting foreign investment to selected locations, again, following the Chinese example. Third, in common with China, Cuba is also a developing country, although with better than average statistical indicators of development, such as education and heath care levels. Income levels may be low generally, and are much less differentiated than in China, reflecting a – still – more egalitarian social-economic system, but Cuba has no legacy of overwhelming foreign debt. However, wildly fluctuating revenue from exporting sugar has affected scope for repayment, and some rescheduling was necessary with the western creditors, although under less favourable conditions than obtained by other Latin American countries (Zimbalist and Eckstein, 1987). The collapse of the Soviet Union brought some perceived repayment relief, as debt owed to the former Soviet Union does not, in Castro's eyes, survive the demise of the creditor (Diaz-Briquets, 2000).

The tensions between ideology and necessary economic pragmatism are reflected in an emerging duality in the economy, defined by the ownership of US dollars. Its recognition as the second official tender, since the mid-1990s, has effectively created an economic demarcation line between different economic prospects. In some instances, this economic divide translates into physical divisions, such as around tourist compounds in the form of fencing and walls (Gunn, 1993). Restricting capitalism to tightly controlled territories has sought to avoid ideological 'contamination' of the rest of the country (Perry et al., 1997). More difficult to control territorially was the acceptance of the US dollar as the second legal tender. The effect here has much more of a social dimension, creating new distinctions between those with and those without access to the currency, and thus economic opportunities.

Tourism has become the new magic word in Cuba for economic growth prospects, but it is also the main arena in which the new social divisions develop along the dividing line between the 'haves' and the 'have nots' when it comes to the possession of foreign currency and thus extra opportunities for consumption. These divisions are just as effective, if less immediately visible, as the physical barriers around the gated tourist compounds (Gunn, 1993), regulating access to the 'western' tourists and the idealised image of Cuba created for them. Tourism was a major economic sector in pre-revolution Cuba, primarily shaped for, and by, visitors from the US, but it came to an abrupt and all but complete standstill in the years immediately after the 1959 revolution. This was the result of, on the one hand, the hostile attitude among the new Cuban leadership towards foreign tourism as a perceived form of capitalist colonialism and, on the other, the economic embargo by the US against 'communist' Cuba. As a consequence, there was no investment in any tourism facilities during the 1960s and 1970s.

It was not until the late 1980s that figures slowly began to increase as a result of the softening of Cuban policies towards dealing with 'capitalism'. By that time, policy makers had realised the potential importance of tourism as a means of economic development by observing other developing countries embarking on the promotion of tourism. However, the main growth began in the mid-1990s, with the opening up of the market to foreign investment, seeing tourist arrivals rise from some three-quarters of a million in 1995, to more than 1.6 million in 1999. At $1.6 million, gross revenues from tourism had by then exceeded those of the traditional main export products, sugar and nickel (Fletcher, 2000). Not unexpectedly, these developments are unevenly distributed across the island in response to identified tourism potential. Eight designated primary regions accounted for some 90 per cent of the rooms at the end of the 1990s, a large share of which are owned by four main joint venture groups between the state and foreign investors, mainly from Spain, Jamaica and France. By 2010, some 7 million visitors are expected (Jedbodsingh, 2001).

The biggest hopes and expectations for economic improvement thus rest on developing the tourism industry, as the fastest and most immediately lucrative source of hard currency (Henken, 2000). Tourism has thus been developed vigorously over the last few years (Crespo and Diaz, 1997; Suddaby, 1997), targeting specifically the mass tourist market, as do the neighbouring islands of the Dominican Republic and Cancun (Crespo and Suddaby, 2000). Revenues from tourism have increased sharply (Perry *et al.*, 1997; Henken, 2000), albeit highly localised. High import costs, however, mean that much of the earnings so far has had to be spent on fuels and food, leaving only about one-third for consumer and investment goods (Gomez, 2001).

But the social-economic divisions between those with and those without access to hard currency are just as effective in separating social groups with different economic prospects. In principle, such a dual currency economy is not new. Under CEE socialism, too, possession of hard currency, usually used for oil imports and the purchase of military hardware, opened up extra opportunities of consumption, allowing one to acquire otherwise unavailable goods (Falk, 1985). In the East German *Intershops*, for instance, any western consumer good could be obtained (ordered), from western cigarettes to cars, all in exchange for western currency (Deutsche Mark). Similar 'foreign currency recovery outlets' (Fabienke, 2001, p. 107) exist in Cuba, and help to bridge shortages in daily goods, even if at prices of up to 20 times those in the state shops. What is different in the Cuban case has, until very recently, been the so public recognition of a western currency as the second legal tender, thus abandoning even the pretence of a unitary national economic space. This new inequality is an obvious retrograde step as far as the Marxist idea of an egalitarian society, which has been at the heart of the Cuban revolution, is concerned. But its meanings go deeper in that they reflect an underlying chasm between a revolutionary identity and idealism held in public discourse, while also aspiring to the 'temptations', and necessities, of western capitalist investment and consumerism.

With communication easy, especially to the Miami-based Cuban exile community, there is a marked awareness among Cubans about the latest 'must haves' on the western markets, just as there was among East Germans following their western compatriots' consumer landscape. Aspirations for 'western' lifestyle impact on the young generation's identity in particular, requiring them to reconcile domestic and foreign influences and paradigms (see also Chapter 9). Restrictive legislation to regulate managerial practices in state companies and joint ventures company (Keiffer, 2001), reflects this ambivalent attitude to marketisation. Government policies have followed the new 'flexible' approach to communist ideology by propagating the rather general Marxist goal of creating a 'better society' as justification for new policies. The ideological focus is thus on unifying Cuban society, despite, or rather because of, the economically driven growing inequalities, while also seeking rapprochement with the neighbouring Latin American countries. Avoiding Marxist rhetoric reduces the threat of a communist invasion, feared by many of them after 40 years of continued propaganda advocating just that (Aguila, 1984).

As a result of these policies, different parallel economies exist in Cuba today. Pastor (2000) distinguishes three distinct coexisting economies: (1) the conventional 'socialist' economy with state-defined conditions; (2) joint ventures between state and foreign (mainly Canadian and western European) investors; and (3) the 'dollar

Table 8.1 Economic development during the 1990s (GDP in constant 1981 prices)

Year	Million pesos	Pesos per capita	Manufacturing (million pesos) out of total
1990	19,008	1,787	4,640
1991	16,975	1,580	—
1992	15,010	1,386	—
1993	12,777	1,172	—
1994	12,868	1,175	—
1995	13,185	1,201	—
1996	14,218	1,290	3,835
1997	14,572	1,317	4,155
1998	14,754	1,327	4,291

Source: Indian Embassy in Cuba, obtainable from: www.indembassyhavana.cu/commercial/cubaccopart3.htm#GDPGrPerCapita (accessed 19 December 2005)

economy' – primarily the tourism enclaves. Fabienke (2001) distingishes between two parallel economies, an 'emerging' segment with private sector input and parastatal organisations, and the 'traditional' segment, containing the state sector economy (ibid., p. 108). Given the changes over the last year, the market has become the main engine of potential growth, albeit under rather restrictive conditions. Thus, for instance, only businesses engaged in food production are allowed to hire assistance, and then only family members (Fabienke, 2001).

In effect, the market has expanded at the expense of the sections that are exclusively state planned, and led to a mixed economy. At the same time, such varied opportunities will facilitate societal divisions (as in China) between those with access to business opportunities and foreign currency, and those without. This includes divisive effects on localities, such as the re-emergence of a 'dual city' in the case of Nicaragua, with dilapidated parts of the city sitting right next to new tourism-related developments in the city centre's 'tourism poles' (Colantonio, 2004). Thus, old buildings are identified as architectural heritage in designated tourism areas and carefully restored, with the help of western capital, while similar buildings elsewhere are left to further decay.

The outcome of this new inequality, especially in relation to the declared (and still officially maintained) communist goal of an egalitarian society, is far from clear (Pastor, 2000), and government institutions find it difficult to adjust to the new entrepreneurs. In fact, it seems that many government officials seem to look for opportunities to obstruct the development of private businesses, despite their official acceptance as part of the economic recovery strategy (Henken 2000). Seeking to avoid any form of 'shock therapy', changes have been more 'glacial' than gradual (Pastor, 2000, p. 35), with strong elements of 'muddling through', rather than a clear strategy for change. This corresponds with the regime's perception of having to 'give in' to capitalism, and thus any concessions will be as much as necessary, while also as little as necessary. The resulting slowness may well discourage much-needed new investment and initiative, both from within and without Cuba.

The highly localised, newly developing tourism economy illustrates the growing contradictions within the Cuban communist system. Adopting the US dollar as the second legal tender, albeit permissible only for foreigners, has institutionalised a divided economic world, effectively establishing a 'fast lane' and a 'slow lane' of economic change and participation therein. While the latter represents the conventional structures of the state-planned economy, the former embodies the new entrepreneurialism needed to drive Cuba's economy. Much of this development is driven by foreign direct investment, effectively operating in a separate economic sphere in collaboration with the state. With Cubans officially barred from possessing dollars, they need to either collude with foreigners to gain access to the hard currency, or obtain and hold them illegally. A legislated 'dollar apartheid' thus separates the foreigners from the indigenous population, a separation often enforced further through physical barriers, such as bouncers outside tourist hotels, or fencing between the state-owned and new public-private owned businesses. Similar divisions exist in business parks between the domestically owned and internationally co-owned businesses – the latter using the US dollar as their currency.

The new, market-influenced business interests take two shapes: indigenous small-scale businesses, usually informal self-employed 'one-man bands', and larger corporate enterprises as part of foreign direct investment. Such divisions affect, in particular, the tourism enclaves, where both forms of entrepreneurialism are concentrated. Given the cheek-by-jowl co-existence of western affluence and Cuban poverty, it is not surprising to find illegal, black-market activities, including prostitution (Trumbull, 2001; Centeno, 2004). These rather less desirable effects are not a uniquely Cuban problem, but are found across the former communist states, as formal employment is limited and expectations are high. Such features reflect a rather less powerful state, with inadequate institutional arrangements to govern the private market/s effectively. Evident economic development is geographically divided, and access to its fruit controlled by an effective demarcation line running through the population, rigorously enforced by the state through employment legislation and selective 'dollarisation' (Perry *et al.*, 1997; Orro 2000). Failing that, hidden transaction costs are not dealt with adequately, and there is an insufficient response to the many, often conflicting, interests in an increasingly diversifying society.

> Into the vacuum left by the state, we do not get a Rousseauian paradise, but the takeover of the society by thugs ... The thuggish possibilities emanating from Miami or from the various drug cartels may some day create a nostalgic longing for the Revolutionary order.
>
> (Centeno, 2004, p. 9)

Unlike in China, there are few attempts at structural reform to facilitate a more entrepreneurially minded administrative-governmental structure. It seems that the Cuban government seeks to maintain a system more in tune with the spirit of its Marxist ideal of an egalitarian, state-directed economy, even at the cost of restricting economic (inward investment) potential. Property rights, and a supportive and predictable legal and administrative framework, are crucial for the successful operation of market principles in general, and luring private inward investment in particular, but they are also at the root of socialist ideology. That is the big challenge faced by those seeking to walk the tightrope between a socialist planned and a market-based economy. This helps to explain why attempts so far to copy the Chinese model, with its Special Economic Zones, have so far had rather limited success. These 'Zones' depend in their credibility on the general underpinning of national economic policy, and there the balance between business and societal interests. But the divisive effect of the duality between those working in the 'dollar enclaves' and those outside, gradually leads to a decreasing sense of solidarity among the population, and, instead, a growing search for individual opportunities to improve one's life under the very unequal conditions offered by the divided economy. 'Correspondingly, there is a widening gulf between much of the regime's socialist ideology and the everyday experiences and practices of the population' (Pickel, 1998, p. 78).

This widening discrepancy between theory and practice threatens to undermine the regime's legitimacy, as it had done in communist Europe during the 1980s. Yet the very fact of tightly controlling the development of, and access to, marketisation reflects the regime's acute awareness of this potential problem (March-Poquet, 2000), especially since the events in the former Soviet Union in 1991. There are thus clear signs of the existing regime adopting reform measures to pre-empt and counteract evident grievances among the population about the poor economic conditions. But these changes are slow, a result of the reluctance of the Cuban regime to succumb to 'the market'. A wholesale replacement of the communist regime *per se*, including the economic and administrative structures, is thus, at the moment, not realistically 'on the cards', although, as history has demonstrated, any assessment of possible future trends in this respect are extremely difficult.

Summarising Cuba's political situation today, after 40 years of development, Azicri (2001, p. 302) observes that 'the different themes and codes of Cuba's political culture have been compressed into a new political culture mix that functionally represents a civil (secular) religion', which is 'rooted in the revolutionary values conforming Cubans' political thought'. This involves developing a new identity internally, selling communism to the people as, first and foremost, 'revolutionary', and thus making it more acceptable across the generations. National sovereignty and independence have thus become the cornerstones of Cuban politics. In so doing, it has had to demonstrate a degree of continued flexibility in defining and re/presenting the socialist values within a set framework, stressing different qualities at different times, and for different audiences – within and outside the country. Thus, regardless of the disappearance of socialism elsewhere, 'Cuba will continue reinventing its own brand of socialism today and tomorrow' (ibid., p. 307). It is this that makes Cuba a special case. It is not merely following the Chinese way of insisting, at least rhetorically, on the continued validity of Marxist–Leninist doctrine and, by implication, one-party rule. But at the same time, a liberal market economy is pursued in a parallel economic 'universe', in complete contradiction to the essence of Marxism. This re-branding and re-focusing of 'communism' uses stronger, more explicit references to the colonial legacy, and liberation from foreign occupation. Concessions to encourage economic development are seen as unfortunate necessities, albeit important ones.

The balancing act between 'market' and 'socialism' is mirrored in the institutional changes to underpin the necessary, if reluctant, economic opening to the outside world. This included the possibility of private ownership, joint ventures with the state in state enterprises, especially within the newly developing tourism industry, decentralisation, and generally a 'dose' of market. But all these changes are very much tentative and piecemeal, particularly in comparison to the sweeping changes in other socialist countries, especially in eastern Europe. In Cuba's case, it is obvious that any changes are *permitted* by the established leadership in response to the external events that so much affect Cuba's livelihood. They are not part of a genuine change to ideological principles, but merely a compromise and reluctant submission to the realities of the new post-Cold War political and economic world order. They seem primarily to serve the survival of the regime (Werlau, 1996), rather than inaugurate a wider shift towards marketisation and democratisation of the kind seen in the CEE countries.

This is one of the main differences from the events in other formerly communist countries. Even though there may be similarities with China at first glance, a second look reveals the quite different political rationale behind the economic policy there. The Chinese leadership subscribes to international trade, foreign investment, consumption and enterprise, as long as the political (and societal) sphere remains under the continuous control of the Communist Party. In Cuba, by contrast, the changes are not the result of a learning process or realisation and admission of the old system's limitations. The strong revolutionary ethos and sense of national identity, combined with a 'sticking-it-out-together-against-all-odds' mentality among the population, have so far allowed Castro to 'get away with' the rather limited concessions to the new international framework, even if it meant tightening belts (even more), but the preparedness to do so may well wear thin and demand a political price.

The continued overemphasis on statist, top-down solutions, seeking to 'direct' and 'order' innovation rather than opening up spaces for entrepreneurship to identify and utilise emerging opportunities, failed to insert some dynamism into the stagnant economies. With ever faster changes in a globalised economy, a *dirigiste* approach will find it difficult to 'keep up with' developments at the time, let alone pre-empt future trends (Amaro, 2000). It is here that China's less strictly centralised model has been successful – providing niches for entrepreneurial reward, and thus helping to establish innovative dynamism, albeit within a controlled environment (Chapter 6). 'The socialist entrepreneur is a normative concept, but so far not the representation of reality' (Brudenius and Gonzalez, 2001, p. 136). A competitive element, with clear rewards for initiative and risk-taking, is an essential ingredient.

An important part of the answer to Cuba's economic difficulties thus seems to rest in the ability to provide a framework for stimulating entrepreneurial capacity and innovation both from within and without the national economy. This means to facilitate indigenous potential, rather than solely rely on outside resources and input. Introverted perspectives, seeking to develop the country under separation from the rest of the world economy, are no longer a realistic option, and the slow establishment of the 'dual' economy reflects that. It is the combination of attracting external funding *and* mobilising indigenous entrepreneurial and innovative capacity, that seems to be the central plank of Cuba's economic development. This makes it different from the Chinese approach of development through foreign direct investment. Whether the new focus on tourism development, driven by the state in conjunction with outside capital, can propel the economy forward, remains to be seen. It is the development of a sizeable, domestic, non-state productive sector that sits at the heart of indigenously driven economic development, including a healthy, diverse service sector.

Post-Soviet socialism and the imagineering of a 'neo-revolutionist' Cuba

The 1990s brought a transition to Cuba of a quite particular kind. The pragmatic and obvious implications of the collapse of the 'Eastern bloc', and especially the Soviet Union, for the Cuban economy, have been discussed already. But there is an interesting second aspect to the post-Soviet era of Cuban socialism, or, perhaps more accurately, Cuban 'neo-revolutionism'. Since the late 1980s, a growing duality between a younger generation's aspirations and expectations, and the established views and ways of doing things held by the older generation in power, is, perhaps, not surprising. As Kapcia (2000) points out, the 1990s have become characterised by competing imagineerings of Cuba and its identity, seen through the history of the revolution. In this context, Che Guevara, the close spiritual companion of Castro during the early phase of the revolution, who later went on his own crusade of spreading socialist revolutionary ideas to other Latin American countries, has regained overwhelming iconic status. He serves as a proxy for criticisms by the younger generation of the realities of Cuban society created by the older generation, measured against the revolutionary ideals. 'Transition' here, therefore, means a re-imaging of the early, idealistic, very much Cuba-centred idealism underpinning the revolution. In this, Che Guevara has become identified with the heroic values of the nationalistic Cuba, depicting it as 'fighting alone in a hostile world' (ibid., p. 189). It is the image of a struggle for the national independence of Cuba that gained influence in public discourse. His ideas, and thus by association those of the younger generation today, are seen as revolving around the personal development of the individual along the new values of communitarianism, and shared values and aspirations. This contrasts with the post-revolutionary reality offered by institutionalised socialism, where the individual is reduced to an unrecognised, faceless, merely functioning member of society.

The collapse of the Soviet Union and Eastern bloc countries is viewed by those Cubans holding the more orthodox, 'revolutionary' ideal, as a blessing, because it is viewed as having liberated Cuba from the imposed, hegemonic requirements to fit in with the Soviet-European model of communism. This, so the argument goes, deprived Cuba of its scope to develop and implement its own, Cubanised interpretation of Marxism–Leninism, with a clear grounding in Latin tradition. In effect, the Sovietised re-interpretation and presentation of the revolution, could be understood as a cultural-political colonisation, working in parallel to the economic hegemony of the former Soviet Union, whether one views Cuba in this asymmetric relationship as 'partner, proxy, puppet, or paladin' (Pastor, 1983). This hegemonic relationship, so it is seen, has, in effect, deprived Cuba of developing its own agenda, and expressing its national self. The Soviet Union's collapse, therefore, has been perceived as an effective removal of outside control. Being seen 'no longer as an advanced arm of a mighty global military force, and ... no longer capable of waging even an ideological guerilla war, because its belief system is so universally rejected and debased' (Rothkopf, 2000, p. 109), Cuba's new independence has made it more acceptable and 'normal' within the Latin American and global community. Transition in Cuba may thus be dubbed as post-Soviet-style communism.

The economic problems of the 'Special Period' of the early 1990s (Henken, 2000; Trumbull, 2000) contributed to a strengthening of national resolve 'against the rest of the world', shifting the perspective decisively to the national self. But this is also the challenge for future development, as economic growth will have to engage the outside world, and thus bears the threat of a new hegemonic dependency. The educational drive had created a more educated young population, with higher aspirations than the realities of a Third World-style economy could offer. With few signs of change visible:

> an increasingly young Cuba found itself frustrated by a system that seemed to have stagnated and become less inclusive, a system still led, moreover, by those who, having come to power in 1959, seemed to age in office, continuing to use an increasingly incredible 'young' revolutionary language that sits uncomfortably with their personal image. In the younger generation's eyes, therefore, the existing political elite was considered out of touch and thus increasingly irrelevant in shaping Cuba's future, hanging on to power too long and speaking a political language that no longer meant much to them.
>
> (Kapcia, 2000, p. 209)

The rise of Gorbachev in the Soviet Union, and his criticism of the era of stagnation under Brezhnev, rang bells for those critical youths, drawing parallels 'between the gerontocratic pre-Gorbachev Soviet Union and the Cuban case' (ibid.), and criticised the absence of such challenging criticism in the Cuban political establishment.

The Cuban leadership has sought to accommodate those concerns by visibly rejuvenating some of the political apparatus, and focusing publicly on the idealism symbolised by Che Guevara. The aim was 'both to reinvigorate the Revolution's ideological thrust and revive the legitimacy of the neglected 1960s, now seen as encapsulating the "essence" of the "real" Revolution' (ibid, p. 210). The associated myth of liberation, communitarianism, and national self-determination and pride, are now being recalled to give the current leadership a new sense of legitimacy and purpose, while seeking to unite the population behind them via the new idealism. 'By clinging to "Che", the young can therefore be "revolutionary" and still distance themselves from the present leadership' (ibid., p212). In effect, the collapse of the

Soviet system has triggered a re-examination and rediscovery of the revolutionary zeal underlying the ambition for self-determination. But the young generation's criticism also emphasises an alternative view to that implemented by the current government.

So far, Cuban-style socialism, not unlike other reforming communist regimes, especially in Asia (Chapter 6), has survived because of two main factors – elements of marketisation and a degree of decentralisation in economic decision-making. Marketisation, as an immediate self-help response in 1990, does not just include the creation of a dollar-based second economy, but also some elements of self-employment within the domestic pesos market, especially in the service sector. In the agricultural sector, the vanguard of 'socialism in practice', several large cooperatives or state farms have been privatised to encourage production. There were even temporarily permitted private farmers' markets, but a public backlash against seemingly extortionate prices led to their abolition not long after their inauguration in the mid-1990s. The second factor, decentralisation, was the some-what inadvertent result of earlier policies. Under the 'rectification' programme (effectively a 'back to basics' campaign) of the late 1980s, the government sought to reinforce the spirit of the revolution, especially egalitarianism, and 'stamp down' on 'out-and-out' consumerism and seeming profiteering by the permitted private market traders (Fabienke, 2001). If nothing else, it at least gave the impression to the population that the government was 'listening' and seeking to maintain the spirit of the revolution.

The economic crisis and ideological uncertainties may be interpreted as pro-ducing collective postmodernism (Kapcia, 2000). This means rejecting the near linear, modernist developmental paradigm of the course of history propagated by Marxism. In Cuba's case, it seems that 'transition' means both drawing on its history, especially the idealism and aspirations that underpinned the revolution in 1959, as a scale against which to measure today's political reality, and also outlining a development strategy into the future. This includes, in particular, the preserva-tion of Cuba's national independence. Dropping its Soviet-style institutionalised doctrine, the Cuban government's ideological-political emphasis has publicly shifted towards the task of preserving national unity and mobilisation *vis-à-vis* the growing challenges of a globalising political economy. But there is the danger of limiting this engagement with the outside world to reluctant 'tinkering on the edges', rather than bold political reform. The big challenge is the continued pressure for more market-driven policies and more interaction with the global economy to develop competitive advantages, while continuing to maintain the idealised perception of societal solidarity and national self. The frequent changes in Cuba's 'stop-and-go' policies of tightening and loosening the state's grip on society and economy, and emphasising the adherence to the principles of 'socialism', allow us to distinguish eight policy periods (see Table 8.2).

Given these particular circumstances underpinning Cuba's immediate history and, especially, its state-building policies under the banner of socialism, there are unique features that help to explain its survival as one of the few still officially communist (Marxist–Leninist) regimes. Looking at the circumstances in more

Table 8.2 Periods of 'stop-and-go' in developing Cuban socialism

Year	Period's political features
1959–60	Anarchic revolution, radicalism
1961–62	Beginning of references to 'socialism'
1962–65	Uncertainty about likely alliances
1966–75	Internal and external radicalisation, exporting communism, challenging Soviet hegemony in the developing 'socialist world'
1975–85	Uneasy Sovietisation
1986–89	'Back to basics' period of rectification
1989–95	Economic turmoil: the 'Special Period'
1995 onwards	Tentative marketisation, neo-revolutionary nationalism, the search for a new role in the global framework

Source: based on Kapcia, 2000

detail, it becomes obvious that Cuba benefits from a unique cocktail of geopolitical location, diverse cultural legacies, the personal characteristics of the leadership (Ennis, 2002), and the ideological legacies of the Cold War. Taken together, these produced the circumstances under which it has been possible to construct Cuba's particular approach to post-colonial nation-building under the unifying ideology of Marxism–Leninism. This 'Cuban way' is now being reinforced, somewhat rejigged, and projected well into the post-Soviet era.

Cuba's particular circumstances could be seen as offering the opportunity of following a different developmental path than that laid down by the Washington Consensus. This means, as discussed in Chapters 4–6:

> adopting a reform strategy that seeks to avoid the high social and econ-
> omic costs of neoliberal radicalism while at the same time embarking on
> meaningful political reform and democratization. If these opportunities
> are acted upon, Cuba may develop an alternative to both the Eastern
> European and the East Asian reform approaches.
>
> (Pickel, 1998, p. 77)

This is a Cuba-specific fusion of *inter alia* part-marketisation, continued state control of some aspects of the economy, spatially restrictive internationalisation, one-party leadership and part-privatisation. It may sound an unlikely combination of models and policies to follow, and so it is not surprising that the Cuban leadership had to adopt a trial-and-error approach, unable to refer to any master plan for this type of multi-dimensional reform approach, defined 'as you go along'. With no clear destination in sight, nor a clearly laid-down path ahead, any new policies were implemented very tentatively, almost gingerly, waiting for the results of each initiative to become apparent, and then being able to respond – if necessary, adjusting and reversing policy initiatives.

In this respect, there are thus obvious parallels to the Chinese approach, albeit, inevitably, at a much reduced scale and with less enthusiasm for marketisation.

Cuba does not have a Hong Kong on its doorstep that it can tap into for its financial and skills resources. Still, it too tries to 'ride two horses' simultaneously, albeit different ones from those chosen by China. Cuba seeks to proceed with a 'dualization of the economy' (ibid.), rather than with a dualisation of the political and economic system, as pursued in China. Cuba seeks to maintain a strong role for state planning as the backbone and general framework for the national economy, within which some market elements are permitted, either in the form of small-scale approved local open markets, or as geographically restricted enclaves of foreign direct investment, evident in the gated tourism enclaves.

However, as pointed out earlier, these changes are not the result of an ideological sea change by the Cuban leadership, but rather a reluctant compromise demanded by the economic crisis of the early 1990s, with a collapse in national output to 60 per cent of the 1989 level. But there are signs of economic rewards for the reform efforts, despite the necessary start from zero. There were no existing international links into global markets, and new markets for Cuban products had to be explored (Pickel, 1998). As a post-colonial country, there was no entrepreneurial tradition to refer back to, a similar situation, for instance, to that found in African countries. Whether this approach can succeed in bringing stable economic development without significant moves towards democratisation remains to be seen (Werlau, 1996).

Conclusions: Cuba, the Third World and 'socialism'

In the countries of the Third World where the bourgeois state was destroyed, the communist movement gained an initial legitimacy in the anti-imperialist struggle, in China, Vietnam, and Cuba. And in spite of similarities, this is what distinguishes them from the bureaucratized societies in Europe.

(Habel, 1991, p. 211)

In other words, it was in those countries where the inherited colonial structure was deconstructed that communism could avoid resistance by the middle classes. These systems faced a credibility crisis at the end of the 1980s and early 1990s, because of the evident shortcomings in improving the living conditions for 'their' people, not only in absolute terms, but (especially) in comparison with the capitalist system. Inevitably, these comparisons were somewhat skewed and generalised, with western qualities of life appearing in a glorified light in the 'east', and projected as superior by western propaganda. That not everything under the capitalist system was automatically superior, and generally better, only emerged later, when the realities were experienced. In addition, the post-colonial revolutions in the Third World countries had raised the stakes by promising more popular involvement through democratisation. Instead, authoritarian regimes installed themselves. Many of the new socialist post-colonial countries, especially in Africa, but also Cuba, thus combined comparative economic underachievement with a lack of popular participation in their countries' governance. From an outsider's perspective, the models of capitalist development adopted by the newly industrialising countries in Latin America and Southeast Asia seem to be more promising in economic, if not democratic, terms, although at the price of rapid and substantial increases in inequalities. This has also been the case with marketisation, as an emergency response to the changing economic realities of the late 1980s/early 1990s. Then, pitching 'market' against 'plan' was perceived as the only way of preserving social-ism, and avoiding a mercantilist future for the communist developing countries. It is at this juncture that Cuba seeks to find its own 'third way', balancing the need to integrate into a global economy, albeit along mercantilist lines, with the desire to retain as much as possible of the ideological legacy of the revolution for independence and subsequent socialist doctrine (Habel, 1991).

Judging the likely path of *post*-communist development in its early stage, in 1990/1, Habel (1991) saw no immediate likelihood of such 'third way' democratic socialism developing in the countries of CEE, because of the attraction of adjacent western Europe, with its market liberalism and parliamentarianism, but also the re-emerging nationalism. Consequently, she concluded, 'it will be hard for demo-cratic socialism to find a space between the liberal racketeering and fundamentalist nationalism which are developing on the compost of rotting Stalinist bureaucracy' (ibid., p. 217). Events in Yugoslavia, and also parts of the former Soviet Union, confirmed her rather pessimistic outlook, although this could only offer a general assessment. In reality, different paths of development have emerged between the 'Second' and 'Third' Worlds, but also within the various global regions, as, for instance, the developments in Russia and the other former Soviet republics illustrate.

The experiences of Third World socialism have been quite different from those in Central and Eastern Europe, with distinctly different histories and experiences. This includes their past experience with capitalism from a colonial dependant's perspective. But the political-economic realities in the early 1990s meant a continuous weakening of the socialist/communist camp. In Cuba's case, the democratisation of Nicaragua in 1990 removed the only remaining 'socialist' ally within Latin America, and the collapse of the Soviet Union thereafter isolated Cuba altogether. The country's strenuous efforts to maintain independence, inclu-ding economic, are driven by its historic background, especially its past political dependencies. To Cuba, as to China, Vietnam and the other few developing countries continuing to proclaim adherence to socialism, westernisation essentially meant reverting to a new version of colonial-style economic dependency on external economic and political control (Walker, 2003). There was no blueprint for post-communist routes of development in non-industrialised post-colonial countries, and thus each country needs to find its own way in post-communist transition, whatever the difficulties.

In Cuba's case, the government's insistence on continued autocratic rule has highlighted Cuba's travel from being an 'outpost' of the Soviet Euro-centric hemisphere to being a part of the post-colonial world, with the communist legacy a mere sub-category, rather than a defining characteristic. The continued insistence on an authoritarian one-party regime shows stronger affinities with the Chinese model of communist-labelled rule than with the post-Soviet model. But there have been some considerable changes to the political economy of post-Soviet Cuba that make Castro's rhetoric appear little more than 'a repetition of well-worn formulae' (Habel, 1991, p. 231). The relationship between government and the people is crucial, and here, Castro's system is in danger of failing to maintain credibility among the public. Repeatedly, it needs to justify itself and the continued exclusion of the people from government decision-making, ultimately being in danger of producing a split between itself and the people.

Keeping the memory of the common revolutionary legacy alive is thus an important part of official political discourse, as it is the memory of the independence struggle that bears the roots of the Cuban government's legitimacy to claim continuous power. It is not surprising, therefore, to see Castro's regime evoking revolutionary spirits and values and idealism in public debates. The economic problems of the early 1990s have heightened the urgency of re-invoking the revolutionary spirit to maintain political-ideological legitimacy. Despite its concerns about losing control, opening up to capitalism and facilitating indigenous entrepreneurialism seem crucial in the attempt to reduce Cuba's 'neo-colonial' economic dependency, despite the official denial of 'giving in' to capitalism. This includes the overt emphasis on tourism as a currency-yielding opportunity, which may be criticised as 'selling out' the revolutionary ideal. But there need to be institutional and structural changes to accommodate and facilitate indigenous enterprising talent and capacity. This does not have to mean 'throwing the baby out with the bath water' and opting for a 'free for all' liberalisation approach. The evidence from Russia and other former communist countries that opted for that radical 'shock therapy' model is of very unpredictable and unequal outcomes with growing social divisions, which may well spell difficulties for the future development of an integrated society and political legitimacy.

The main 'special factor' allowing the regime to maintain 'socialism' as its ideology, appears to be Cuban nationalism – rekindled by the government post 1990. This, combined with the charisma of the popular leader Fidel Castro, who maintains and cultivates the image as revolutionary leader, links the current government back to the struggle for national independence and an end to colonial dependency. This is not dissimilar to the roles of Mao Zedong and Deng Xiaoping in China, whose leading positions in the revolution gave them authority and recognition among the population, even if the governmental bureaucracy was an object of public complaint and disaffection. The revolutionary leaders stood above such day-to-day problems and criticisms. In contrast to China, the ongoing stand-off with the US provides the Cuban government with an important propagandist image of an enemy and threat, *against* which national identity and the resolve to protect independence can be projected and authoritarian government measures justified. National security is a powerful argument in governmental attempts to expand control, whatever the ideological underpinnings.

Nevertheless, despite politically successful manoeuvring, the economic situation is still in need of further reform. Pickel (1998) judged the reforms so far to be too narrow in focus, and too technocratic, lacking a vision about future scenarios of development. This includes, in particular, the absence of democratisation, allowing for a merely nascent civil society, 'kept at bay' by an overbearing state. But whether all three elements of reform, economic and political, and cultural renewal, need to go together in order to produce a successful outcome, as suggested by him, remains to be seen. Certainly, there are now signs of concern about likely societal instability, emanating from the rapidly growing disparities in China as a result of its economic-reform-only approach – causing considerable concern among the leadership about a possible loss of popular support, and thus claim to power, and Cuba's path of a dual economy with dollar apartheid points in a similar direction of creating and manifesting considerable economic inequalities among its population.

The main challenge for the Cuban regimes is maintaining a sense of popular shared purpose and national benefit from the continued conditions, more difficult and austere than promised by capitalism. Evidence that some are doing much better than others, and without so much obvious struggle, is likely to weaken public resolve to continue the route of the 'Cuban way' of socialist post-colonial development. If the revolutionary legacy is reduced to little more than idealised romanticism, however, popular expectations may be unrealistically high and disappointment lead to a fading away of support for the socialist path of development, however implemented on the ground. The Nicaraguan experience of the 1990 election of a neo-liberal government to replace the incumbent socialist Sandinistas (Smith, 1998) demonstrates the importance of rhetoric meeting lived realities 'on the ground'. But it also shows the crucial importance of external (US) influences, both politically and economically, on Latin American states (Walker, 2003). Yet, it seems, the economic crisis generated by the collapse of statist communism in Europe 'is helping to shatter old dogmas and create intellectual and political openings for new, although still tentative, thinking about democracy in a socialist society' (McCaughan, 1997, p. 75).

Yet the idea of socialism as the most effective and promising developmental strategy to overcome inequalities and dependencies, seen not just from an ideological-political, but also academic perspective at the time (Huberman and Sweezy, 1969), still lives on in the background. And the new, ever starker contrasts between those benefiting from liberalisation and internationalisation, and those that are not, may continue to provide Cuba's ongoing insistence on the principles of socialism with new support.

9

IDENTITY UNDER COMMUNISM AND POST-COMMUNISM

Craig Young and Kathrin Hörschelmann

Introduction

This chapter considers the key processes shaping the particular politics of identity that have emerged under post-socialism. The issue of identity was of particular concern to communist regimes. In addition to efforts to reform economic and political systems, such regimes were concerned with the identities that their 'citizens' constructed for themselves. This was part of the much broader project of 'building' new socialist (here referring to the ideology of Marxism) societies. Within this project each individual had a particular idealised role to play. At an ideological level, at least, this political project envisaged nothing less than a complete remodelling of society. This was important to maintain the legitimacy of one-party rule, but it was also about attempts to create a new society. The first basis for this was the attempt to establish new sources of identity that people constructed in relation to the Communist Party and to notions of 'Communist society'. This involved state-sponsored attempts to remove previous sources of identities and allegiances, including efforts to suppress or modify national identities, pre-Communist gender identities and religious adherence. This was first experimented with in the USSR post-1917 before being spread in various forms to other communist countries post-1945, where the nature and depth of state-driven identity projects varied considerably.

With the transformation of communist societies in the different regions, a variety of post-communist identity processes has emerged. Democratisation, political pluralism, freedom of expression, and the re-emergence of civil society and personal choice to varying degrees in the contexts of 'Europeanisation' and globalisation have created a variety of new contexts in which post-communist identity formation can take place. These varied processes of identity formation are outlined in this chapter, drawing on examples from the major global post-communist areas. After considering the re-emergence of national and ethnic distinctions and the role of religion, identity changes at various micro-scales are considered in relation to work and consumption. This chapter will focus on the key themes only, while moving between different global regions. It examines the reconfiguration of national identities; religion; ethnic exclusion; the transformation of work and of social relations; and changing consumer practices and identity. These issues are examined both more generally and through examples from Russia, Central Asia, China and Central and Eastern Europe, including the Balkans, thus showing how post-communist transformation is lived and made sense of in specific locales.

The reconfiguration of national identities

One of the major political consequences of the transformation of former communist regimes has been the dissolution and reconstitution of national states and a renewed interest in, as well as problematisation of, ethnicity. Political forces in both new and persisting states have sought to find different schemes for the formulation of national identity. In many cases, this has meant the search for an 'uncontaminated' history, as though, by eradicating signs of the Communist period, time could be reversed and a less ambiguous, unifying identity recovered (Verdery, 1999). The pre-Communist past is actualised and more or less intentionally mobilised as a way of providing orientation in the present. In the nation-building projects of post-communist politicians, the past is selectively reinterpreted in the service of current political interests, and Communism is presented as an aberration rather than as part of the nation's 'normal' historical development. Yet, anthropologists have found that some ideologies and practices adopted under Communism have left more permanent traces in the lives of people and continue to offer representational schemes for the evaluation of transformation experiences (Humphrey 1995; Burawoy and Verdery 1999; Berdahl 2000; Lampland 2000; Hann 2002).

A second feature of nation-building processes in Central and Eastern Europe and the former Soviet states has been their reliance on ethnic criteria for the formulation of citizenship laws. While the clear-cut distinction drawn by Hans Kohn (1944, 1946) between an Eastern 'ethnic' and a Western 'civic' nationalism has been rightly critiqued (see Kuzio 2002; Shulman 2002), few states have sought to define their citizenry using multicultural principles. With different degrees of severity, ethnic minorities have found themselves not only excluded from equal citizenship rights, but also made the target of aggressive anti-minority discourses and practices. They have become one of the most obvious victims of nationalist policy-making, which seeks to forge national unity out of contrast with ethnic 'others' (and note here the continued salience of notions of identity based on constructions of 'them' and 'us'), especially in a situation where unambiguous identity markers cannot be easily reclaimed from the past. One response, in turn, has been a strong movement for ethnic separatism in several of the new states as well as in Russia, where the principle of autonomy has been discredited by its ineffectiveness under the Soviet central regime.

In many eastern European and post-Soviet states outside of Russia identities have become formulated in opposition to communism as a foreign, colonialist import. This option has not been available to Russian politicians, who have looked instead towards unearthing allegedly primordial characteristics and towards religion in the form of the Orthodox Christian Church in order to build a distinct national identity. Urban (1994) argues that in Russia, the collapse of the USSR has triggered an acute crisis of identity, which is now becoming redefined through two key processes: the recovery of identity markers, and the purging of the nation of markers from the Communist past. As far as possible, national identity is constructed through a dis-association with Communism. Contemporary political actors instead look towards the pre-Communist past as a store of cultural materials for (re)constructing national identity. Between them, however, despite significant similarity of objectives, a blame-game forecloses possibilities for dialogue: 'by unmasking others as "communists", they present themselves as defenders of the nation, as bearers of the national interest' (ibid., p. 738–9). Other political players are thus demonised and made culpable for the discredited Communist past.

Urban (1998) sees politicians as major players in the production of would-be national ideas. Liberal democrats, communist-patriots and the state under Yeltsin's presidency competed for the 'soul' of the nation and through their discourses on national identity effectively summoned particular forms of Russian 'culture' into existence. Liberal democrats presented themselves as the most vehement anti-communists and individualists. In devaluing the Soviet past, however, they also devalued the lives of people under the Soviet regime, and became discredited by the 'reforms' enacted by the Yeltsin government. As communism collapsed, they quickly ran out of an 'other' against whom to define themselves. As a result, some neo-liberal politicians looked towards a combination of their ideas with nationalist thought, seeking to construct a synthesis of definitive national traits with free market ideology. Urban gives the example of Chubais, who sought to assemble components of a new national idea largely from Russia's past. He separated features deemed obsolete (like *sobornost'* and *kollektivnost'*) from enduring ideas (like Orthodox religion), but maintained a staunchly anti-communist position.

Communist-patriotic politicians equally invoked particular constructions of the past in order to 'reinvent' the nation. Their discourse evolved around a fairy-tale story, from which the Russian nation emerged as a hero-victim. External and internal forces were blamed in this construction for a series of misfortunes visited upon the nation. Communist-patriots share with their political opponents a messianic conception of Russian culture, which:

> is portrayed as embodying those traits and values uniquely suited to the flourishing of human life on earth. Principal among these are 'spirituality', a selflessness enabling the individual to search for the true and the good; *sobornost'*, a mystical notion of harmonious communion of the people based on the Orthodox faith; and certain 'instincts' that this nation has to form and support a great power state (*derzhava*) that rules a temporal order corresponding to the true nature of this nation.
>
> (Urban 1998, p. 980)

There are numerous similarities between the Russian situation and that of the New Independent States regarding the process of constructing national identity. To paraphrase Gellner (1964), they do not awaken nations to self-consciousness but invent them where they do not exist. Yet, in countries such as Uzbekistan, the question is even more complex and the answers found by policy makers are still more ambiguous. Uzbekistan was first founded as a state by Stalin. None the less, the current autocratic ruler, President Karimov, has sought to establish historical continuity in his writing by selectively combining figures and events from the past that can be woven into a narrative of long-term fights for independence and a proud history of intellectual Islamic development. As March (2002) explains, he and his 'court intellectuals' have discursively produced a new myth of nationhood by inventing traditions and national continuities that at close inspection seem both arbitrary and contradictory, given that figures as diverse as the medieval Islamic scholar Tamerlane and the Soviet dissident Rashidov have been included along the same lines.

All intellectual activity since Tamerlane and the great Islamic medieval thinkers is presented as a single phenomenon directed uninterruptedly towards a common set of values and the idea of national independence, although 'Uzbekistan is precisely a case of the achievement of statehood and nationhood through historical accident and Karimov's project is precisely an effort to create a national ideology *ex post facto*' (ibid., p. 374). Public rituals, space and architecture are appropriated for the purposes of making the new identity 'real' and former Soviet signifiers are simply supplanted with new symbolising structures, as in the case of Lenin statues exchanged for monuments to Tamerlane. Karimov utilises the narrative of national independence struggle for legitimising his own position of power by identifying new threats to the nation, such as Islamic radicalism or fundamentalism, and by presenting his authoritarian style of government as a necessary, transitory prerequisite for democracy. The ruling ideology, which is taught in schools and propagated through the government-dominated media, is thus made to seem as in line with the interests of a vulnerable nation, whose peace and stability is endangered by radical forces inside and outside the country (ibid., p. 382).

The distinction between Russian foreign oppressors and subjugated nationals is not always as easily drawn. Thus in Ukraine, Wanner (1998) has identified a conflict between nationalists who define the historical relationship between Ukrainians and the Russian/Soviet states in terms of cultural subjugation, economic exploitation, forced annihilation and genocide, and the large number of Russified Ukrainians, who see themselves as freely assimilated to Russian culture. Shulman (2002, p. 19) equally notes a major distinction between Ethnic Ukrainians and Eastern Slavonics, where 'the Eastern Slavic nationalist discourse promotes a vision of Ukraine as consisting of two organically related and equally native ethnic groups sharing a common cultural and historical space' and 'the strong presence of Russian culture and language is portrayed as something to be valued, celebrated and preserved – in contrast to the discourse of most Ethnic Ukrainian nationalists'. Kuzio (2002), however, warns that the division between a Catholic, nationalist West and a Russian-speaking, Orthodox, pro-Russian East, much repeated in western media representations of the recent election protests, is too simplistic and ignores the large group of people who use Russian and Ukrainian interchangeably and whose identity is very much in flux. Both the national democratic opposition under Yuvchenko (now in government) and the centre-right ex-government under Kuchma, however, shared an interest in forging a new, ethnically based sense of belonging, which links individuals to the state via national identity (Wanner 1998; Kuzio 2002). History is used by these political elites once more as a reservoir for the crafting of post-Soviet national culture.

Wanner (1998, p. xxv) focuses on key sites that make the new national identity 'material' or 'real', in particular schools, festivals, the state calendar and monuments in public space, thus arguing that the new identities are not just imagined but actualised through material objects and practices. Shulman (2002) has also identified a range of political practices aimed at building an Ethnic Ukrainian national identity, such as the 1989 Language Law, which made Ukrainian the sole state language; the 1992 directive for bringing the language of instruction in public schools in line with the national composition of population in each region, which weakened the rights of parents to choose the language of instruction for their children; and further changes in education policies whereby Ukrainian-language schools are no longer required to teach Russian and higher-education entrance exams have to be taken in Ukrainian. Russia has become an 'other' for the definition of self-identity in Ukrainian nationalist politics. It is presented as part of an Asiatic empire built on oppression and subjugation, as sinister, brutal and uncouth, whereas Ukrainians are constructed as civilised, peace-loving and European (Wanner 1998). Holy (1996, p. 77) notes a similar construction of other and self in discourses of Czech national identity since the Velvet Revolution in 1989 and the separation from Slovakia in 1993, in which the Czech nation is represented in such discourses as 'highly cultured and well-educated' and as Protestant, democratic and civilised compared with the 'backward, Catholic traditionalist' Slovaks.

Cultural closeness to Europe was also one of the arguments used by the leaders of the former Yugoslav republics of Slovenia and Croatia in their drive for independence and by Serbian and Croatian forced nationalism in the Bosnian war (Sells, 1996). In founding independent states based on national collectivities, they appealed to modern European principles of nation-building and presented themselves as defenders of 'European culture', yet most drastically inverted the myth of European tolerance, as Hayden (2002) explains. While western media reports and academic commentators such as Meštrovic (1994) and Hammel (1993) have sought to explain the violent dissolution of former Yugoslavia via reference to the resurfacing of deep-rooted ethnic conflicts that were glossed over but not abolished by the communist federal state, others have constructed more careful and complex interpretations, which show, among other things, that the assumption of 'deep-rooted ethnic conflict' is not only grossly misleading and oversimplified, but also in turn supports the very segregationist policies it seeks to critique. It is motivated most often by an orientalist construction of Balkan 'others', that Croatian, Serbian and Slovenian nationalists have ironically also applied to justify their own nationalist and anti-Muslim policies.

As Bringa (1995), Hayden (1996, 2002), Salecl (1994) and Sells (1996) argue, even the response of European and American political forces to the conflicts in former Yugoslavia was coloured by assumptions about ethnic difference mirrored in the propaganda of nationalist leaders in Slovenia, Croatia, Serbia and Bosnia. Indeed, the division of Bosnia-Herzegovina into ethnically based zones by the

Vance-Owen Plan (1993) and later the Dayton Accord (1995) could be seen as an acquiescence to, and a *post factum* justification for, ethnic cleansing by the nationalist forces involved in the conflict (Bringa 1995; Sells 1996; Hayden 2002).

Hayden (1996) understands the wars in former Yugoslavia as arising not from deep ethnic divisions but from a clash between prescribed and lived (mixed) cultures that nationalist governments sought to eradicate as part of a process of consolidating their power in an otherwise unstable political and economic situation. He points out that the wars erupted almost entirely in the most 'mixed' regions: 'the wars have been about forced unmixing of peoples whose continuing coexistence was counter to the political ideologies that won the free elections of 1990' (ibid., p. 783). Hayden (1996, p. 784) defines ethnic cleansing as 'the removal of specific kinds of human matter from particular places', produced by a 'lack of congruence between the present reality of life as lived and the objectification of life as it suddenly must be lived', and includes in this process not just violent manifestations of war and conflict, but also bureaucratic measures such as the new constitutions of states through which nationalist ideologies are manifested and institutionalised. Hayden emphasises that a Yugoslav community had in fact existed in many parts of the country and had become increasingly intermingled, but the mixed territories were perceived as anomalous and threatening, 'since they served as living disproof of the nationalist ideologies' (ibid., p. 788).

New citizenship laws across most of the former Yugoslavia have created a new regime of inclusion and exclusion. Nationals living in other states have been granted citizenship, while those from different nationalities who resided within the state suddenly became foreigners of their own republics. Ron (2000) has shown the extent to which the drawing of new boundaries has formed part of a process of state crafting. He points out that Serbian leaders chose to attack Muslims in Bosnia because it was a 'foreign' land, thus less costly than an attack on Muslims internally. Serbian nationalist historiography claims Kosovo as a significant site by constructing the 1389 Battle of Kosovo as the beginning of centuries of oppression. Denitsch (1996, p. 113) explains that the myth of national suffering was a key feature of Serb nationalist propaganda. For Sells (1996) and Bringa (1995), the war in Bosnia-Herzegovina was not only motivated by a desire for ethnic 'un-mixing', but more substantially aimed at eradicating all traces of Muslim life and culture. Duijzings (2000) has shown that the reference to ancient ethnic hatred is also misleading for Kosovo, where ethnically mixed pilgrimages and contacts across ethnic and religious boundaries were widespread. However, in Letnica, where he conducted his research, strong pressures existed to protect a family's integrity and avenge infringements upon its reputation, and this was a source of hostilities that was exploited in the subsequent conflict (also see Bax, 2000).

While the authors cited thus far have given a thorough insight into the symbolic construction of national identities and ethnic difference in former Yugoslavia, and have pointed towards the role played by governmental propaganda and intellectual discourses, Gordy (1999) goes further to examine the reasons why the Serbian government under Milosevic did not meet with more resistance, despite its failure to ever gain an absolute majority in national elections. He argues that the use of nationalist rhetoric and the appeal to nationalist movements was a key strategy used to augment the power of Milosevic's Socialist Party of Serbia, but that the regime was particularly able to produce feelings of defeat, exhaustion and hopelessness that led to political apathy and resignation. Milosevic formed an insecure coalition of opposites by supporting nationalist Serbs in Kosovo, while opposing the Communist old guard. He capitalised on the fears of Serbs in other republics by building an image as their protector and as defender of the national interest, and used the media to present his party as a relatively moderate force compared with the extreme nationalist opposition.

Media propaganda helped significantly in discrediting opposition to the regime and in sowing nationalist panic in response to the politics of independence and nationalist revival pursued in Slovenia and Croatia. There was thus little room in the media for the presentation of alternatives, especially as independent outlets were closed and marginalised. In addition, Gordy (1999, p. 197) shows the effect of economic instability and hyperinflation in destroying sociability and forcing people to concentrate on the pursuit of everyday needs: 'The cumulative effect of instability, impoverishment, and restriction was that "normality" became a scarce commodity. Deprived of this fundamental backdrop for social activity, autonomous means of social exchange, both economic and interpersonal, broke down.' The dependence of most people on the regime deepened, and space for autonomous action and decision became restricted with profound social consequences. Highlighting these political and economic conditions, however, does not answer all questions. It fails to provide a general explanation for the enormous extent of violence and the personal responsibility of the perpetrators of genocide in former Yugoslavia. Those answers can only be given through an examination of specific cases, not in general and from above.

Religion

Religion was treated in different ways under communism, with outright suppression of religious practice relatively rare. However, the re-establishment of religious freedom after the fall of communism in CEE and the FSU, and its continued presence in other 'socialist' contexts, has meant that religion has come to play an important role in the formation of post-communist identities. In some contexts, such as Poland, the Catholic church has played a strong role in defining the emergent civil society, especially in areas such as gender roles and identities. The Orthodox church has re-emerged as an important force in society, while other contexts have witnessed a return to other religious beliefs, such as the 'new shamanism' in Siberian cities (see Humphrey, 2002). Alongside the return of traditional religion, there has also been a significant growth in new cults, which have proved particularly appealing to those members of society who are struggling to adjust to the changes ongoing in post-communist societies (see Ramet, 1995). Across the post-communist world, religion is interwoven with a number of other sources of identity to create new hybrid forms of identity.

In the Yugoslav conflicts, for example, ethnic identities were articulated primarily on the basis of religious affiliation. In the case of Croats and Serbs, this was indeed the major way of claiming difference, since other characteristics, such as language, were very much shared. Religious difference was emphasised during the conflicts so that it took on more political significance than was previously the case. Leutloff (2000, p. 1) argues that for Croatian Serbs in the Krajina, politics, religion and remembering the past became tightly connected: 'The church supported the upcoming national politics by justifying and stirring up national differences and sometimes even national hatred.' Serbian national identity has become very much bound up with the Orthodox confession, while being Orthodox is less bound to religious practice than to symbolising belonging to the Serb nation. As Leutloff (2000, p. 11) explains:

> As Krajina-Serbs lost trust in political institutions and did not have their own political representatives, the church functioned as the only institution which gave Serbs some structure and which had the power to unite them. In this way, the church had a we-group establishing function among Krajina-Serb refugees.

On a level of communal practice, however, the Church can also have an integrative function. Thus Leutloff cites the example of a Village-Slava, the biggest Orthodox festival, in which local Croats again came to participate after the war, since it was seen as a community festival more than a solely religious event.

Lederer (2001) focuses on the role of Islam in the former Eastern Europe. He also finds that the connection between nationalism and Islam was previously weak and is still not shared by many. In Kosovo, anti-Serb protests in the 1980s had practically no religious connotation, since they were started by leftist and patriotic students. Serbian propaganda, however, manufactured such a link in claiming that Iranian intelligence needed Kosovo as a beachhead for the expansion of militant Islam in Europe. Equally, in Bosnia-Herzegovina, religiosity was low and the link between Islam, Islamic tradition and ethnicity must have been indirect. Yet, during the Bosnian war, 'the ideologists of the Party of Democratic Action (SDA – *Stranka Demokratske Akcije*) introduced and emphasized the concept of Boshnyak nation, which meant the Bosnian people of Muslim extraction, so the Orthodox and the Catholics became, by definition, reduced to minority status in the country where they lived, not being part of the Nation' (ibid., p. 8). Islamic orientations were thus pronounced as central to national identity.

Religion and national identity have also become more tightly connected in Russia, although in contrast to Poland, for instance, practical religiosity remains at low levels (Agadjanian, 2001). For nationalist politicians, the Orthodox Church represents a barrier against the destruction of Russian statehood and an antidote to Western values and religious cults. Orthodox religious belief is also linked to messianistic notions of 'Slavic civilisation' and to the 'Eurasian' project of opposing the 'Atlantic' civilisation of modernity with a 'social alliance' between Islam and the Orthodox Church. The links between Russian culture and Orthodox Christianity are seen to consist of a similar sense of world mission (spiritual, linked to suffering), collectivism (built on Christian togetherness), and authoritarianism (State Church). At the time of economic and social instability following the White House coup, the church was furthermore perceived:

> as a fresh spiritual and emotional compensation for the shocking breakup of the social system, as well as a repository of cultural arguments, collective memories, and the symbolic strength needed to build new national, group, and individual identities. Religious arguments, among others, were instrumentalized to create a new Russian nationalism ... the striking disparity between the low level of religious practice and the high level of discussion about religion is explained by the fact that religion in late twentieth-century Russia functioned not so much as a source of beliefs, values, and social identities, but as *public religion* par excellence. Religion is both the subject and the object of contestation and debate. In Russia the national church is in search of the state, and vice versa, for the state is in search of religious legitimation.
>
> (Agadjanian, 2001)

While both church and state thus see benefits in forging links between religion and national identity, spirituality (rather than religiosity) has gained new significance for different reasons. Kaneff (2002) notes a rising belief in mysticism and superstition in the Bulgarian village where she conducted ethnographic research in the 1990s. She sees this very much as an attempt by villagers 'to make sense of their disrupted lives under post-socialism' (ibid., p. 102). Death, in the example she analyses, has become reinterpreted as not occurring 'naturally' any longer, indicating that culturally codified beliefs about the ties between persons, society and 'nature' have become destabilised in the period of transformation:

> The perceived upheaval in village life over recent years of political and economic change is accompanied by a breakdown of the total set of social relations that constituted the socialist world ... The 'natural order', what is 'culturally significant', beliefs associated with birth, weddings, and death, the ways individuals position themselves and relate to official state structures, are all under negotiation ....
>
> (Ibid.)

Ethnic exclusion

Under post-communism, ethnic identity, never entirely suppressed under communism, has also become the focus of new identity politics. The conflicts in the former Yugoslavia, for example, were motivated to a significant extent by efforts to create ethnically homogenous national territories (Storey, 2002). Similar motives were the driving force of violent conflicts elsewhere in the post-communist world (e.g. in Armenia, Azerbaijan, Georgia, Chechnya, Moldova and Uzbekistan).

Exclusion and discrimination on ethnic grounds has become a common feature in many post-communist states, even those that do not define citizenship on the basis of ethnic belonging. One particularly marginalised group in Central and Eastern Europe today are the Roma, who had experienced a phase of relative equality under state socialism. Nedelsky (2003) shows how, even in the civically defined Czech nation, they are subject to discrimination and abuse. Constitutionally, the requirement of having a clean criminal record for five years (independent of the severity of the crime) and residency in the Czech Republic for two years, imposes a serious barrier to Roma access to citizenship. Those who were deemed Slovak by the 1996 law are considered foreigners and have to apply for citizenship. Since nearly all Czech Roma were murdered during the holocaust, most of them were either born in Slovakia or have Slovak parents. The residency requirement is further complicated by housing shortages. The Czech government has also been slow to take effective action against racist violence targeted at Roma (ibid., p. 98).

Non-violent discrimination occurs at many levels, showing a gap between constitutional rights and localised policies. Thus, in the town of Jirkov, Roma were evicted from apartments deemed unsanitary or overcrowded after the 'velvet divorce' between Slovakia and the Czech Republic. Requirements of guest registration were introduced, whereby a family was only allowed to have visitors twice a year, while in another town, Kladno, the mayor banned Roma children from using the public swimming pool after a hepatitis outbreak. Nedelsky (2003) also notes cases of government employment offices marking Roma files with an 'R' as a signal to potential employers, a practice that was, however, quickly prohibited. There are also assumptions of Roma children having more learning disabilities, which are unfounded, but lead to them being sent to specialist schools. Often, the real reason is an insufficient knowledge of the Czech language. Since tests are language-oriented, Roma children become classified as less able. In schools, Roma history is rarely taught. Education is instead based on 'white history', 'white music' and 'white customs'. Although there are many civic organisations to promote Roma interests, Roma parties are unlikely to ever jump the 5 per cent barrier needed to gain representation in parliament. Thus, in national politics, they fail to have a significant voice.

Ethnic discrimination frequently entails a territorial dimension and involves the claiming of certain places and objects in the built or natural landscape as property of one ethnic group. Hann (1999) illustrates this with the case of the Ukrainian minority in southeast Poland and their struggle to (re)claim a Greek Catholic Church that had become Roman Catholic after the migratory movements of the Second World War. He shows that, although Poland has been almost entirely free of ethnic violence, the potential for nationalism is deeply rooted and had continued to grow under socialism. In Poland, nationalism is closely linked to religious belief, since the Roman Catholic Church promoted the nation as a basic principle of cultural ordering. Contemporary national discourses in Poland refer to the inter-war period as a period of positive development in the life of the nation that was rudely interrupted by war and then socialism. In the strive to assert the 'purity' of the nation today in all areas of social life, Ukrainians are an 'other' who make apparent the false premise of ethnic and cultural homogeneity. In claiming the Cathedral as an important site for religious practice and heritage preservation, they manifest a material-symbolic presence in the landscape that goes against the grain of nationalist assertions of overlapping political and cultural boundaries.

In Russia, the search for an 'other' against whom national identity can be asserted has led to a bourgeoning of racism, especially towards Caucasians. Roman (2002) highlights the ways in which people of colour are racialised through official municipal discourse and the local registration system in Moscow. The city is reimagined as non-communist, white, Slavic and 'under siege', especially, following recent Chechen terror attacks. While before 1990, non-Russians in the Soviet Union were regarded as 'little brothers' with distinct nationalities, they are now reduced to a single criminal, uncivilised, black race. Caucasians are frequently represented in local media and political discourse as 'capitalist exploiters' associated with crime, while their families are construed as a burden on the already strained social system. The strict registration rules, which contravene international Human Rights legislation, themselves do much to encourage criminal activity, while enabling local police authorities to earn significant income from fines and taxes. Caucasians are heavily targeted for identity checks and subjected to dehumanising abuse, which Roman (2002) argues desensitises Russian audiences, who begin to accept that dark-skinned people are indeed criminals.

Although for many, ethnic minorities in post-communist states the situation has become worse since 1990, there are numerous examples of groups that suffered severe repression under the communist regime and have gained new cultural as well as political rights as a consequence of transformation. John Pickles (2001) thus discusses the situation of the Turkish minority in Bulgaria, who suffered much discrimination and violence at the hands of the state pre-1989, leading to mass migrations. Between June and August 1989, over 360,000 Turkish and Roma Bulgarians left the country following intense pressures for assimilation and outright physical attacks by the nationalist Zhivkov government. Even before this date, however, Muslim minorities had been under pressure to integrate and assimilate, through closures of ethnic minority schools, the Bulgarianisation of syllabi, restrictions enforcing the use of Bulgarian in newspapers, the closure of Turkish-language cultural institutions such as theatres, and campaigns to re-name places. The Bulgarian communist state portrayed Turkish residents in Bulgaria as a 'fifth column' of Turkey, as alien and disloyal. After the ousting of Zhivkov in November 1989 many of their civic rights were restored, and a high number (c.120,000) of those who had been expelled or who had emigrated returned. However, Pickles argues that they now experience new forms of social injustice as a result of economic liberalisation and political deregulation, which hit the rural areas in which the majority of Turkish Bulgarian live particularly hard. Decollectivisation was followed by rapid rises in unemployment and increasing household income inequalities. The region of Kurdjali thus lost 70 per cent of jobs between 1990 and 1993, compared with a national average of 21 per cent (ibid., p. 8).

In China, minority representation in nationally distributed culture and media is still largely monopolized by the Han majority and the state (Baranovitch, 2001).

However, Baranovitch traces how minorities in present-day China are achieving more of a voice in the construction of their own identities, and how these narratives are increasingly informing constructions of their identity in the larger public sphere.

The transformation of work and of micro-social relations

The political-economic transformation of former communist countries has also affected people's identities at the level of everyday social interactions and micro-social relations. Not only were national identities and ethnic affiliations re-evaluated, but people's social relations were also subject to change because of different interpretations of people's role in the communist regime, their political affiliations and their ability or failure to prosper in the new system. Anthropologists have noted that the transformation of work, and the rise in inequalities, have had particularly drastic effects on the organisation of societies at the micro-level. They have identified a rise in insecurities about people's present social roles and their future opportunities as well as an increase in the level of distrust towards others, which leads on the one hand to a greater focus on 'the local', e.g. consumption practices (see Humphrey 1995), and on the other hand to a breakdown in established family and friendship networks, as demonstrated by West (2002). Work was a defining feature of a person's worth under socialism. As it became restructured and increasingly scarce in most former communist societies, people's sense of self-worth was radically altered.

Buchovski describes how in rural western Poland the disruption of the previously direct and traceable link between physical labour effort and profit has led to a depreciation of farming identities:

> For farmers, diligent work is a major idiom of their identity that also helps them to distance themselves from the other classes. However, mystified commodification of labour, strengthened through the obscure effects of market forces leaves them much confused about the traditionally absolute value of their ethos.
>
> (Buchovski 2003, 26–27)

Attitudes to work have changed only gradually for farmers, but the social value of their labour has significantly declined, thus causing a spasm between self- and other-identification. Farmers continue to see office work as 'artificial' and over-paid, with top administrators earning unmerited high incomes, while salaries for farm labour are regarded as unjustifiably low. Since there are few jobs for young people, they often fall outside of the value-categories applied by farm workers, who regard them as 'lazy and unable to learn'. If the value of a person is seen to depend on his or her ability to work diligently and efficiently, unemployed youths occupy a bottom rank in social esteem.

Kaneff's research in Bulgaria (2000) has further shown a strong link between land and identity in rural communities, where (communal) work via the cooperative provided connections between local residents and village history through buildings that were constructed in communal labour (also see Pine (1998) on Poland and Verdery (1999) on Romania). While collectivisation had been a painful experience for many farmers, Kaneff's respondents reflected positively on the resulting improvement in living standards. The current process of decollectivisation and land privatisation was seen as a threat to local identities and social relations. It challenged the assumption that property ownership is legitimised by work and that knowledge about ownership is local and connected to identity, rather than legalistic. Private, individual farms were rejected for this reason. Cooperative farms had formed an integral part of communal life, underpinned by complex notions of responsibility and reciprocal obligations between the state and the individual or community. Through the cooperative, public buildings in the village (as well as many private houses) had been constructed with 'hard' physical labour and local resources. The work lent them their value, while social hierarchies were constructed around participation in the labour process. The buildings formed sources of pride as visual, concrete symbols of joint effort. Their privatisation, and the dissolution of the collective by government decree, were resented as outside decisions that ignored their symbolic importance for the social organisation of the village and for work-centred rural identities.

While much anthropological work has focused on the situation of rural communities, Kideckel (2002) has argued for greater attention to the plight of industrial workers and the working class. He critiques the assumption that post-communist societies are moving too slowly towards capitalism. Rather, 'the region's problematic is not too slow a movement to capitalism (as "transition" would have it) but too fast; not too little capitalism, but too much' (ibid., p. 115). The 'neo-capitalist' systems emerging in the 'transition economies' are even more unequal, Kideckel claims, than Western models, and the pace and extent of class differentiation exceeds Western experience. There is no middle class to speak of, while public resources have been appropriated by narrow elites without transparent, equitable distribution. Workers have become the 'other' in this system. Their conditions and identities have been 'shaped by rapidly diminished access to resources' (ibid., p. 116), as mass unemployment and major increases in the cost of living have eroded their standards of living. Yet, in public discourses, the meaning of workers' lives and concerns are dismissed and their very category is made almost invisible (ibid., p.122). This is confirmed by Oushakine (2000) and Savchenko (2002), who document the gap between the majority Russian population and the 'new' Russian elite's evaluations of the new system.

The sense of vulnerability emerging from these changes has been exacerbated by the collapse of social security arrangements. Together with the general sense of rapid transformation and future unpredictability, it has led to 'ontological insecurity' (Giddens, 1990) about the self and its conditions of existence. Ilkhamov (2001) has noted a similar development in Uzbekistan. He argues that the new insecurities are leading to the emergence of a new 'poor identity' and a strengthening of religion, in this case Islam. Ilkhamov warns about the possibility of social explosion and the possible rise of neo-radical ideologies. Respondents in his research identified both positive and negative aspects of the Soviet regime, but valued particularly their previous ability to plan ahead and to lead coherent, stable lives.

Husain Jurayev, speaking for the rural Muslim community in Uzbekistan (see Ilkhamov, 2001), is opposed to social injustice, an attitude that is shared in traditional Muslim belief. Islam provides him with a new sense of moral order. It is similarly conservative to the Soviet regime in valuing communal concerns, etatism and universalism over individualism and finds support particularly among the impoverished rural poor, to whom it offers a voice. As Ilkhamov cautions, 'the consciousness of the masses still echoes with the slogans and calls of communist ideology. Views, opinions and judgements that draw upon socialist lexicons combined with national-patriotic buzzwords can often be heard in the streets of Uzbekistan ... ' (ibid., p. 40–1). If charismatic local leaders adopt a combination of Islamic belief and radical left-wing ideologies, Ilhamov argues, they are likely to find a significant following among the rural poor.

While new political leaders often seek to distance themselves from communism, its ideologies and symbolic constructions continue to provide a framework for the interpretation of present developments among many 'ordinary' people, as the example above clearly shows. Oushakine (2000, p. 1010) suggests that these engagements with the past are also due to the loss of a meta-language for describing the current situation, leaving the 'post-Soviet' as 'an empty space, a non-existence, devoid of its subjectifying force, its own signifier, and its own meaning effect'. The post-Soviet person does not have language to describe his/her situation. Communism has yet to be replaced by new normative ideals, as the wordless Russian anthem exemplifies. In his research with young Russians, Oushakine found that the students were unable to find a proper symbol, a proper signifier to represent their post-Soviet location. Oushakine argues that this reflects 'a more fundamental tendency of individual and collective inability either to put into words normative ideals and desired goals of the post-communist period or to express the changes that have already happened in Russia.' (ibid., p. 997). He adopts Naumova's term of the 'speechless culture' (1999), 'in which silence is a reaction of post-Soviet people to the threatening instability of social system' (ibid., p. 993). The students in this research showed little sense of group belonging. They distrusted the social system and, instead of finding specific terms for the description of post-Soviet society and subjects, they referred back to known signifiers from the communist past. Thus, there is as yet no conceptual framework to place 'things' in the present. Symbolic vocabulary of the previous era is re-used as an intermediary, which enables slow disinvestment and disattachment, while allowing subjects to locate themselves within a larger cultural context.

Changing practices of consumption and identity

While Oushakine's research in many ways confirms the assumption of a vacuum in the moral and symbolic space of post-communist societies, Savchenko (2002) delivers a more nuanced interpretation. In her research on young Russians' consumption practices, she demonstrates how individuals in fact construct individual identities out of the necessity of coping with uncertainty. Humphrey (1995) and Ransig (2002) have also found consumption a revealing area for the examination of identity change. This is partly due to the fact that lack of consumer goods and material shortages in communist societies had caused much social unrest and were a key area for concern during *perestroika*. The arrival of Western goods on private market stalls was one of the most obvious signs of transformation in everyday life. For some, it meant the beginning of a more liberal society of choice. For others, however, as both Humphreys and Ransig demonstrate, consumer society formed the basis of new social divisions and inequalities. The following excerpt from an article by the Russian journalist Svetlana Graudt, published in the British newspaper *The Observer* (6 March 2005) makes this contradiction particularly clear:

> We may dance to Scissor Sisters and Robbie Williams at their Moscow gigs, but we also frequent Soviet-themed restaurants such as Petrovich and Zhiguli for their simple Russian food and dancing to Seventies pop. We like change and the transformation is everywhere. The old food shop, over the road from the 17-storey block of flats where – at 27 – I still live with my parents, in the past two years has been transformed into an expensive supermarket complete with casino, gym and restaurant. Fifteen years ago I queued there all day for sugar with a number written with a ballpoint pen on my hand … We are living it up in Moscow, but our lifestyles hardly reflect the true hardships many Russians find. The society remains vastly polarised: the rich live well and the rest are just trying to survive.

In the years after the failed Coup of 1990, consumers in Russia were faced with a bewildering growth in economic transactions and the dissolution of old laws, while new laws failed to keep pace with economic life on the street (Humphrey, 1995). Simultaneously, shoppers' financial resources were drastically curbed as wages failed to keep up with inflation and were often only paid after lengthy delays. Inflation also eroded savings, and privatised enterprises were beginning to lay off labour. As a consequence, their consumer behaviour was to a large extent determined by economic want. While markets failed to provide cheap basic products and thus made the search for affordable local products a necessity, consumers also reacted with suspicion to the new traders and their goods. For Humphrey, this was related both to the legacy of Soviet propaganda, where private entrepreneurs were regarded as 'speculators' and 'profiteers', and to new insecurities. She recounts the stories of several research respondents who saw non-Russian goods as somehow contaminated and deceitful and therefore actively avoided purchasing them. This sense of undeserved privileges and untrustworthiness was equally extended to 'new Russians'.

A similar sense of suspicion has been noted by Ransig (2002) in Estonia, where the availability of Western products was on the one hand welcomed as a sign of the country's 'return to Europe' and to 'normality'. On the other hand, however, residents of the small northern town where she conducted her research reacted with ambivalence to the social differentiation that it made transparent. While many had supported Estonian independence and the country's inclusion in the European Community as a way of being accepted as 'western' rather than 'Russian', they experienced the resulting changes as a move from order to disorder and from certainty to uncertainty, not just in terms of personal opportunities, but also in relation to the social organisation of local communities. As Ransig explains, those 'most opposed to the old system . . . were also the people who were in many ways the least happy about the new one' (ibid., p. 130), since consumer choice for them had little to do with the freedom of opinion and expression that they had demanded under *perestroika*.

Changes in economic and political structures have also had effects on the organisation of consumption in China. From 1979 onwards, the Chinese government under Deng Xiaoping had relaxed the grip of the state on the economy by launching policies of decollectivisation and relegitimising private entrepreneurship (Davis, 2000). As a consequence, financial inequalities, especially between rural and urban populations, began to re-emerge, while private market trading and consumer choice increased and sites of production were disconnected gradually from sites of consumption. Davis (2000) notes that widened opportunities for personal consumer choice can provide a greater social space for urban residents to invest in non-official initiatives, whether this leads to outright resistance to the Party State or not. Grocery shopping has become an avenue for developing new cultural distinctions, such as how financial resources are funnelled towards children in one-child households, thus changing the character of childhood, or how the home has become a space for increased personal expression, and how family rituals have become more elaborate as an expression of increased opportunities to invest in the family and the home. Major changes have also occurred in the construction of social relations outside the home and the organisation of leisure time. There are now more commercial leisure sites, such as bowling alleys, dance halls and restaurants, including global fast-food chains that provide a space for the performance of personal identity as well as for the establishment of new socio-cultural hierarchies.

Conclusion: post-communist identities – hybrid of 'old' and 'new'

Post-communist change extends to encompass a fundamental reordering of people's identities, a process that illustrates the complexities of societies experiencing either the end of formal communist structures or their more gradual reformation. This chapter has illustrated the many different and intersecting sources of identity formation and politics that have come into play under conditions of post-communism. There is no simple account or explanation of changing identities in such situations. It is too simplistic to explain new identities as a re-emergence of identities that pre-dated, and were suppressed under, communist systems. While various socialist (communist) societies are often stereotyped as suppressing people's sources of identity, notably national, religious, gender and sexual identities, these sources of identity were treated in a variety of ways in different socialist times and places. Neither were the populations of such regimes always simply the passive recipients of state-sponsored identity-creation programmes.

Nevertheless, the activities of communist states to alter senses of identity did have a significant impact, whether people adopted officially promoted identities or defined themselves in opposition to them. What is important is that attempts to influence identity under socialism have left important legacies that combine in complex ways with new influences in post-communist identity formation. The simplistic 'us' and 'them' conception of identity fostered under communism in some contexts leads into equally simplistic post-communist conceptions of 'self' and 'other', particularly with regard to ethnic and national minorities. In those countries where communism has come to an end, the communist era has left a multitude of resources that contemporary identity cultures can rework and manipulate as a part of producing new post-communist identities. For, example, Yurchak (1999) analyses how post-Soviet youth melded the symbolism of the USSR's space race and key figures such as Yury Gargarin with Western house and techno music and clubbing to create new post-communist youth cultures.

In some contexts, the very process of post-communist transformation itself creates the conditions for producing new identities. Verdery (1996), for example, shows how privatisation of farmland in Romania led to new post-communist conflict based on national differences between Romanian villagers and ethnically German villagers who felt discriminated against in the privatisation process. Yurchak (1999) relates how Russian youth were able to exploit the slow collapse of the Soviet state by using derelict and unsupervised properties as new spaces for creating youth cultures beyond the influence of both the Communist state and the encroaching market. The emergence of new independent post-communist nations with political plurality opens up new arenas in which elites compete to establish particular constructions of national identity for political ends. In other contexts, where communist regimes continue to hold power but the economic context is rapidly changing, such as Cuba and China, the state's influence over identities is also changing, and people are finding relatively more scope for expressing their own identities based on ethnicity, religion, gender and other factors. Rapid restructuring in production, consumption, mass media and labour markets has opened up new spaces for the creation of different forms of contested identities, particularly associated with increasingly (if still relatively limited) globalisation. Newly emerging identities are hybrid, combining new influences from globalised Western culture with pre-communist traditions and the legacies of socialism, shaped in the new contexts produced by post-communist transformation.

10

SUMMARY AND
CONCLUSIONS

From communisms to post-communisms:
global patterns of transformation

Since the collapse of the Iron Curtain, the debate on post-socialisms has been shaped by two distinct features: a strong Europe-/Soviet-centred view, and a fuzziness about the meaning of 'socialism' *per se*. Fundamentally in the shadow of the Iron Curtain, simplistic propaganda of 'East' versus 'West' and communism versus democracy, has shaped the view of, and response to, the collapse of the Soviet Union, and thus the historic epicentre of the 'communist East'. The Cold War propaganda and the subsequent euphoria about the demise of the communist regimes meant that 'western' values, societal arrangements and state structures were deemed the 'natural' choice for the disintegrating CEE states and the Soviet Union. This has directly contributed to the third main characteristic of post-'socialist' transition, that of defining 'transition' as a universal, essentially uniform, process, that requires equally uniform and prescriptive policy responses. The issue of policy and value transfer has thus been a key concern in the process of 'transition' from socialism to post-socialism (however defined). But the realities of these changes have been much more diverse and unpredictable than had been suggested, and expected, in public discourse at the time; this includes geographical variations across the spatial scales national to local. It also includes the realisation that there is more to 'socialism' and 'post-socialism' than the European version, which dominated by far the debate and thus appeared as the embodiment of post-socialist developments *per se*.

This book has set out to expand this geographically limited view to a truly global perspective in recognition of the many different versions of 'socialism', and, indeed, 'communism', and the subsequent 'post' versions, following the demise of the Soviet Union as the second superpower. The emergent 'Chinese model', publicised by the country's spectacular economic rise to public consciousness, highlighted the need for a broader perspective. This includes opening up to the effect and aftermath of 'socialism' as a political and societal-economic factor in the developing world, closely intertwined with the processes and challenges of post-colonial development. The main exception here has been Cuba with its special status as something of a cause célèbre, largely because of its standoff with the US as a David versus Goliath competition.

This broader view has revealed considerable variations in the notions of 'socialism' and its application as an 'actually existing' regime. Thus, while socialism has positive connotations in the post-colonial developing countries, with associations of freedom fights, liberation from foreign domination and control, independence and national autonomy, the situation was quite the opposite in Eastern Europe. There, 'socialism' or 'communism' were associated with foreign (Soviet) domination and control, effectively perceived as a form of colonialism, not just economically and politically, but also culturally and in terms of identity and sense of belonging in the wider European context. Indeed, the very terms 'socialism' and 'communism' vary in their usage, reflecting different conceptual and ideological backgrounds and objectives. While 'socialism' tends to be more aimed at the Marxist philosophy of class struggle towards an egalitarian society with, ultimately, no longer a need for a state to hold up the divisions between the 'haves' and 'have nots', 'communism' is aimed at the Leninist application of the idea through a strong (rather than diminishing) authoritarian state. Communism is thus closely associated with Sovietisation, that is the spread of the Soviet interpretation and application of communism (differing, for instance, from the Chinese version). For eastern European countries, subjugated to Soviet domination, communism, Sovietisation and external domination are thus closely intertwined. And this has fundamentally shaped the nature and development of post-*communism*.

These issues became particularly evident in the Baltic States, for instance, very visibly and intentionally turning their backs to Russia in favour of approaching (western/northern) Europe. And in Poland, the great efforts invested in restoring the destroyed old town centres to their pre-war appearance, rather than building the 'socialist' or 'Soviet' city, can also be read as an act of resistance, seeking to emphasise the European connections and legacies *vis-à-vis* the Russian/Soviet hegemony. As Chapter 2 discussed (see excurse by Zarycki), the very usage of the terms 'socialism' and 'communism' may well be reflecting such programmatic, politically driven rationales, with the term 'communism' used to emphasise a Soviet-style state-dictatorial regime and unwanted domination of other nations. But both terms, while differing considerably in their ideological and practical rationales, draw conceptually on nineteenth-century European industrial society and its particular problems. Outside their geographical reference areas of north-western Europe, they become less immediately relevant. Neither the essentially feudal conditions in early twentieth-century Russia, nor those of non-industrialised, former colonial countries, were taken into consideration. Not surprisingly, therefore, 'socialism', or 'communism' (and the usage of the two terms is not consistent), had to be 'adjusted', whether leading to Leninist Sovietism or African Marxism, Chinese Maoism, or Cuban Castroism. Each brought into the equation their own cultural-historic legacies, as well as the personal ambitions and idealisms of the respective leadership. The simplistic profiling of 'socialism' under the Cold War could thus not represent the whole story. So it is not surprising that the move away from that seemingly standard format 'communist regime' to a 'post-communist' status was portrayed as similarly standardised and predictable. The one-size-fits-all application of the Washington Consensus paradigm, proposing a shift from communist authoritarianism to a liberal market democracy as 'automatic' and universal, illustrates this way of thought.

The notion of 'transition' is much less clear-cut than the wide use of the term may suggest. This does not only apply to the nature of 'transition' itself, that is the process of changes, but also the direction and ultimate end point. The use of the term 'transformation' offers an alternative option, being less heavily loaded with specific meanings and expectations, and thus permitting a less prescriptive under-standing. An alternative to that is the distinction between two phases of 'transition': the initial period of at times fundamental change, followed by a (longer) period of consolidation. If referring to 'democratisation', as frequently it does, it is the second phase that marks the crucial period of establishing a genuine, generally practised, rather than merely formal, government-propagated form of governance. There are thus many different dimensions to the meanings of both 'socialism' and 'transi-tion', and their interpretation and analysis requires a much broader, while also more detailed, view of underlying rationales, historic backgrounds and geographic variations in economic and political relationships and opportunities. There is certainly more to the notions of 'socialism', 'communism' and 'transition' than the Eurocentric, Soviet-focused perspective that dominated the times of the Cold War and has since shaped responses and expectations under 'post-socialism' in a con-tinued ideological-conceptual hegemony.

Despite this seeming universality of post-communist 'transition', past legacies in national identity, state building, economic competitiveness and personalities have had a major impact on the ways in which changes occurred, with different starting and end points. The histories of communism and subsequent 'post-communism' varied considerably over time and place. Post-soviet Russia and the Central Asian Republics, China, Mozambique and Cuba all followed their own paths towards communism, however defined. The departure points for further developments thus varied, and thus post-communism took on many different faces and outcomes. In Central and Eastern Europe, the more than four decades of communism were experienced as a foreign occupation of national territories by the Soviet Union, and the visible presence of Soviet forces across the region made that situation clear to all residents. Inevitably, such outside interference in well-established, historically based nation states was resented, especially as it also meant severing long-established links with the rest of Europe in favour of a colonial-style dependency on Moscow. Overcoming that dependency and breaking out of the Soviet-controlled sphere was thus seen as an act of liberation and re-connection with national histories, identities and Europeanness. The bottom-up nature of the challenges to the communist, Moscow-supported regimes was thus not surprising *per se*, as it reflected underlying resentment and disaffection with national governments that were con-sidered as not 'theirs' but rather Moscow's puppet regimes. The various uprisings in 1953 (East Germany), 1956 (Hungary), 1968 (Czechoslovakia) and 1981 (Poland) reflected the underlying tensions, and only Moscow's forceful intervention could retain the status quo.

The obvious reluctance by the Soviet Union after 1985 to use force as a means of maintaining the empire changed that, and was seen as a green light for public expressions of popular disaffection with the Soviet-backed regimes. Calls for democracy became more audible and confident, and, with western encouragement, succeeded in making it clear to the communist governments *in situ* that they had no legitimacy and no support from either within or without 'their' countries. These bottom-up pressures and public displays of civic interests and demands can well be interpreted as evidence of a (re-)emerging civil society. Indeed, the frequent reference to pre-communist times, including past, if at times only tentative, democratic experiences, is presented as more of a re-emerging, rather than newly emerging, civil society. This is one of the main differences from events elsewhere, and thus the nature of post-communist developments in Europe.

But rather than continuing to embark on an indigenously defined path of post-communist transformation, developments after the initial collapse of the communist regimes became part of the dichotomic Cold War-era divisions between the images of a democratic, liberal West and a communist, dictatorial East. With the latter gone, so the general assumption was, the western model was the only 'show in town' left, and democratisation, liberalisation and marketisation, as symbolised by the Washington Consensus, became the new (and only) realistic paradigm. Early attempts at pursuing a 'third way' of societal-economic arrangement were soon brushed aside. The post-unification developments in eastern Germany illustrate this process of 'westernisation', and despite the fact that the new head of government is from eastern Germany, an outcome also driven by tactical considerations, the whole machinery is clearly western-biased. The continued economic imbalances, requiring considerable ongoing financial transfers from west to east, underpin the effective eastern dependency on the western part of the country.

Post-communism in Central and Eastern Europe has thus been driven by external (western) ambitions to 'bring in' the previously separated East, with the implicit understanding of a 'natural' transition towards western conditions, i.e. emulating the west by leaving behind the apparently failed communist model and everything associated with it. This effective *tabula rasa* approach ignored any underlying differences, especially economic prospects, and projected a rose-coloured image of the seemingly miraculous powers of the liberal market economy. References to the 'economic miracle' in western Europe after the Second World War were used in support of this assumption. Inevitably, this resulted in a correspondingly simplistic transfer of policy measures and governmental structures, again, seeking to extend western European practices, values and ways of doing things to 'the New Europe', and expecting outcomes as seen in the west. From within eastern Europe, this was seen as the logical and most promising, that is the quickest, way to achieve western forms and qualities of life, lifestyles and associated political-economic conditions. Following western ways of doing things thus seemed to many the obvious route to take. It was only when the predicted quick solutions failed to materialised in their expected form and quantity, that questions emerged about particular communist or local/regional legacies, 'disturbing' the simplistically envisaged input–output calculations. The harsh realities of a competitive global market, the dominant power of western capital, and the geographically and socially very uneven distribution of 'winners' and 'losers', right down to the intra-local level, have come to pose major political challenges not just for further economic development, but also for the viability of national consensus and support for the democratic, liberal market model.

Looking at eastern Germany, the issues of policy transfer and associated linear predictions of outcomes become quite obvious, especially in economic development. Under the impact of globalisation, governance structures are judged by their effectiveness in manoeuvring a territory (nation or locality) successfully onto the global stage and attracting a share of presumed footloose capital. Pressure exists, therefore, to employ similar, apparently successful or 'proven' strategies, and it was a liberal, free-market economy that was seen as the quickest and most promising way forward at the end of the 1980s. The importance of key international institutions, such as the World Bank or the IMF, helped to push this agenda truly globally, wherever financial assistance through them was sought. But each formerly communist-run region brings with it specific mixes of constitutional, geographic, historic or cultural legacies, which define *actual* qualities and thus competitiveness, and set the framework for policy-making. This makes it much more difficult to argue for simple policy transfers. The soon evident stark territorial variations in economic development prospects posed a formidable challenge to the utility and credibility of 'western' policy paradigms. These experiences gradually fed back into 'mainstream' discussions of global economic processes and required appropriate policy responses, although the dominant view still seems focused on structure, rather than process, that is political structures, rather than defining and making policies. Given the very dynamic nature of the transformation process, such attempts increasingly look questionable.

The legacies of the socialist period have had an important impact on the further economic and regulative development of the Central and Eastern European countries. While showing basic similarities in the main features of the planned economic geography created under the communist regime, in eastern Germany the situation was somewhat special because of the exceptionally rapid and wholesale change of state regulation. This included seeking to airbrush that period out of history altogether as much as possible. Thus, for instance, the boundaries of the new formal planning regions in the five new *Länder* were in several instances cut in such a way as to obscure their association with their socialist-era predecessors (Herrschel, 2001). References to the former regional capital cities as switchboards of the party machinery's local control were also avoided by giving regions more 'egalitarian' names based on geographic landscapes. In many instances, these are little known outside the area and thus cannot serve as a recognition factor for the region.

The role and rationale of the geography of planning regions was thus guided by 'post-communist' ideological concerns, rather than the pragmatic necessities of competitive economic development policy and management in response to internationally defined opportunities. All this happened *vis-à-vis* the rapidly growing challenges of economic restructuring and resulting geographical unevenness in its course. But there was little immediate concern for that when the new territorial governance was established. Instead, distinctly intra-regional, introspective agendas prevailed, which, particularly in the early years of post-socialism, concentrated on dealing with new-found powers and responsibilities and the disappearance of the old state. Global competition and its differently scaled perspectives did not feature much on the political radar at that time, allowing locally centred interests to prevail. The reason was partly an unawareness of the largely inexperienced, newly empowered local and regional administrations of the new necessities to actively engage in global competition for economic development, and partly petty, historically driven localism following decentralisation and an unfamiliarity with local policy-making autonomy gained as a result of the 'westernisation' of government.

This initial simplistic and somewhat naive understanding was, however, increasingly challenged in the second half of the 1990s, when the fundamental problems of developing a new economic structure became more apparent to the, by then, more 'seasoned' policy makers. The particularly stark differences in development opportunities after the end of communism, have exposed the limitations of the simply imported conventional western models. This contributed to the emergence of an array of different understandings and interpretations of policy requirements and 'best practices'. This was made possible *inter alia* by the growing familiarisation of the new democratic institutions and their policy makers with the post-unification state structures, policy-making mechanisms and, importantly, economic realities. These included the workings of globally operating capitalism. In many places, conventional, 'safe' (western) strategies, with their technocratic and planning-focused perspective, no longer seemed appropriate and adequate. These lessons have had to be learned quickly, but there are variations in this process. The scale and force of transformation has, at least in some quarters, opened up established rigidities and fixed ways of doing things, and that also applies to the international scale, such as European Union policies and funding.

Inequalities in social, economic and state-administrative respect are even more pronounced in the former Soviet Union, where clear divisions emerged between the European- and Asian-oriented parts of the country, overlaying strong urban–rural contrasts. But there is also the way in which the end of communism came about and, therefore, post-communism began in the FSU. In contrast to events in Central and Eastern Europe, top-down implementation of policy changes 'accidentally' led to the collapse of the whole system, rather than its reform. Both the beginning and the end of communism were initiated as an elite-directed process from the top down. There was no public display of demanding change, suggesting a weak civil society. Instead, a strong state overshadowed society (and it still does). The merely marginal attention given to political parties *vis-à-vis* an increasingly powerful presidency reflects the asymmetric relationship between people and government. This, and the sheer duration of communist rule in the former Soviet Union, contributed to the emergence of a particular, Soviet form of communism, and, after its end, post-communism. Multi-ethnicity, multi-nationality with not always clear territoriality, and imperial-style dependencies on Russia as the centre of the Soviet empire, have all been hallmarks of Soviet communism and post-communism. Using the two-phase transition model (see Chapter 3), Russia/the FSU underwent rather dramatic, if protracted, changes immediately after the end of the communist regime, while the subsequent 'consolidation' period shows a rather haphazard and 'messy' picture as far as the presumed processes of democratisation and liberalisation are concerned. This applies in particular to the Central Asian Republics, where autocratic regimes continue to be the norm, often with the former communist leaders, who now sail under the banner of nationalism, still in office.

China and the other communist Asian countries nearby developed a quite different model of communism in its own right and, even more significantly, embarked on their own forms of post-communist development. This includes a rurally peasant-based development of the communist state, rather than an urban industrial focus, and a much lesser degree of centralisation in economic management, and marks a major difference from the Soviet model with its all-embracing central control. China's transformation process is particularly interesting, in so far as it continues to claim adherence to communist principles. But these apply only to the political-governmental sphere, while the economic sphere is opened up to market forces. This systemic duality of 'communism' and 'marketisation' is ideologically rather incompatible, but a strict separation between the two has, so far, allowed communist principles to be combined with quite amazing economic growth. China's (and Vietnam's) transformation thus differs fundamentally from the recession-based changes in eastern Europe and the FSU.

How long this dual track approach can be maintained is, however, not quite clear. Essentially, although still claiming a communist identity, China follows the state-developmental model of the 'Asian Tiger' countries, such as the Philippines or Taiwan. But rather than proclaiming western values, especially with regard to democratic principles, China continues to insist on 'communism' as the official ideology. In effect, therefore, transformation in China is two-track and two-speed, with the political-societal track held at standstill, while the economic track speeds ahead. China's transformation thus includes two dimensions – transformation as a systemic change, as in the economic sphere, and the growing gap between the new economic paradigm and the political status quo. The resulting tensions pose a major potential challenge to the regime.

Africa's post-communist transition, again, looks quite different from both Europe's and Asia's, although it shares some elements of a colonial legacy with the latter. Socialism, and later communism, in Africa, was closely associated with post-colonial independence. Nationalism, African identity and independence from global economic forces were the immediate objectives associated with socialism. But internal divisions along ethnic lines and a strong role of tribal identities and loyalties meant a much reduced scope for nationalism to act as a rallying point and common reference point for determined national development projects, as, for instance, in Latin America. Promoting a national cause as unifying driver in the cause of developing a socialist, egalitarian society was thus much less effective in galvanising people and making them put up with economic hardship in Africa, than it was in stronger nation-based cultures as found in Asia or Latin America.

Increasingly, therefore, with a less euphoric and more realistic assessment of conditions and scope for developing a strong cause, the idealist 'African socialism' gave way to a more party-focused, Soviet-oriented, functionalist 'communist' approach. This was driven less by idealistic ambitions than by a pragmatic use of communist (Leninist) ideology to justify a dictatorial regime, often with military leaders who came to power through army coups. Reference to communism offered a convenient justification for the maintaining of an authoritarian, centrally controlled state. It also offered the opportunity to gain access to Soviet economic and military aid as part of the East–West competition for influence in Africa. The collapse of the Soviet Union removed this rationale, leaving adherence to the western paradigm of a liberal market democracy the only available avenue to international support. There was no more 'mileage' to be gained from claiming to be 'communist' – quite the opposite. Consequently, with a handful of exceptions where communism was pursued in a more ideologically based way, the label 'communist' or 'socialist' was dropped without much resistance, although that did not mean an automatic shift towards democratisation. As in the Central Asian republics, nationalism, with distinct ethnic undertones, was the new justification for dictatorial leaders to retain power, albeit with some democratic 'noises' to satisfy western aid agencies and governments. Economic pressures and the absence of external support eventually also brought down the more diehard communist regimes, such as Angola or Ethiopia. The post-communist period in Africa has thus been much less dramatic than in Central and Eastern Europe or, indeed, the Soviet Union. Regime change has been limited, with most of the established political elites carrying on, irrespective of the 're-labelling' of state names and party names. Socially and economically, there was little change either, with a continuation of post-colonial economic dependency and 'underdevelopment'.

Finally, Latin America, and here first and foremost Cuba, almost as a cause célèbre, offers different variations on the implementation of socialism/communism, and the adjustment to the end of the Cold War and disappearance of the Soviet Union. Latin America's geographic position, next to the United States, has meant that the US has viewed Latin America effectively as its 'backyard' and thus taken a hegemonic interest in political and economic developments there. Colonial-style economic (and political) dependencies, with varying degrees of visibility, are thus the usual pattern. Cuba's resistance to this hegemonic arrangement, claiming adherence to communism right under the nose of the US government, has until today added a particular dimension to its situation. Other attempts at following a socialist path, such as in Nicaragua, have failed not least through external involvement. Indeed, projecting itself as David in the fight against the US Goliath has added credibility and support to Fidel Castro's regime until today, especially also among many Latin Americans who resent US domination of their affairs. The combination of anti-colonial struggle, defence of national autonomy and independence, and a Latin preference for strong men in leadership, often with a military background, but also a distinct personal charisma of the leader, all combined to retain internal support for a regime that has struggled to maintain, let alone improve, the quality of people's daily lives.

The fact that the revolutionary generation is still in power gives the regime added legitimacy, especially in the eyes of the generation of that time. This is different from the situation in Africa, where a new generation of leaders had accessed power, with no connection to the anti-colonial struggles. In Cuba, communism has thus a strong revolutionary, nationalistic undertone, and this is now increasingly emphasised by the Cuban government, rather than the virtues of communism *per se*. The dramatic economic problems immediately after the end of the Soviet Union as Cuba's main economic sponsor has made such shift in emphasis necessary. Promising a 'better' life was less credible under such conditions than 'defending freedom'.

New economic realities have meant that concessions to capitalism, as the only realistic source of urgently needed investment, were needed. With one eye on China, Cuba has adopted a dual economy approach, encouraging, while limiting, foreign investment to clearly demarcated areas, especially for tourism developments. Such enclaves, effectively gated communities, accept the US dollar as legal tender, thus highlighting their separate, almost exterritorial nature. But this dollarisation of parts of the economy has also created new social divisions between those with access to foreign currency, and thus much improved scope for obtaining consumer goods, and those without. This division challenges one of the main pillars of Cuban communism, that of an egalitarian society. Reducing that in importance and visibility will inevitably challenge the credibility and legitimacy of communism as official state doctrine. Cuba's new geopolitical position, with the predominant East–West axis, which made Cuba appear as Moscow's western outpost, having given way to a stronger emphasis on the North–South relationship, has meant a greater emphasis on its Latin American belonging as part of a shift from Cuban 'communism' to 'nationalism'. By re-emphasising its Latin belonging, Cuba also strengthens its iconic role for other Latin American states wishing to emulate its success of maintaining independence and national autonomy *vis-à-vis* the North American hegemon. Effectively, this provides Cuba with a protective umbrella against external attempts at facilitating regime change. This may be one explanation why there has been no obvious attempt at actively changing the regime.

Looking across all countries claiming to have followed a socialist or communist agenda and now undergoing a market-oriented transformation, then, the issue of inequality and unevenness in the outcome of the changes is the most prevailing, be it based on territorial, societal or social factors such as identity or social status. In fact, two main sets of factors may be identified as underpinning and shaping the phenomenon of post-communist (post-socialist) transition (Table 10.1): structure and territory on the one hand, and the role of processes and relationships on the other. The former set includes territory both as actual, physical and as 'virtual' imagined space of belonging and engagement, separated by borders and dividing lines, again, both real and imagined. The issue of identity and sense of belonging, including claims to territorial representation and ownership, are part of this (see also Chapter 9). This includes ideological and idealistic objectives and programmes, such as Africanism or Cubanism.

Past histories, including the experience of colonialism, and the ways in which the communist regime was established, have had a considerable impact on the nature and progress of post-communist transition. Imposed by an external, occupying

Table 10.1 Key determinants of shape and outcome of post-socialist transition

Role of territory and structure	Role of process and relationships
– economic inequality: ('Special Economic Zones', tourism enclaves, urban rural contrasts)	– democratisation (Washington Consensus) – liberalisation
– territorial reference of identity (national to local)	– personalities in leaderships, and their networks
– previous experience with democracy, legacies of governmental practice	– policy transfer and degree of 'adjustment' to the new conditions (learning process)

force, such as in eastern and central Europe or parts of the former Soviet Union, it led to an association of communism with neo-colonial features, challenges to national identity and autonomy. This Sovietisation resulted in a completely negative view of socialism as something unwanted and alien to established political culture and national ways of doing things. Referring to the term 'communism', rather than 'socialism', has become an expression of rejection of this unwanted Soviet interference with national life, and the subsequent experience of political, economic and cultural impotence and inadequacies.

Post-communism, therefore, was meant to be the very opposite – national self-affirmation, often leading to quite explicit nationalism, a free market in contrast to the planned market, and generally as little control and planning as possible. In fact, the very word planning, including development planning for settlements, infra-structures and so on, had become a 'dirty' word in the early 1990s. Only now, the pendulum has swung back from that extreme to a more intermediary, conciliatory position. Anything reminiscent of the communist-era practices of governance was rejected – even the names of administrative territories, such as regions and, in some cases, cities (such as former Karl-Marx-Stadt in eastern Germany, or Leningrad, in Russia). In some of the Central and Eastern European countries, such as the Czech Republic, the whole administrative scale of 'regions' was abolished as representing too much of communist-era structures. Only the requirements of the European Union led to a re-introduction of regions, albeit in a heavily revised form to emphasise the break with the communist past.

The other main set of factors in the 'territory and structure' category includes economic inequality and unevenness in the outcomes of marketisation, whether the deliberate result of policies, as in Cuba or China, or incidental results of selective investment by (western) capital, as in eastern Europe. Competitive pressures by a globalising capital market have affected all countries with a legacy of a planned (socialist) economy. The need to open up to these pressures and come out of the protective, state-controlled environment of the communist era has meant *de facto* abandoning the notion of an egalitarian society, even if clinging to its portrayal in public political discourse, as in China and Cuba with their effective duality between society and economy, and within the economy, respectively.

Elsewhere, all communist-era rhetoric has been abandoned, be it because of it being despised as a reflection of imposed alien structure, as in eastern Europe, or for pragmatic reasons, as in Africa, as the only source of development aid is now western organisations, and being 'communist' reduces the chances of gaining support from there. These economic inequalities stretch across all scales, national to local, and are particularly visible and stark in contrast at the local level, whether in eastern European cities promoting a westernised, 'fairytale' city centre to incoming tourists and business people, or state-imposed enclaves of gated tourist compounds and 'islands' of capitalism, as in Cuba and China. How far, and for how long, these visibly different qualities of life can be kept apart without political repercussions remains to be seen.

The second main set of transition-shaping factors is that of processes and relationships. Here sit the key pillars of the Washington Consensus, democratisation, liberalisation and marketisation, and, as part of that, the whole issue of policy transfer. But there are also the roles of personalities, their ambitions and networks. They all have been crucial in shaping the transformation processes and their outcomes so far. These factors have come into effect after the initial abandonment of the communist regime – either in whole or part. They are thus situated in the second, the consolidation phase of transformation, especially with regard to democratisation. While market reforms can be implemented very quickly, and new rules of making business put in place, it is much more difficult for a sense of popular ownership of the state and public policy to emerge. In some instances, this is actively inhibited, such as in China or Cuba, which seek to maintain the political status quo while allowing only the economy to change. Elsewhere, as in Russia, earlier established, albeit tentative and incomplete, practices of a democratic society are slowly being rolled back. Most progress in terms of democratisation has been made in those countries where the collapse of communism was initiated by popular pressure in the first place, as in eastern Europe. The transformation process was thus much more 'owned' by the people than in those countries where change was initiated and managed from 'above' without popular involvement.

But despite these key differences, the dominance of the ideological juggernaut of the Washington Consensus has had a fundamental impact on post-communist developments, whether as a direct or indirect result. The absence of a credible alternative to the western model of market economy has meant that all countries had to embark on emulating market principles, albeit in various forms. But global capitalism set the rules. Central and Eastern European countries embraced this new capitalist structure, not least as a sign of rejecting communism, but also as part of a 'returning to the pre-communist days' move. And a free market seemed to offer all those things previously only heard of as hearsay. African communist countries had little choice. Their continued economic and military dependency meant that marketisaton was the only option, as it was a condition imposed by western international agencies. There was no more alternative source in Moscow. The result has been a complete abandonment of all references to communism, unlike in the few countries seeking to retain some of the communist spirit by separating marketisation from the political-societal sphere. This is particularly the case in China and some neighbouring countries seeking to emulate China's policies, such as Vietnam. In Cuba, this strategy operates at a much smaller scale, effectively creating western 'capitalist ghettos' in the setting of a struggling centrally planned economy. This is accompanied by ideological campaigns and reasoning to explain and justify the coexistence of such conceptually mutually exclusive systems of market and state control, with the former generating and emphasising inequalities, and the latter seeking to achieve an egalitarian society.

How long these contradictions and uneven results can continue to be explained and justified by the incumbent regimes, remains to be seen. Certainly, there is no one model of transition, transformation, post-socialism or post-communism, as Table 10.2 shows. The five global regions investigated here show distinct features shaping their respective legacies of communist values, whether positive in conjunction with post-colonial liberation and national emancipation (Africa, Latin America, Asia), or negative, as part of a neo-colonial dependency and external domination (CEE). Not surprisingly, this has resulted in differing modes of initiation of, and engagement with, the end of communist regimes. The degree of popular ownership of these developments emerged as a key factor; that is, the distinction between top-down and bottom-up causation of the shift towards post-communism and the subsequent popular involvement with the ensuing developments.

Processes of change vary between places and over time, and there is certainly no end to this process as yet. While formalities are easy to change and adjust, it is the underlying values, 'hang-ups' and concerns that continue to impact on the transformation process of a continued approximation of conditions, whether genuinely desired or accepted as the inescapable *sine qua non* of the current world

Table 10.2 Patterns of post-socialist transition (overview)

Global region of post-socialist 'transition'	Type of transition	Typical features
Central and Eastern Europe	– all-out shift towards liberal market economy as antidote to communist era – varying speeds between 'shock therapy' and 'gradualism' – notion of 'catching up' and re-joining Europe – clear agenda of leaving communist era behind – end of communist regimes facilitated through bottom-up pressures (civil society), people were stakeholders of changes	– attempt at airbrushing communist period out of history – referring to pre-communist history of civil society and democratisation (Central Europe) – emphasising European tradition *vis-à-vis* Russia (looking west) – socialist period viewed as imposed 'anomaly' under a form of colonial dependency – socialism holds decidedly negative connotations
Former Soviet Union	– hesitant (stagnant) change without clear agenda – collapse of Communist regime was 'accidental' through reforms 'gone wrong' – changes initiated top-down, not much involvement of public – differing outomes across former SU (democratisation, authoritarianism)	– trauma of loss of empire (post-imperial 'blues' and unclear identity) – new nationalisms and identities among former Soviet states – question marks over democratisation – re-emerging ethnic identities challenge inherited territoriality *Continued*

237

Table 10.2 Continued

Global region of post-socialist 'transition'	Type of transition	Typical features
Asia and Latin America (especially China, Vietnam, Cuba)	– strictly centrally controlled partial reforms (*economic* system only) – search for a 'third way' between marketisation and maintaining the government-societal status quo – applying dual economy with selective dollarisation (Cuba) or designated Special Economic Zones (China) – shifting emphasis: nationalism, rather than socialism in Cuba – socialism has positive connotations of post-colonial revolutionary liberation	– balancing liberal free market and socialist principles in two parallel worlds – inevitably creating conditions for social inequality (access to resources) – seeking to maintain alternative position/ paradigm in unipolar (western) world – adopting modified version of developmental policy of 'Asian Tiger' states (China, Vietnam)
Africa	– socialism has credibility as 'revolutionary' struggle against colonialism – initial post-colonial euphoria about forging pan-Africanism with socialist values soon proved too idealistic ('African Socialism') – socialism increasingly merely a convenient label for authoritarian regimes. – 'de-based' socialism (end of bloc affiliation) was simply abandoned by regimes seeking to gain western economic support – post-colonial dependency issues shared with other African countries with or without past socialist ambitions.	– colonialist inheritance presents socialism as antidote to colonialist capitalism – socialism as expression of national empowerment and identities. – socialism not rejected for ideological but purely pragmatic (economic) reasons – ethnic divisions outweigh sense of national belonging, as state territories cut through ethnic lines, and make national mobilisation for joint goal difficult

order. The outcome, however, is far from clear, and projecting them as a linear extension of today's or past conditions is unlikely to match reality.

This multitude of 'socialisms' and 'post-socialisms' may be seen as the third phase in the development of the understanding of the 'post-socialist condition' (Herrschel 2001). This follows the two previous phases – with the first phase of the early 1990s dominated by a rather simplistic, universalist view of the nature of post-socialism as a standardised process (epitomised by the Washington Consensus doctrine), and the second phase, in the later 1990s, when geography was added by admitting differing paths of post-socialist transformation (to use the term less prescriptive than 'transition'). Increasingly, therefore, the initial all-embracing neo-liberal discourse had to be revised as the continued impact of territorially and societally specific legacies of the applied forms of Marxism–Leninism became apparent. This has added to the challenges of studying post-socialism, because it is not merely restricted to the formal features of systemic and regime change, but stretches to social, political and economic values, held by the people in these areas. This is much more difficult to examine and interpret and much less clearly identifiable and predictable in its likely outcome than the structural-institutional characteristics.

BIBLIOGRAPHY

Adler, G. and Webster, E. (1995) Challenging transition theory: the Labour Movement, radical reform and the transition to democracy in South Africa. *Politics and Society*, 23 (1): 75–106.

Agadjanian, A. (2001) Public religion and the quest for national ideology: Russia's media discourse. *Journal of the Scientific Study of Religion*, 40 (3): 351–65.

Agarwal, C. (2004) Cuba's path to a market economy: Washington Consensus, Doi Moi, or Reforma à la Cubana? *ASCE (Association for the Study of the Cuban Economy) Annual Proceedings*, 14: 312–24. Available at: http://lanic.utexas.edu/project/asce/publications/proceedings/

Âgh, A. (1998) *The Politics of Central Europe*. London: Sage.

Âgh, A. (1999) Processes of democratisation in the East Central European and Balkan states: sovereignty-related conflicts in the context of Europeanisation. *Communist and Post-Communist Studies*, 32: 263–79.

Aguila, J. M. del (1984) *Cuba: Dilemmas of a Revolution*. London: Westview Press.

Ahl, R. (1999) Society and transition in post-Soviet Russia. *Communist and Post-Communist Studies*, 32: 175–39.

Akaev, A. (1998) Jibek jolynyn diplomatiyasï (Kïrgïz Respublikasïnïn Prezidenti Askar Akaevdin doktrinasï) [Silk Road Diplomacy (The doctrine of the president of the Kyrgyz Republic, Askar Akaev)]. *Kïrgïz Tuusu*, 27–29/10/1998.

Albright, D. (1980a) Moscow's African policy in the 1970s. In: D. Albright (ed.), *Communism in Africa*, pp. 35–66. London: Indiana University Press.

Albright, D. (ed.) (1980b) *Communism in Africa*. London: Indiana University Press.

Aligica, P. (2003) *Neoclassical Economics and the Challenge of Transition: Lessons and Implications of the Eastern European Economic Reform Experience*. IWM (Institut für die Wissenschaft vom Menschen/Institute for Human Sciences), Vienna. Working Paper No. 3.

Allen, R. (1992) *Farm to Factory: A Reinterpretation of Soviet Industrial Revolution*. Princeton, NJ: Princeton University Press.

Altvater, E. (1993) *The Future of the Market: An Essay on the Regulation of Money and Nature after the Collapse of 'Actually Existing Socialism'*. London: Verso.

Amaro, N. (1996) Decentralization, local government and citizen participation in Cuba. *ASCE (Association for the Study of the Cuban Economy) Annual Proceedings*, 6: 262–82. Available at: http://lanic.utexas.edu/project/asce/publications/proceedings/

Amaro, N. (2000) Models of development and globalization in Cuba: Cuba in transition. *ASCE 2000*: 277–88.

Anderson, J. (1997) *The International Politics of Central Asia: Regional International Politics*. Manchester: Manchester University Press.

Anderson, L. (1999a) Introduction. In: L. Anderson (ed.), *Transitions to Democracy*, pp. 1–13. New York: Columbia University Press.

Anderson, L. (ed.) (1999b) *Transitions to Democracy*. New York: Columbia University Press.

Anheier, H.-K., Priller, E. and Zimmer, A. (2000) Civil society in transition: the East German third sector ten years after unification. *Civil Society Working Paper 15*, available at: www.lse. ac.uk/collections/CCS/pdf/CSWP_15.pdf (accessed 8 May 2005).

Appel, H. (2001) Corruption and the collapse of the Czech transition miracle. *East European Politics and Societies*, 15(3): 528–53.

Artisien-Maksimenko, P. (ed.) (2000) *Multinationals in Eastern Europe*. Basingstoke: Macmillan.

Ash, T. G. (2005) How the dreaded superstate became a commonwealth. *The Guardian*, 6 October 2005: 27.

Ashwin, Sarah (2002) 'A woman is everything.' The reproduction of Soviet ideals of womanhood in post-communist Russia. In: A. Rainnie, A. Smith and A. Swain (eds), *Work, Employment and Transition: Restructuring Livelihoods in Post-Communism*, pp. 117–33. London: Routledge.

Aydingdün, A. (2002) Creating, recreating and redefining ethnic identity: Ahiska/ Meskhetian Turks in Soviet and post-Soviet contexts. *Central Asian Survey*, 21 (2): 185–97.

Azicri, M. (2001) *Cuba Today and Tomorrow. Reinventing Socialism*. Tampa: University of Florida Press.

Babu, A. M. (1981) *African Socialism or Socialist Africa?* London: Zed Press.

Bachtler, J., Downes, R. and Gorzelak, G. (2000a) Introduction: challenges of transition for regional development. In: J. Bachtler, R. Downes and G. Gorzelak (eds), *Transition, Cohesion and Regional Policy in Central and Eastern Europe*, pp. 1–10. Series EPRC Studies in European Policy. Aldershot: Ashgate.

Bachtler, J., Downes, R. and Gorzelak, G. (eds) (2000b) *Transition, Cohesion and Regional Policy in Central and Eastern Europe*. Series EPRC Studies in European Policy. Aldershot: Ashgate.

Baer, M. (1993) Profiles in transition in Latin America and the Caribbean. *Annals of AAPSS*, 526 (March): 47.

Baer, M. D. and Weintraub, S. (eds) (1994) *The NAFTA Debate: Grappling with Unconventional Trade Issues*, 3–33. London: Lynne Rienner Publishers.

Baev, P. (1996) A new look at Russia in transition. *Journal of Peace Research*, 33 (3): 371–6.

Balcerowicz, L. (1991) Problems with the definition of socialism in today's world. In: O. Sik (ed.), *Socialism Today? The Changing Meaning of Socialism*, pp. 65–74. New York: St Martin's Press.

Balcerowicz, L. (1995) *Socialism, Capitalism, Transformation*, pp. 19–27. Budapest: Central European University Press.

Banchoff, T. (1999) National identity and EU legitimacy in France and Germany. In: T. Banchoff and M. P. Smith (eds), *Legitimacy and the European Union*, pp. 180–98. London, Routledge.

Baranovitch, N. (2001) Between alterity and identity: new voices of minority people in China. *Modern China*, 27: 359–401.

Barany, Z. and Volgyes, I. (1995) *The Legacies of Communism in Eastern Europe*. Baltimore, MD: Johns Hopkins University Press.

Barker, A. M. (1999) The culture factory: theorizing the popular in the Old and New Russia. In: A. M. Barker (ed.), *Consuming Russia: Popular Culture, Sex and Society since Gorbachev*. Durham, NC: Duke University Press.

Bartlett, D. (2001) Economic development in the newly independent states: the case for regionalism. *European Journal of Development Research*, 13 (1): 135–53.

Bastian, J. (ed.) (1998) *The Political Economy of Transition in Central and Eastern Europe*. Aldershot: Ashgate.

Bates, R., Rui, J., de Figureiredo, J. and Weingast, B. (1998) The politics of interpretation: rationality, culture and transition. *Politics and Society*, 26 (4): 603–42.

Batt, J. (2001) European identity and national identity in Central and Eastern Europe. In: H. Wallace (ed.), *Interlocking Dimensions of European Integration*, pp. 247–62. Basingstoke: Palgrave.

Bax, Mart (2000) Planned policy or primitive Balkanism? A local contribution to the ethnography of the war in Bosnia-Herzegovina. *ETHNOS*, 65 (3): 317–40.

Bell, D. (1995) Pleasure and danger: the paradoxical spaces of sexual citizenship. *Political Geography*, 14(2): 139–53.

Berdahl, D. (2000) Introduction. In: D. Berdahl, M. Bunzl and M. Lampland (eds), *Altering States: Ethnographies of Transition in Eastern Europe and the Former Soviet Union*, pp. 1–13. Ann Arbor, MI: University of Michigan Press.

Berg, E. and Oras, S. (2000) Writing post-Soviet Estonia on to the world map. *Political Geography*, 19: 601–25.

Berki, R. N. (1988) *The Genesis of Marxism: Four Lectures*. London: Dent

Betancourt, R. (2004) *The Role of the State in a 'Democratic' Transition: Cuba*. Cuba Transition Project. Miami, FL: University of Miami.

Bicanic, Ivan (1995) The economic causes of new state formation during transition. *East European Politics and Society*, 9 (1): 2–21.

Billig, Michael (1995) *Banal Nationalism*. London: Sage

Black, G. (1981) *Triumph of the People: The Sandinista Revolution in Nicaragua*. London: Zed Books.

Blanchard, Olivier (1997) *The Economics of Post-Communist Transition*. Oxford: Clarendon Press.

Blecher, Marc (2003) *China against the Tides: Restructuring through Revolution, Radicalism and Reform*. London: Continuum.

Bojkov, V.D. (2004) Neither here, nor there: Bulgaria and Romania in current European politics. *Communist and Post-Communist Studies*, 37: 509–22.

Bönker, F., Müller, K. and Pickel, A. (2002a) Cross-disciplinary approaches to postcommunist transformation: context and agenda. In: F. Bönker, K. Müller, and A. Pickel (eds), *Postcommunist Transformation and the Social Sciences: Cross-Disciplinary Approaches*, pp. 1–38. Oxford: Rowman and Littlefield.

Bönker, F., Müller, K. and Pickel, A. (eds) (2002b) *Postcommunist Transformation and the Social Sciences: Cross-Disciplinary Approaches*. Oxford: Rowman and Littlefield.

Boukhalov, O. and Ivannikov, S. (1995) Ukrainian local politics after independence. *Annals of the American Academy, AAPSS*, 540 (July): 125–36.

Boycko, M., Shleifer, A. and Vishny, R. (1996) *Privatizing Russia* (2nd edn). London: MIT Press.

Bozóki, A., Körösényi, A. and Schöpflin, G. (eds) (1992) *Post-Communist Transition. Emerging Pluralism in Hungary*. London: St Martin's Press.

Bradshaw, M. and Prendergrast, J. (2005) The Russian heartland revisited: an assessment of Russia's transformation. *Eurasian Geography and Economics*, 46 (2): 83–122.

Bradshaw, M. and Stenning, A (2000) The progress of transition in East Central Europe. In: J. Bachtler, R. Downes and G. Gorzelak (eds), *Transition, Cohesion and Regional Policy in Central and Eastern Europe*, pp. 11–32. Series EPRC Studies in European Policy. Aldershot: Ashgate.

Bradshaw, M. and Stenning, A. (2004) Introduction: transformation and development. In: M. Bradshaw and A. Stenning (eds), *East Central Europe and the Former Soviet Union: The Post-Socialist States*, pp. 1–32. Harlow, Pearson.

Bringa, Tom (1995) *Being Muslim the Bosnian Way. Identity and Community in a Central Bosnian Village*. Princeton, NJ: Princeton University Press.

Brown, J. (1994) *Hopes and Shadows. Eastern Europe after Communism*. Durham, NC: Duke University Press.

Brown, J. D. and Earle, J. (2003) *Economic Reforms and Productivity-Enhancing Reallocation in the Post-Soviet Transition*. Centre for Economic Reform and Transformation, Discussion Paper 2004/4. Edinburgh: Herriot Watt University. Available at: www.sml.hw.ac.uk/cert

Brudny, Y. (1997) Neoliberal economic reform and the consolidation of democracy in Russia: or why institutions and ideas might matter more than economics. In: K. Dawisha (ed.), T*he International Dimension of Post-Communist Transitions in Russia and the New States of Eurasia*, pp. 278–97. London: ME Sharpe.

Brundenius, C. (1981) Growth with equity: the Cuban experience. *World Development*, 9 (11/12): 1083–96.

Brundenius, C. and Gonzales, P. M. (2001): The future of the Cuban model: a longer view. In: C. Brundenius and J. Weeks (eds), *Globalization and Third World Socialism*, pp. 129–52. Basingstoke: Palgrave.

Brundenius, C. and Weeks, J. (eds) (2001) *Globalization and Third World Socialism*. Basingstoke: Palgrave.

Brus, W. (1991): Socialism: the very concept under scrutiny. In: O. Sik (ed.), *Socialism Today? The Changing Meaning of Socialism*, pp. 47–57. New York: St Martin's Press.

Bruszt, L. (1992) Transformative politics: social cost and social peace in East Central Europe. *East European Politics and Societies*, 6 (1): 55–72.

Bruton, M., Bruton, S. and Li, Y. (2005) Shenzhen: coping with uncertainties in planning. *Habitat International*, 29: 227–43.

Buchovski, M. (2003) Redefining social relations through work in a rural community in Poland, *Max-Planck-Institute for Social Anthropology Halle*, Working Paper No. 58. Available at: www.eth.mpg.de/pubs/wps/pdf/mpi-eth-working-paper-0058.pdf (accessed December 2004).

Buckley, M. (1997) Victims and agents: gender in post-Soviet states. In: M. Buckley (ed.), *Post-Soviet Women: From the Baltic to Central Asia*, pp. 3–16. Cambridge: Cambridge University Press.

Bunce, V. (1995) Should transitologists be grounded? *Slavic Review*, 54 (1): 111–27.

Burawoy, M. (1994) Why coupon socialism never stood a chance in Russia: the political conditions of economic transition. *Politics and Society*, 22 (4): 585–94.

Burawoy, M. (2001) Transition without transformation: Russia's involuntary road to capitalism. *East European Politics and Societies*, 15 (2): 269–90.

Burawoy, M. and Verdery, K. (1999) *Uncertain Transition: Ethnographies of Change in the Postsocialist World*. Lanham, MD: Rowman & Littlefield.

Carothers, T. (2002) The end of the transition paradigm. *Journal of Democracy*, 13 (1): 5–21.

Carson, C. (1990) *Basic Communism: Its Rise, Spread and Debacle in the 20th Century*. Wadley, AL: American Textbook Committee.

Carter, F. W. and Maik, W. (eds) (1999) *Shock-Shift in an Enlarged Europe: The Geography of Socio-Economic Change in East-Central Europe After 1989*. Aldershot: Ashgate.

Cartier, C. (2001) *Globalizing South China*. Oxford: Blackwell.

Castillo, O. and Gaspar, N. (2002) Tourism development. Locomotive for the Cuban economy. *Revista: Harvard Review of Latin America*, Winter: 76–8. Available at: www.fas. harvard.edu/~drclas/publications/revista/Tourism/castillo.htm

Centeno, M. (2004) The return of Cuba to Latin America. Keynote speech, Annual Conference of the Society of Latin American Studies, Leiden, The Netherlands, 2–4 April 2004.

Cha-He, R., Wen-Yan, L., Linge, G. and Forbes, D. (1997) Linking the regions: a continuing challenge. In: G. Linge (ed.), *China's New Spatial Economy: Heading Towards 2020*, pp. 46–72. Oxford: Oxford University Press.

Chandler, A. (1998) *Institutions of Isolation: Border Controls in the Soviet Union and Its Successor States, 1917–1993*. London: McGill-Queen's University Press.

Chazan, N., Lewis, P., Mortimer, R., Rothschild, D. and Stedman, S. (1999) *Politics and Society in Contemporary Africa* (3rd edn). Boulder, CO: Lynne Riener Publishers.

Chen, K., Jefferson, G. H. and Singh, I. (1992) Lessons from China's economic reform. *Journal of Comparative Economics*, 16, (2): 201–25.

Chen, P. (1993) China's challenge to economic orthodoxy: Asian reform as an evolutionary, self-organizing process. *China Economic Review*, 4 (2): 137–42.

Chirot, D. (ed.) (1989) *The Origins of Backwardness in Eastern Europe*. Berkeley: University of California Press.

Christensen, P. (1998) Socialism after communism: the socioeconomic and cultural foundations of Left politics in post Soviet Russia. *Communist and Post-Communist Studies*, 31 (4): 345–7.

Cirtautas, A. (1995) The post-Leninist State: a conceptual and empircal examination. *Communist and Post-Communist Studies*, 28 (4): 379–92.

Clapham, C. (1992) The socialist experience in Ethiopia and its demise. *Journal of Communist Studies*, 8 (2): 105–25.

Cliffe, L. (1972) Tanzania – socialist transformation and party development. In: L. Cliffe and S. Saul (eds), *Socialism in Tanzania: An Interdisciplinary Reader. Vol. 1 Politics*, pp. 266–76. Nairobi: East African Publishing House.

Cliffe, L. and Saul, S. (eds) (1972) *Socialism in Tanzania: An Interdisciplinary Reader. Vol. 1 Politics*. Nairobi: East African Publishing House.

Colantonio, A. (2004) Tourism in Havana during the Special Period: impacts, residents' perceptions, and planning issues. *ASCE (Association for the Study of the Cuban Economy) Annual Proceedings*, 14: 20–42. Available at: http://lanic.utexas.edu/project/asce/publications/proceedings/

Comisso, E. (1991) Property rights, liberalism, and the transition from 'actually existing' socialism. *East European Politics and Societies*, 5 (1): 162–88.

Cornell, Svate E. (1999) The devaluation of the concept of autonomy: national minorities in the former Soviet Union. *Central Asian Survey*, 18 (2): 185–96.

Corr, E. (1995) Societal transformation for peace in El Salvador. *Annals of the AAPSS*, 541: 144–56.

Coulson, A. (1979a) Introduction. In: A. Coulson (ed.), *African Socialism in Practice: The Tanzanian Experience*, pp. 1–15. Nottingham: Spokesman.

Coulson, A. (ed.) (1979b) *African Socialism in Practice: The Tanzanian Experience*. Nottingham: Spokesman.

Crespo, N. and Díaz, S. (1997) Cuban tourism in 2007: economic impact. *ASCE (Association for the Study of the Cuban Economy) Annual Proceedings*, 7: 150–1. Available at: http://lanic. utexas.edu/project/asce/publications/proceedings/

Crespo, N. and Suddaby, C. (2000) A comparison of Cuba's tourism industry with the Dominican Republic and Cancún, 1988–1999. *ASCE (Association for the Study of the Cuban*

Economy) Annual Proceedings, 10: 321–36. Available at: http://lanic.utexas.edu/project/asce/publications/proceedings/

Crotty, J. (2003) Managing civil society: democratisation and the environmental movement in a Russian region. *Communist and Post-Communist Studies*, 36: 448–508.

Curtis, G. (1998) *Cambodia Reborn: The Transition to Democracy and Development*. Washington, DC: Brookings Institute Press.

Dabrowski, M. and Antczak, R. (1997) Economic transition in Russia, Ukraine, and Belarus: a comparative analyis. In: B. Kaminski (ed.), *Economic Transition in Russia and the New States of Eurasia*, pp. 42–80. London: ME Sharpe.

Dacosta, M. and Carroll, W. (2001) Township and village enterprises, openness and regional economic growth in China. *Post-Communist Economies*, 13 (2): 229–41.

Daniel, C. and Reid, J. (1998) Observations on the relations between the transition of Eastern European/Central Asian economies, economic growth, and direct foreign investment. *Social Science Journal*, 35 (3): 455–9.

Davis, Deborah S. (2000) Introduction: a revolution in consumption. In: Deborah S. Davis (ed.) *The Consumer Revolution in Urban China*. Berkeley, CA: University of California Press.

Davis, H. and Scase, R. (1985) *Western Capitalism and State Socialism: An Introduction*. Cambridge: Basil Blackwell.

Dawisha, K. and Parrott, B. (eds) (1997) *The Consolidation of Democracy in East-Central Europe*. Cambridge: Cambridge University Press.

Dawisha, K. and Turner, M. (1997) The interaction between internal and external agency in post-communist transitions. In: K. Dawisha (ed.), *The International Dimension of Post-communist Transitions in Russia and the New States of Eurasia*. London: ME Sharpe.

Dawson, J. (2001) Latvia's Russian minority: balancing the imperatives of regional development and environmental justice. *Political Geography*, 20: 787–815.

de Melo, M., Denizer, C. and Gelb, A. (1996) *From Plan to Market: Patterns of Transition*. World Bank Policy Research Working Paper No. 1564. Available at ssrn:http://ssrn.com/abstract=604909

Démurger, S., Sachs, J., Woo, T., Bao, S. and Chang, G. (2002) Geography, economic policy, and regional development in China. *Asian Economic Papers*, 1:1 (MIT): 146–97.

Denich, Bette (1994) Dismembering Yugoslavia: nationalist ideologies and the symbolic revival of genocide. *American Ethnologist*, 21 (2): 367–90.

Denitch, Bogdan (1996) *Ethnic Nationalism: The Tragic Death of Yugoslavia*. Minneapolis, MN: University of Minneapolis Press.

Diaz-Briquets, S. (2000) Role of the United States and international lending institutions in Cuba's transition. *Studies in International Comparative Development*, 34 (4): 73–86.

Dingsdale, A. (1999) Redefining 'Eastern Europe': a new regional geography of post-socialist Europe? *Geography*, 84 (3): 204–22.

Dingsdale, A. (2002) *Mapping Modernities: Geographies of East and Central Europe, 1920–2000*. London: Routledge.

Dryzek, J. and Holmes, L. (2002) *Post-Communist Democratization: Political Discourse across Thirteen Countries*. Cambridge: Cambridge University Press.

Duijzings, Ger (2000) *Religion and the Politics of Identity in Kosovo*. London: Hurst and Company.

Duke, V. and Grime, K. (1997) Inequality in post-communism. *Regional Studies*, 31: 883–90.

Dunford, M. and Smith, A. (2004) Economic restructuring and employment change. In: M. Bradshaw and A. Stenning (eds), *East Central Europe and the Former Soviet Union: The Post-socialist States*, pp. 33–58. Harlow: Pearson.

Dyczok, M. (2000) *Ukraine: Movement without Change, Change without Movement*. Amsterdam: Harwood.

EC (European Commission) (2001) *Perceptions of the European Union, 2001*. Available at: www. europa.org (accessed October 2001).

Eggertsson T. (1998) Limits to institutional reforms. *Scandinavian Journal of Economics*, 100 (1): 335–57.

Eimer, D. (2005): Democracy withers away as China's leaders gather. *The Independent*, 8 October.

Einhorn, Barbara (1993) *Cinderella Goes to Market: Women's Movements in Central and Eastern Europe*. London: Verso.

Eke, S.M. and Kuzio, T. (2000) Sultanism in Eastern Europe: the socio-political roots of authoritarian populism in Belarus. *Europe-Asia Studies*, 52: 523–47.

Ekiert, G. (2001) *The State after State Socialism: Poland in Comparative Perspective*. IWM Institute for Human Sciences Working Papers, Vienna, Working Paper No. 6.

Elster, J., Offe, C. and Preuss, U. (1998) *Institutional Design in Postcommunist Societies: Rebuilding the Ship at Sea*. Cambridge: Cambridge University Press.

Ennis, E. (2002) Revolutionary leaders, ideology and change. *ASCE (Association for the Study of the Cuban Economy) Annual Proceedings*, 12: 312–24. Available at: http://lanic.utexas.edu/project/asce/publications/proceedings/

Espino, M. D. (2000a) International tourism in Cuba: an economic development strategy? *Cuba in Transition*, 1. Available at: http://lanic.utexasedu/la/cb/cuba/asce/cuba1/espino.html

Espino, M. D. (2000b) Cuban tourism during the Special Period. *ASCE (Association for the Study of the Cuban Economy) Annual Proceedings*, 10: 360–73. Available at: http://lanic.utexas.edu/project/asce/publications/proceedings/

Evans, D. (1993) *Sexual Citizenship: The Material Construction of Sexualities*. London: Routledge.

Fabienke, R. (2001) Labour markets and income distribution during crisis and reform. In: C. Brundenius and J. Weeks (eds), *Globalization and Third World Socialism*, pp. 102–28. Basingstoke: Palgrave.

Falk, P. (1985) The cost of Cuba's trade with socialist countries. In P. Falk, *Cuban Foreign Policy*. Lexington, MA: Lexington Books. Reprinted in: P. Brenner *et al.* (eds) (1987), *The Cuba Reader: The Making of a Revolutionary Society*, pp. 307–15. New York: Grove Press.

Fang, C., Zhang, X. and Fan, S. (2002) Emergence of urban poverty and inequality in China: evidence from household survey. *China Economic Review*, 13: 430–3.

Feng, Y. and Zak, P. J. (1999) The determinants of democratic transitions. *Journal of Conflict Resolution*, 43 (2): 162–77.

Ferry, M. (2003) The EU and recent regional reform in Poland. *Europe-Asia Studies*, 55: 1097–116.

Ferry, W. and Kanet, R. (eds) (1998) *Post-Communist States in the World Community*. Basingstoke: Macmillan.

Fish, M. S. (2001) The Inner Asia anomaly: Mongolia's democratization in comparative perspective. *Communist and Post-Communist Studies*, 34: 323–38.

Fitzgerald, F. (1994) *The Cuban Revolution in Crisis. From Managing Socialism to Managing Survival*. New York: Monthly Review Press.

Fletcher, P. (2000) Cuba's tourism still booming. *Financial Times*, 17 May.

Forsyth, T. (2004) *Encyclopedia of International Development*. London: Routledge,

Fowkes, B. (1999) *The Post-Communist Era: Change and Continuity in Eastern Europe*. Basingstoke: Palgrave.

Friedheim, D. V. (1993): Bringing society back into democratic transition theory after 1989: pact making and regime change. *East European Politics and Societies*, 7 (3): 482–512.

Friedland, W. and Rosberg, C. (1964a) Introduction: the anatomy of African socialism.

In: W. Friedland and C. Rosberg (eds), *African Socialism*, pp. 1–11. Stanford, CA: Hoover Institution.

Friedland, W. and Rosberg, C. (eds) (1964b) *African Socialism*. Stanford, CA: Hoover Institution.

Friedmann, J. (2005) *China's Urban Transition*. London: University of Minnesota Press.

Friis, L. and Murphy, A. (1999) The European Union and Central and Eastern Europe: governance and boundaries. *Journal of Common Market Studies*, 37: 211–32.

Frydman, R., Rapaczynski, A., Earle, J. *et al.* (1993) *The Privatization Process in Central Europe*. Budapest: Central European Press.

Fu, Z. (2002) The state, capital, and urban restructuring in post-reform Shanghai. In: J. Logan (ed.), *The New Chinese City: Globalization and Market Reform*, pp. 106–20. Oxford: Blackwell.

Fukuyama, F. (1989) The end of history? *The National Interest*, Summer: 9.

Fukuyama, F. (1992) *The End of History and the Last Man*. London: Hamish Hamilton.

Gadzey, A. (1992) The state and capitalist transformation in Sub-Saharan Africa: a development model. *Comparative Political Studies*, 24 (4): 455–87.

Gar-on, Yeh, A. and Li, X. (1999) Economic development and agricultural land loss in the Pearl River Delta, China. *Habitat International*, 23 (3): 373–90.

Gar-on Yeh, A. and Wu, F. (1999) The transformation of the urban planning system in China from a centrally-planned to transitional economy. *Progress in Planning*, 51 (3): 167–252.

Gati, C. (1996) If not democracy, what? Leaders, laggards, and losers in the postcommunist world. In: M. Mandelbaum (ed.) *Post-Communism – Four Perspectives*, pp. 168–98. New York: Council of Foreign Relations Book.

Gelb, A., Jefferson, G. and Singh, I. (1993) Can communist economies transform incrementally? The experience of China. *Economies of Transition*, 1(4): 401–35.

Gellner, Ernst (1964) *Thought and Change*. London: Weidenfeld and Nicolson.

Gel'man, V. (2003) In search of local autonomy: the politics of big cities in Russa's transition. *International Journal of Urban and Regional Research*, 27 (1): 48–61.

Gentile, M. (2002) Residential segregation in a medium-sized post-Soviet city: Ust'-Kamenogorsk, Kazakhstan. *Tijdschrift vor Economische en Sociale Geografie*, 94 (3): 589–605.

Gibb, R. and Michalak, W. (1993) The European Community and Central Europe: prospects for integration. *Geography*, 78: 16–30.

Giddens, A. (1990) *The Consequences of Modernity*. Cambridge: Polity Press.

Glasman, M. (1994) The great deformation: Polanyi, Poland and the terrors of planned spontaneity. In: C. Bryant and E. Mokrzycki (eds), *The New Great Transformation. Change and Continuity in East-Central Europe*. London: Routledge.

Gomez, A. (2001) Crisis, economic restructuring and international reinsertion [of Cuba]. In: C. Brundenius and J. Weeks (eds), *Globalization and Third World Socialism*, pp. 61–70. Basingstoke: Palgrave.

Gong, S. and Li, B. (2003) *Social Inequalities and Wage, Housing and Pension Reforms in Urban China*. Asia Programme Working Paper No. 3. London: Royal Institute of International Affairs. Available at: www.chathamhouse.org.uk/pdf/briefing_papers/Li%2520and %2520Gong.pdf

Gonzalez, E. (1980) Cuba, the Soviet Union, Africa. In: D. Albright (ed.), *Communism in Africa*, pp. 145–168. London: Indiana University Press.

Gonzalez, E. and McCarthy, K. (2001) *Cuba after Castro. Legacies, Challenges, and Impediments*. Report prepared for the National Defense Research Institute. Pittsburgh: Rand Corporation.

Gordy, Eric D. (1999) *The Culture of Power in Serbia. Nationalism and the Destruction of Alternatives.* University Park, PA: The Pennsylvania State University Press.

Gorzelak, G. and Jałowiecki, B. (2002) European boundaries: unity or division of the continent? *Regional Studies*, 36: 409–19.

Gowan, P. (1995) Neo-liberal theory and practice for Eastern Europe. *New Left Review*, 1: 3–60.

Grabher, G. and Stark, D. (eds) (1997) *Restructuring Networks in Post-Socialism.* Oxford: Oxford University Press.

Graudt, Svetlana (2005) How we said goodbye to Lenin, hello to Cosmo. *The Observer*, 6 March: 22.

Gray, J. and White, G. (eds) (1982) *China's New Development Strategy.* London: Academic Press.

Greskovitz, B. (2002) The path-dependence of transitology. In: F. Bönker, K. Müller and A. Pickel (eds), *Postcommunist Transformation and the Social Sciences: Cross-Disciplinary Approaches*, pp. 219–51. Oxford: Rowman and Littlefield.

Grichtchenko, J. and Gritsanov, A. (1995) The local political elite in the democratic transformation of Belarus. *Annals of AAPSS*, 540 (July): 118–25.

Grzymała-Busse, A. (2002) *Redeeming the Communist Past: The Regeneration of Communist Parties in East Central Europe.* Cambridge, Cambridge University Press.

Grzymała-Busse, A. and Jones Luong, P. (2002) Reconceptualising the state: lessons from post-communism. *Politics and Society*, 30 (4): 529–54.

Gunn, G. (1993) The sociological impact of rising foreign investment. Cuba Briefing Paper Series, No. 1. Available at www.strategicstudiesinstitute.army.mil/pdffiles/PUB48.pdf.

Habel, J. (1991) *Cuba: The Revolution in Peril.* London: Verso.

Haggard, S. and Kaufman, R. (1999) The political economy of democratic transitions. In: L. Anderson (ed.), *Transitions to Democracy*, pp. 72–96. New York: Columbia University Press.

Hall, D. R. (1984) Foreign tourism under socialism. The Albanian 'Stalinist' model. *Annals of Tourism Research*, 11: 539–55.

Hall, D. R. (1990) Stalinism and tourism: a study of Albania and North Korea. *Annals of Tourism Research*, 17: 36–54.

Hammel, E. A. (1993) Demography and the origins of the Yugoslav civil war. *Anthropology Today*, 9 (1): 4–9.

Han, S. and Yan, Z. (1999) China's coastal cities: development, planning and challenges. *Habitat*, 23 (2): 217–29.

Hann, C. (1999) Postsocialist nationalism: rediscovering the past in southeastern Poland. *Slavic Review*, 57 (4): 840–63.

Hann, C. (ed.) (2002) *Postsocialism: Ideals, Ideologies and Practices in Eurasia.* London: Routledge.

Hardy, J. (1998) Cathedrals in the desert? Transnationals, corporate strategy and locality in Wrocław. *Regional Studies*, 32: 639–52.

Hausner, J. (1995) Contradictions and dilemmas in the development of post-socialist societies. In: M. Mendell and K. Nielsen (eds), *Europe – Central and East*, pp. 56–73. London: Black Horse Books.

Hausner, J., Kudeusz, T. and Szlachta, J. (1997) Regional and local factors in the restructuring of south-eastern Poland. In: G. Grabher and D. Stark (eds), *Restructuring Networks in Post-Socialism: Legacies, Linkages and Localities*, pp. 190–208. Oxford: Oxford University Press.

Hayden, R. M. (1996) Imagined communities and real victims: self-determination and ethnic cleansing in Yugoslavia. *American Ethnologist*, 23 (4): 783–801.

Hayden, R. M. (2002): Antagonistic tolerance: competitive sharing of religious sites in South Asia and the Balkans. *Current Anthropology*, 43: 205–31.

Haynes, J. (1992) One-party state, no-party state, multi-party state? 35 years of democracy, authoritarianism and development in Ghana. *Journal of Communist Studies*, 8 (2): 41–62.

He, S. and Wu, F. (2005) Property-led redevelopment in post-reform China: a case study of Xintiandi redevelopment project in Shanghai. *Journal of Urban Affairs*, 27 (1): 1–23.

Henderson, K. (1999) *Back to Europe. Central and Eastern Europe and the European Union*. London: UCL Press.

Henken, T. (2000) The last resort or bridge to the future? Tourism and workers in Cuba's second economy. *ASCE (Association for the Study of the Cuban Economy) Annual Proceedings*, 10: 321–36. Available at: http://lanic.utexas.edu/project/asce/publications/proceedings/

Hernández-Catá, E. (2000) The fall and recovery of the Cuban economy in the 1990s: mirage or reality. *ASCE (Association for the Study of the Cuban Economy) Annual Proceedings*, 10: 24–36. Available at: http://lanic.utexas.edu/project/asce/publications/proceedings/

Herr, H. and Priewe, J. (1999) High growth in China: transition without a transition crisis? *Intereconomics*, 34 (6): 303–16.

Herrschel, T. (2001) Environment and the post-socialist 'condition'. *Environment and Planning A*, 7 (1): 569–72.

Herrschel, T. (2005) 'Competing Regionalization' through territory and cluster networks – experiences from post-socialist eastern Germany. *GeoJournal*, 62(1): 59–70.

Herrschel, T. and Forsyth, T. (2001) Constructing a new understanding of 'environment' under post-socialism. *Environment and Planning A*, 33 (4): 573–87.

Ho, S. P. S. and Lin, G. C. S. (2003) Emerging land markets in rural and urban China: policy and practices. *The China Quarterly*, 175: 681–707.

Hoffmann, B. (2001) Transformation and continuity in Cuba. *Review of Radical Political Economics*, 33: 1–20.

Holmes, L. (1996) Cultural legacies or state collapse? Probing the post-communist dilemma. In: M. Mandelbaum (ed.), *Post-Communism – Four Perspectives*, pp. 22–76. New York: Council of Foreign Relations Book.

Holmes, L. (1997) *Post-Communism – an Introduction*. Cambridge: Polity Press.

Holy, Ladislav (1996) *The Little and the Great Czech Nation: National Identity and the Post-Communist Social Transformation*. Cambridge: Cambridge University Press.

Hopf, T. (2002) Making the future inevitable: legitimizing, naturalizing and stabilizing. The transition in Estonia, Ukraine and Uzbekistan. *European Journal of International Relations*, 8 (3): 403–36.

Hörschelmann, Kathrin (2000) 'Go east, young man . . .': a gendered representations of identity in television dramas about 'east Germany'. In: Patrick Stevenson and John Theobald (eds), *Relocating Germanness. Discursive Disunity in Unified Germany*, pp. 43–59. Basingstoke: Macmillan.

Hörschelmann, Kathrin and van Hoven, Bettina (2003) Experiencing displacement: the transformation of women's spaces in (former) East Germany. *Antipode*, 33 (4): 742–60.

Huang, J.-T., Kuo, C.-C. and Kao, A.-P. (2003) *Journal of Chinese Economic and Business Studies*, 1 (5): 273–85.

Huberman, L. and Sweezy, P. (1969) *Socialism in Cuba*. London: Modern Reader Paperbacks.

Hughes, A. (1992) The appeal of Marxism to Africa. *Journal of Communist Studies*, 8 (2), Special Issue 'Marxism's Retreat from Africa': 5–19.

Humphrey, Caroline (1995) Creating a culture of disillusionment: consumption in Moscow, a chronicle of changing times. In: Daniel Miller (ed.), *Worlds Apart: Modernity Through the Prism of the Local*, pp. 43–68. London: Routledge.

Humphrey, Caroline (2002) *The Un-Making of Soviet Life. Everyday Economies after Socialism.* Ithaca, NY: Cornell University Press.

Hunter, A. and Sexton, J. (1999) *Contemporary China.* Basingstoke: Macmillan.

ILGA Europe (2001) *Equality for Lesbians and Gay Men: A Relevant Issue in the EU Accession Process.* Brussels: ILGA-Europe.

Ilkhamov, Alisher (2001) Impoverishment of the masses in the transition period: signs of an emerging 'new poor' identity in Uzbekistan. *Central Asian Survey*, 20 (1): 33–54.

Ishiyama, J.T. and Kennedy, R. (2001) Superpresidentialism and political party development in Russia, Ukraine, Armenia and Kyrgyzstan. *Europe-Asia Studies*, 53: 1177–91.

Ismagambetov, T. (2002) Structuring Central Asia's new geopolitical space: regional characteristics and prospects. *Central Asia and the Caucasus*, 2 (14): 7–15.

Jackson, L. (2004) Nationality, citizenship and identity. In: M. Bradshaw and A. Stenning (eds), *East Central Europe and the Former Soviet Union: The Post-Socialist States*, pp. 187–218. Harlow: Pearson.

Jackson, Robert H. (1990) *Quasi-States: Sovereignty, International Relations and the Third World.* Cambridge: Cambridge University Press.

Jameson, K. and Wilber, C. (1981) Socialism and development: Editors' introduction. *World Development*, 9, (9/10): 803–11.

Janos, A. C. (1996) What was Communism? A retrospective in comparative analysis. *Communist and Post-Communist Studies*, 29 (1): 1–24.

Jebodsingh, L. (2001) *Cuba: A Market Profile. Arthur Andersen.* Available at: www.hotel-online. com/Trends/Andersen/2001_CubaProfile.html

Jenkins, P. (2001) Strengthening access to land for housing for the poor in Maputo, Mozambique. *International Journal of Urban and Regional Research*, 25 (3): 629–47.

Jones, D., Cheng, L. and Owen, A. (2003) Growth and regional inequality in China during the reform era. *China Economic Reform*, 14: 186–200.

Joseph, R. (1999) Democratization in Africa after 1989: comparative and theoretical perspectives. In: L. Anderson (ed.), *Transitions to Democracy*, pp. 237–60. New York: Columbia University Press.

Kaminski, B. (ed.) (1996a) *Economic Transition in Russia and the New States of Eurasia.* London: ME Sharpe.

Kaminski, N. (1996b) Introduction. In: B. Kaminski (ed.), *Economic Transition in Russia and the New States of Eurasia*, pp. 3–10. London: ME Sharp.

Kaneff, D. (2002) Why people don't die 'naturally' any more: changing relations between 'the individual' and 'the state' in post-socialist Bulgaria. *Journal of the Royal Anthropological Institute* 8: 89–105.

Kaneff, D. (2000) Property, work and local identity. *Max-Planck-Institute for Social Anthropology Halle*, Working Paper No. 15. Available at: www.eth.mpg.de/pubs/wps/pdf/mpi-eth-working-paper-0015.pdf (accessed December 2004).

Kapcia, A. (2000) *Cuba: Islands of Dreams.* Oxford: Berg.

Karl, T. and Schmitter, P. C. (1992) The types of democracy emerging in southern and eastern Europe and south and central America. In: P. Volten (ed.), *Bound to Change: Consolidating Democracy in Central Europe*, pp. 42–68. New York: Institute for East–West Security Studies.

Karl, T. L. and Schmitter, P. C. (1995) From an iron curtain to a paper curtain: grounding transitologists or students of post-communism? *Slavic Studies* (54): 965–78.

Kautsky, J.H. (1998) Centralization in the Marxist and in the Leninist tradition. *Communist and Post-Communist Studies*, 30 (4): 379–400.

Kay, Rebecca (2005) Heroes or villains? Russian media representations of men and masculinity in the post-Soviet era. Paper presented at the Annual Conference of the American Association of Geographers, Denver, 4–9 April.

Keiffer, A. (2001) Perfeccionamiento empresarial: Entrepreneurial perfectionism and the Cuban tourism industry. *ASCE (Association for the Study of the Cuban Economy) Annual Proceedings*, 11: 38–46. Available at: http://lanic.utexas.edu/project/asce/publications/proceedings/

Khakimov, A. (1996) Prospects of federalism in Russia: a view from Tartarstan. *Security Dialogue*, 27, (1): 69–80.

Khan, A. R. and Riskin, C. (2000) *China: Income Distribution and Poverty in the Age of Globalization*. London: Oxford University Press.

Kideckel, David A. (2002) The unmaking of an East-Central European working class. In: Chris Hann (ed.), *Post-Socialism: Ideals, Ideologies and Practices in Eurasia*, pp. 114–32. London: Routledge

Kihlgren, A. (2003) Small business in Russia – factors that slowed its development – an analysis. *Communist and Post-Communist Studies*, 36: 193–207.

Klinghoffer, A. (1998) High fidelity: getting attuned to Cuba's new market beat. *Management Decision*, 36 (3): 175–9.

Kloos, H. and Adugna, A. (1989) The Ethiopian population: growth and distribution. *Geographical Journal*, 155 (1): 33–51.

Kohn, Hans (1944) *The Idea of Nationalism: A Study in its Origins and Background*. New York: Macmillan.

Kohn, Hans (1946) *Prophets and People: Studies in Nineteenth Century Nationalism*. New York: Macmillan.

Kolko, G. (1995) *Vietnam: Anatomy of a Peace*. London: Routledge.

Kolodko, G. (1999) Transition to a market economy and sustained growth. Implications for the post-Washington consensus. *Communist and Post-Communist Studies*, 32: 233–61.

Kolodko, G. (2000a) Transition to a market and entrepreneurship: the systemic factors and policy options. *Communist and Post-Communist Studies*, 33: 271–93.

Kolodko, G. (2000b) *From Shock to Therapy: The Political Economy of Postsocialist Transformation*. Oxford: Oxford University Press.

Kolodko, G. (2001) Globalization and catching-up: from recession to growth in transition economies. *Communist and Post-Communist Studies*, 34: 279–322.

Kopytoff, I. (1964) Socialism and traditional African societies. In: W. Friedland and C. Rosberg (eds), *African Socialism*, pp. 53–62. Stanford, CA: Hoover Institution.

Kornai, J. (1990) *Vision and Reality, Market and State: Contradictions and Dilemmas Revisited*. London: Routledge.

Kornai, J. (1992) *The Socialist System: The Political Economy of Communism*. Princeton, NJ: Princeton University Press.

Kornai, J. (1994) Transformational Recession: the Main Causes. *Journal of Comparative Economics*, 19: 39–63.

Kornai, J. (2000) What the change of system from socialism to capitalism does and does not mean. *Journal of Economic Perspectives*, 14 (1): 27–42.

Kostovicova, D. and Young, C. (2003) Competing geographies of identity in post-socialist Europe: an introduction. Paper presented at the Association of American Geographers Annual Conference, New Orleans, USA.

Kovacs, J. (1991) The regulated market – the future of socialism? In: O. Sik (ed.), *Socialism Today? The Changing Meaning of Socialism*, pp. 109–21. New York: St Martin's Press.

Kubicek, P. (2005) The European Union and democratization in Ukraine. *Communist and Post-Communist Studies*, 38: 269–92.

Kucia, M. (1999) Public opinion in Central Europe on EU Accession: the Czech Republic and Poland. *Journal of Common Market Studies*, 37: 143–52.

Kurtz, M. J. and Barnes, A. (2002) The political foundations of post-communist regimes: marketization, agrarian legacies, or international influences. *Comparative Political Studies*, 35 (5): 524–53.

Kuus, M. (2002) Sovereignty for security? The discourse of sovereignty in Estonia. *Political Geography*, 21: 393–412.

Kuzio, T. (2002) Nationalism in Ukraine: towards a new theoretical and comparative framework. *Journal of Political Ideologies*, 7 (2): 133–61.

Kuzio, T. (2005) Regime type and politics in Ukraine under Kuchma. *Communist and Post-Communist Studies*, 38: 167–90.

Lagerspetz, M. (2002) From 'parallel polis' to 'the time of the tribes': post-socialism, social self-organization and post-modernity. *Journal of Communist Studies and Transition Politics*, 17 (2): 1–18.

Lampland, M. (2000) Afterword. In: D. Berdahl, M. Bunzl and M. Lampland (eds), *Altering States: Ethnographies of Transition in Eastern Europe and the Former Soviet Union*, pp. 209–18. Ann Arbor, MI: University of Michigan Press.

Lane, D. (1996) *The Rise and Fall of State Socialism*. London: Polity Press.

Laski, K. and Bhaduri, A. (1997) Lessons to be drawn from main mistakes in the transition strategy. In: S. Zecchini (ed.), *Lessons from the Economic Transition: Central and Eastern Europe in the 1990s*, pp. 103–21. Dordrecht: Kluwer.

Lavigne, M. (1999) *The Economics of Transition. From Socialist Economy to Market Economy*, 2nd edn. Basingstoke: Macmillan Press.

Lavigne, M, (2000) Ten years of transition: a review article. *Communist and Post-Communist Studies*, 33: 475–83.

Leaf, M. (2002) A tale of two villages: globalization and the peri-urban change in China and Vietnam. *Cities*, 19 (1): 23–31.

Lederer, Gyorgy (2001) Islam in East Europe. *Central Asian Survey*, 20 (1): 5–32.

Lee, J. (2000) Change in the source of China's regional inequality. *China Economic Review*, 232–45.

Legum, C. (1980) African outlooks toward the USSR. In: D. Albright (ed.), *Communism in Africa*, pp. 7–34. London: Indiana University Press.

LeoGrande, W. (1980) *Cuba's Policy in Africa, 1959–1980*. Berkeley: Institute of International Studies, University of California. Reprinted in: P. Brenner *et al.* (eds) (1987) *The Cuba Reader: The Making of a Revolutionary Society*, pp. 375–395. New York: Grove Press.

Leutloff, C. (2000) Politics, religion and remembering the past: the case of Croatian Serbs in the 1990s. *Max-Planck-Institute for Social Anthropology Halle*, Working Paper No. 17. Available at: www.eth.mpg.de/pubs/wps/pdf/mpi-eth-working-paper-0017.pdf (accessed December 2004).

Lewin, M. (1991) *The Gorbachev Phenomenon: A Historical Interpretation*. Berkley: University of California Press.

Lewis, J. and Litai, X. (2003) Social change and political reform in China: meeting the challenges of success. *China Quarterly*, 927–42.

Light, D. and Phinnemore, D. (1998) Teaching 'transition' in Central and Eastern Europe through fieldwork. *Journal of Geography in Higher Education*, 22: 185–99.

Light, M. (1992) Moscow's retreat from Africa. *Journal of Communist Studies*, 8 (2): 21–40.

Lin, G. (2002) Region-based urbanization in post-reform China: spatial restructuring in the Pearl River Delta. In: J. Logan (ed.), *The New Chinese City. Globalization and Market Reform*, pp. 245–57. Oxford: Blackwell.

Lin, J. (2004) *Lessons of China's Transition from a Planned Economy to a Market Economy*. China Centre for Economic Research Working Paper Series, no. E2004001, 2 February, 2004.

Lin, J. Y. and Liu, Z. (2000) Fiscal decentralization and economic growth in China. *Economic Development and Cultural Change*, 49: 1–21.

Lin, J. Y. and Yao, Y. (1999) Chinese rural industrialization in the context of the East Asian miracle. Working paper. Beijing: China Center for Economic Research, Beijing University. Available at: www.ccer.edu.cn/en/ReadNews.asp?NewsID=656.

Lin, J. Y., Cai, F. C. and Li, Z. (2003) *The China Miracle: Development Strategy and Economic Reform*. Hong Kong: The Chinese University Press.

Linge, G. (ed.) (1997) *China's New Spatial Economy. Heading Towards 2020*. Oxford: Oxford University Press.

Linz, J. and Stepan, A. (1996) *Problems of Democratic Transition and Consolidation: Southern Europe, South America and Post-Communist Europe*. London: Johns Hopkins University Press.

Lipton, D. and Sachs, J. (1990) Creating a Market Economy in Eastern Europe. *Brookings Papers on Economic Activity, no 1*, pp. 75–133.

Lock, P. (1994) Review essay: Russia and the World Economy. Problems of Integration, by Alan Smith. *Journal of Peace Research*, 31 (3): 351–7.

Logan, J. (2002a) Three challenges for the Chinese city: globalization, migration, and market reform. In: J. Logan (ed.), *The New Chinese City. Globalization and Market Reform*, pp. 3–21. Oxford: Blackwell.

Logan, J. (ed.) (2002b) *The New Chinese City: Globalization and Market Reform*. Oxford: Blackwell.

Luntley, M. (1989) *The Meaning of Socialism*. London: Duckworth.

McCaughan, E. (1997) *Reinventing Revolution*. Oxford: Westview Press.

McDonald, J. (1993) Transition to Utopia: a reinterpretation of economics, ideas, and politics in Hungary, 1984 to 1990. *East European Politics and Societies*, 7 (2): 203–39.

MacEwan, A. (1981) *Revolution and Economic Development in Cuba*. London: Macmillan.

McFaul, P. (2001) *Russia's Unfinished Revolution. Political Change from Gorbachev to Putin*. London: Cornell University Press.

McFaul, P. (2004) *Russia's Transition to Democrcay and US-Relations: Unfinished Business*. Cambridge, MA: Hoover Institution.

McIntyre, R. (1992) Industrial reform in Bulgaria. In: I. Jeffries (ed.), *Industrial Reform in Socialist Countries*, pp. 62–73. London: Edward Elgar.

Mandel, D. (1992) Post-perestroika: revolution from above versus revolution from below. In: S. White, A. Pravda and Z. Gitelman (eds), *Developments in Soviet and Post-Soviet Politics*, pp. 300-31. Durham, NC: Duke University Press,

Mandel, E. (1978) *From Stalinism to Eurocommunism: The Bitter Fruits of Socialism in One Country*. London: New Left Books.

Mandelbaum, M. (1996) Introduction. In: M. Mandelbaum (ed.), *Post-Communism – Four Perspectives*, pp. 1–21. New York: Council of Foreign Relations Book.

March, Andrew (2002) 'The use and abuse of history: "national ideology" as transcendental object' in Islam Karimov's 'ideology of national independence'. *Central Asian Survey*, 21 (4): 371–84.

March-Poquet, J. (2000) What type of transition is Cuba undergoing? *Post-Communist Economies*, 12 (1): 92–117.

Marton, A. (2000) *China's Spatial Economic Development. Restless Landscapes in the Lower Yangzi Delta*. London: Routledge.

Massey, D. (1984) *Spatial Divisions of Labour*. London: Hutchinson.

Massey, D. (1988) *Global Restructuring, Local Responses*. Worcester, MA: Graduate School of Geography, Clark University.

Massey, D. and Allen, J. (1988) *Uneven Re-development: Cities and Regions in Transition: A Reader*. London: Hodder and Stoughton in association with the Open University.

Matei, Sorin (2004) The emergent Romanian post-communist ethos: from nationalism to privatism. *Problems of Post-Communism*, 51 (2): 40–7.

Mayhew, A. (1998) *Recreating Europe: The EU's Policy towards Central and Eastern Europe*. Cambridge: Cambridge University Press.

Mendell, M. and Nielsen, K. (1995a) Introduction. In: M. Mendell and K. Nielsen (eds), *Europe: Central and East*. London: Black Horse Books.

Mendell, M. and Nielsen, K. (eds) (1995b) *Europe: Central and East*. London: Black Horse Books.

Mesa-Lago, C. (1998) Assessing economic and social performance in the Cuban transition of the 1990s. *World Development*, 26 (5): 857–76.

Meštrovic, G. (1994) *The Balkanisation of the West: The Confluence of Postmodernism and Post-communism*. London: Routledge.

Meurs, M. (1992) Popular participation and central planning in Cuban socialism: the experience of agriculture in the 1980s. *World Development*, 20 (2): 229–40.

Meurs, M. and Djankov, S. (1998) Economic strategies of surviving post-socialism: changing household economies and gender divisions of labour in the Bulgarian transition. In: A. Rainnie, A. Smith and A. Swain (eds), *Work, Employment and Transition*, pp. 213–26. London, Routledge.

Midlarsky, M. and Roberts, K. (1985) Class, state, and revolution in Central America. *Journal of Conflict Resolution*, 29, (2): 163–93.

Miliband, R. (1977) *Marxism and Politics*. Oxford: Oxford University Press.

Miliband, R. and Saville, J. (eds) (1974) *The Socialist Register 1973*. London: The Merlin Press.

Mitchell, D. (2000) *Cultural Geography: A Critical Introduction*. Oxford: Blackwell.

Mohiddin, A. (1972) Ujamaa na Kujitegemea. In: L. Cliffe and S. Saul (eds), *Socialism in Tanzania: An Interdisciplinary Reader. Vol 1 Politics*, pp. 165–77. Nairobi: East African Publishing House.

Mohiddin, A. (1981), *African Socialism in Two Countries*: London: Croom Helm.

Moisio, S. (2002) EU eligibility, Central Europe, and the invention of applicant state narrative. *Geopolitics*, 7: 89–116.

Moss, R. (1973) *Chile's Marxist Experiment*. Newton Abbott: David & Charles.

Mueller, J. (1996) Democracy, capitalism, and the end of transition. In: M. Mandelbaum (ed.), *Post-Communism – Four Perspectives*, pp. 102–67. New York: Council of Foreign Relations Book.

Mulikita, N. (2003) A false dawn? Africa's post-1990 democratization waves. *Security Review*, 12 (4).

Munck, G. L. and Leff, C. S. (1999) Modes of transition and democratization: South America and Eastern Europe in comparative perspective. In: Anderson, L. (ed.), *Transitions to Democracy*, pp. 193–216. New York: Columbia University Press.

Murray, P. and Szelenyi, I. (1984) The city in the transition to socialism. *International Journal of Urban and Regional Research*, 8: 90–107.

Murrell, P. (1992) Conservative political philosophy and the strategy of economic transition. *East European Politics and Societies*, 6 (1): 3–16.

Myrna, A. (2005) Cuba – Governance and Social Justice. Cuba's Transition – Lessons from Other Countries. Paper presented to Mexico City conference, 21–22 April 2005, Canadian Foundation for the Americas.

Nanto, D. and Sinha, R. (2001) China: a major economic power? *Post-Communist Economies*, 13 (3): 345–70.

Nathan, A. (1997) *China's Transition*. New York: Columbia University Press

Naumova, N. (1999) *Retsidiviruyushchay a modernizatsiy a v Rossii: beda, vina ili resurs chelovechestv a*. Moscow: Editorial URSS.

Nedelsky, Nadya (2003) Civic nationhood and the challenges of minority inclusion. *Ethnicities*, 3 (1): 85–114.

Ng, M. (2002) Sustainable development in the rapidly growing socialist market economy of Shenzhen. *DISP*, 151: 42–50.

Ng, M. and Xu, J. (2000) Development control in post-reform China.: the case of Liuha Lake Park, Guangzhou. *Cities*, 17 (6): 409–18.

Ngai, P. (1999) Becoming Dagongmei (working girls): the politics of identity and difference in Reform China. *The China Journal*, 42.

Nove, A (1991): Socialism – Why? In: O. Sik (ed.), *Socialism Today? The Changing Meaning of Socialism*, pp. 75–88. New York: St Martin's Press.

Nunez, R., Brown, J. and Smolka, Martim (2000) Using land value to promote development in Cuba. *Land Lines*, 12 (2).

O'Donnell, G. and Schmitter, P. (1986) *Transitions from Authoritarian Rule*. London: Johns Hopkins University Press.

O'Donnell, G. (1994) Delegative Democracy. *Journal of Democracy*, 5 (1): 55–69.

Offe, C. (1994) *Der Tunnel am Ende des Lichts. Erkundungen der politischen. Transformation im neuen Osten*. Frankfurt: Campus.

Offe, C. (1996) *The Varieties of Transition: The East European and East German Experience*. Cambridge, MA: MIT Press.

Okoko, K. (1987) *Socialism and Self-Reliance in Tanzania*. London: KPI.

Orro, R. (2000) Has Cuba definitely found the path to economic growth? *ASCE (Association for the Study of the Cuban Economy) Annual Proceedings*, 10: 39–47. Available at: http://lanic. utexas.edu/project/asce/publications/proceedings/

Osborn, A. (2005) The 5-minute briefing: the Mikhail Khodorkovsy verdict. *The Independent*, 15 May.

Ost, S. (1992) *Shaping a New Politics in Europe: Interests and Politics in Post-Communist East Europe*. Progam on Central and Eastern Europe Working Paper Series No. 8. Cambridge, MA: Center for European Studies, Harvard University, p 12. Quoted in: M. Mendell and K. Nielsen (eds), *Europe: Central and East*. London: Black Horse Books.

Otayek, R. (1992) The democratic 'Rectification' in Burkina Faso. *Journal of Communist Studies*, 8 (2): 83.

Ottaway, M. (1980) The theory and practice of Marxism-Leninism in Mozambique and Ethiopia. In: D. Albright (ed.), *Communism in Africa*, pp. 118–44. London: Indiana University Press.

Ottaway, M. (1999) *Africa's New Leaders: Democracy or State Reconstruction?* Washington, DC: Carnegie Endowment for International Peace.

Ottaway, M. (2003) *Democracy Challenged: The Rise of Semi-authoritarianism*. Washington, DC: Carnegie Endowment for International Peace.

Ottaway, M. and Ottaway, D. (1986) *Afro-Communism* (2nd edn). London: Africana Publishing Company.

Ó Tuathail, Gearóid and Dahlman, Carl (2004) The effort to reverse ethnic cleansing in Bosnia-Herzegovina: the limits of return. *Eurasian Geography and Economics*, 45 (6): 439–64.

Oushakine, Serguei (2000) Third Europe-Asia Lecture. In the state of post-Soviet aphasia: symbolic development in contemporary Russia. *Europe-Asia Studies* 52 (6): 991–1016.

Paasi, A. (2001) Europe as a social process and discourse: considerations of place, boundaries and identity. *European Urban and Regional Studies*, 8: 7–28.

Parkin, R. (2002) Administrative reform, cross-border relations, and regional identity in western Poland. *Max-Planck-Institute for Social Anthropology Halle*, Working Paper No. 47. Available at: www.eth.mpg.de/pubs/wps/pdf/mpi-eth-working-paper-0047.pdf (accessed December 2004).

Pastor, R. (1983) Cuba and the Soviet Union: does Cuba act alone. In: B. Levine (ed.), *The New Cuban Presence in the Caribbean*. Boulder, CO: Westview. Reprinted in: P. Brenner *et al.* (eds) (1987) *The Cuba Reader: The Making of a Revolutionary Society*, pp. 296–307. New York: Grove Press.

Pastor, R., Jr (2000) After the deluge? Cuba's potential as a market economy. In: S. Kaufmann Purcell and D. Rothkopf (eds), *Cuba: The Contours of Change*, pp. 31–56. London: Lynne Rienner Publishers.

Paulson, J. (ed.) (1999) *African Economies in Transition. Vol 1: The Changing Role of the State*. Basingstoke: Macmillan.

Paulson, J. and Gavin, M. (1999) The changing role of the state in formerly-socialist economies of Africa. In: J. Paulson (ed.), *African Economies in Transition. Vol 1: The Changing Role of the State*, pp. 11–65. Basingstoke: Macmillan.

Pearce, J. (1993) Foreword. In: H. Smith, *Nicaragua: self-determination and Survival*. London: Pluto Press.

Pearson, R. (2002) *The Rise and Fall of the Soviet Empire*, 2nd edn. Basingstoke: Palgrave.

Peck, J. (2001) *Workfare States*. New York: Guilford

Pei, M. (1994) *From Reform to Revolution. The Demise of Communism in China and the Soviet Union*. London: Harvard University Press.

Pei, M. (1996) Microfoundations of state-socialism and patterns of economic transformation. *Communist and Post-Communist Studies*, 29: 131–45.

Peréz, L. (1988) *Cuba – Between Reform and Revolution*. Oxford: Oxford University Press.

Perry, J., Steagall, J. and Woods, L. (1997) Cuban tourism, economic growth, and the welfare of the Cuban worker. *ASCE (Association for the Study of the Cuban Economy) Annual Proceedings*, 7: 141–9. Available at: http://lanic.utexas.edu/project/asce/publications/proceedings/

Philo, S. and Kearns, G. (1993) Culture, history, capital: a critical introduction to the selling of places. In: G. Kearns and S. Philo (eds), *Selling Places: The City as Cultural Capital, Past and Present*. Oxford: Pergamon.

Pickel, A. (1998) Is Cuba different? Regime stability, social change and the problem of reform strategy. *Communist and Post-Communist Studies*, 31 (1): 75–90.

Pickel, A. (2002) Transformation theory: scientific or political? *Communist and Post-Communist Studies*, 35: 105–14.

Pickles, John (2001) 'There are no Turks in Bulgaria': violence, ethnicity, and economic practice in the border regions and Muslim communities of post-socialist Bulgaria. *Max-Planck-Institute for Social Anthropology Halle*, Working Paper No. 25. Available at: www.eth.mpg.de/pubs/wps/pdf/mpi-eth-working-paper-0025.pdf (accessed December 2004).

Pickles, J. and Smith, A. (eds) (1998) *Theorising Transition*. London: Routledge.

Pickles, J. and Smith, A. (2005) Technologies of transition: foreign investment and the (re)articulation of East Central Europe into the global economy. In: D. Turnock (ed.), *Foreign Direct Investment and Regional Development in East Central Europe and the Former Soviet Union*, pp. 21–38. Aldershot: Ashgate.

Pickles, J. and Unwin, T. (2004) Transition in context: theory in post-socialist transformation. In: van Hoven, Bettina (ed.), *Europe: Lives in Transition*, pp. 9–28. London: Prentice Hall.

Pierson, C. (1995) *Socialism after Communism. The New Market Socialism*. London: Polity Press.

Pilkington, H. (1998) *Migration, Displacement and Identity in Post-Soviet Russia*. London: Routledge.

Pine, Frances (1998) 'Dealing with fragmentation. The consequences of privatisation for rural women in central and southern Poland'. In: Sue Bridger and Frances Pine (eds), *Surviving Post-Socialism: Local Strategies and Regional Responses in Eastern Europe and the Former Soviet Union*, pp. 106–23. London: Routledge.

Pine, Frances (2002) Retreat to the household? Gendered domains in post-socialist Poland. In: Chris Hann, Chris (ed.), *Post-Socialism: Ideals, Ideologies and Practices in Eurasia*, pp. 95–113. London: Routledge.

Pipes, R. (2001) *Communism: A Brief History*, London: Weidenfeld and Nicholson.

Pipes, R. (2002) *Communism. A History of the Intellectual and Political Movement*. London: Phoenix.

Pollis, A. (1981) Human rights, Third World socialism and Cuba. *World Development*, 9 (9/10): 1005–17.

Pomfret, R. (2001) Reintegration of formerly centrally planned economies into the global trading system. *ASEAN Economic Bulletin*, 18 (1): 35–47.

Post, K. and Wright, P. (1989) *Socialism and Underdevelopment*. Routledge: London.

Poznanski, K. (1999) Recounting transition. *East European Politics and Societies*, 13 (2): 328–44.

Poznanski, K. (2001) Transition and its dissenters: an introduction. *East European Politics and Societies*, 15 (2): 207–20.

Poznanski, K. (2002) Transformation as a subject of economic theory. In: F. Bönker, K. Müller, and A. Pickel (eds), *Postcommunist Transformation and the Social Sciences: Cross-Disciplinary Approaches*, pp. 55–76. Oxford: Rowman and Littlefield.

Pridham, G. (2002) EU enlargement and consolidating democracy in post-communist states: formality and reality. *Journal of Common Market Studies*, 40: 953–73.

Pridham, G. and Âgh, A. (eds) (2001) *Democratic Transition and Consolidation in East-Central Europe*. Manchester: Manchester University Press.

Primbetov, S. (1996) Central Asia: prospects for regional integration. In: B. Kaminski (ed.), *Economic Transition in Russia and the New States of Eurasia*, pp. 159–70. London: ME Sharpe.

Protsyk, O. (2003) Domestic political institutions in Ukraine and Russia and their responses to EU enlargement. *Communist and Post-Communist Studies* 36: 427–42.

Qian, Y. and Weingast, B. (1996) China's transition to markets: market-preserving federalism, Chinese style. *Journal of Policy Reform*, 1: 149–85.

Radu, M. (1995) Cuba's transition: institutional lessons from Eastern Europe. *Journal of Interamerican Studies and World Affairs*, 37 (2): 83–112.

Rahman, M. (2000) *Sexuality and Democracy: Identities and Strategies in Lesbian and Gay Politics*. Edinburgh: Edinburgh University Press.

Ramet, S. R. (1995) *Social Currents in Eastern Europe* (2nd edn). Durham, NC: Duke University Press.

Ransig, Sigrid (2002) Re-constructing the 'normal': identity and the consumption of western goods in Estonia. In: Ruth Mandel and Caroline Humphrey (eds), *Markets and Moralities. Ethnographies of Postsocialism*, pp. 127–42. Oxford: Berg.

Richardson, D. (1998) Sexuality and citizenship. *Sociology*, 32 (1): 83–100.

Roberts, D. (2001) *Political Transition in Cambodia 1991–99: Power, Elitism and Democracy*. Richmond: Curzon

Roman, Meredith L. (2002) Making Caucasians black: Moscow since the fall of communism and the racialisation of non-Russians. *Journal of Communist Studies and Transition Politics*, 18 (2): 1–27.

Ron, James (2000) Boundaries and violence: repertoires of state action along the Bosnia/ Yugoslavia divide. *Theory and Society*, 29: 609–49.

Róna-Tas, Á. (1998) Path dependence and capitalist theory: sociology of the post-communist economic transformation. *East European Politics and Societies*, 12 (1): 107–31.

Rothkopf, D. (2000) A call for a post-Cold War Cuba policy . . . then years after the end of the Cold War. In: S. Kaufmann Purcell and D. Rothkopf (eds), *Cuba: The Contours of Change*, pp. 105–26. London: Lynne Rienner Publishers.

Round, John (2005) Coping with post-socialism: changing middle-aged masculinities in post-Soviet Russia. Paper presented at the Annual Conference of the American Association of Geographers, Denver, 4–9 April.

Ruccio, D. (1987) The State and planning in Nicaraga. In: R. J. Spalding (ed.), *The Political Ecoomy of Revolutionary Nicaragua*, pp. 61–84. London: Allen and Unwin.

Rudolph, R. and Brade, I. (2005) Moscow: processes of restructuring in the post-Soviet metropolitan periphery. *Cities*, 22 (2): 135–50.

Rustow, D. (1970) Transitions to democracy: toward a dynamic model. *Comparative Politics*, 2. Reprinted in: L. Anderson (ed.), *Transitions to Democracy*, pp. 14–41. New York: Columbia University Press.

Šabič, Z. and Brglez, M. (2002) The national identity of post-communist small states in the process of accession to the European Union: the case of Slovenia. *Communist and Post-Communist Studies*, 35: 67–84.

Sachs, J. (1993) *Poland's Jump to the Market Economy*. Cambridge, MA: MIT Press.

Sachs, J. and Woo, W. (2001) *China's Transition Experience, Reexamined*. World Bank Research Report

Saich, T. (2001) *Governance and Politics of China*. Basingstoke: Palgrave

Sakwa, R. (1999) *Postcommunism*. Buckingham: Open University Press.

Sakwa, R. (2003) *Russian Politics and Society*, 3rd edn. London: Routledge.

Salecl, R. (1994) *The Spoils of Freedom: Psychoanalysis and Feminism after the Fall of Socialism*. London: Routledge.

Sandle, M. (1999) *A Short History of Soviet Socialism*. London: UCL Press

Saul, J. (1985a) The content: a transition to socialism. In: J. Saul (ed.), *A Difficult Road: The Transition to Socialism in Mozambique*, pp. 75–154. New York: Monthly Review Press.

Saul, J. (ed.) (1985b) *A Difficult Road: The Transition to Socialism in Mozambique*. New York: Monthly Review Press.

Savchenko, A. (2002) Toward capitalism or away from Russia? Early stage of post-Soviet economic reforms in Belarus and the Baltics. *American Journal of Economics and Sociology*, 61 (1): 233–57.

Schroeder, G. (1996) Economic transformation in the post-Soviet republics – an overview. In: B. Kaminski (ed.), *Economic Transition in Russia and the New States of Eurasia*, pp. 11–41. London: ME Sharpe.

Seleny, A. (1994) Constructing the discourse of transformation: Hungary, 1979–82. *East European Politics and Societies*, 8 (3): 439–66.

Sells, Michael A. (1996) *The Bridge Betrayed: Religion and Genocide in Bosnia*. Berkeley, CA: University of California Press.

Semetko, H. and Krasnoboka, N. (2003) The political role of the Internet in societies in transition: Russia and Ukraine compared. *Party Politics*, 9 (1): 77–104.

Share, D. (1987) Transitions to democracy and transition through transaction. *Comparative Political Studies*, 19 (4): 515–48.

Shaw, D. (1999) *Russia in the Modern World: A New Geography*. Oxford: Blackwell.

She, X. (1999) Spatial inequality of rural industrial development in China, 1989–1994. *Journal of Rural Studies*, 15 (2): 179–99.

Shen, J. (1998) China's future population and development challenges. *The Geographical Journal*, 164 (1): 32–40.

Shulman, Stephen (2002) Sources of civic and ethnic nationalism in Ukraine. *Journal of Communist Studies and Transition Politics*, 18 (4): 1–30.

Sidaway, J. and Power, M. (1995) Socio-spatial transformations in the 'postsocialist' periphery: the case of Maputo, Mozambique. *Environment and Planning A*, 27: 1463–91.

Sik, O. (1967) *Plan und Markt im Sozialismus*. Vienna: Verlag Fritz Molden.

Sik, O. (1976) *Das Kommunistische Machtsystem*. Hamburg: Hoffmann und Campe.

Sik, O. (1978) *Plan and Market Under Socialism*. New York : John Wiley & Sons.

Sik, O. (1991a) Socialism – theory and practice. In: O. Sik (ed.), *Socialism Today? The Changing Meaning of Socialism*, pp. 1–29. New York: St Martin's Press.

Sik, O. (ed.) (1991b) *Socialism Today? The Changing Meaning of Socialism*. New York: St Martin's Press.

Sim, S. (1998) *Post-Marxism: A Reader*. Edinburgh: Edinburgh University Press.

Simon, F. (1995) Tourism development in transition economies: the Cuba case. *Columbia Journal of World Business*, Spring: 26–40.

Skidelsky, R. (1995) *The World After Communism: A Polemic for our Times*. Basingstoke: Macmillan.

Skidelsky, R. (1996) The State and economy: reflections on the transition from communism to capitalism in Russia. In: M. Mandelbaum (ed.), *Post-Communism: Four Perspectives*, pp. 77–101. New York: Council of Foreign Relations Book.

Skidmore, Thomas E. and Smith, Peter H. (2001) *Modern Latin America*. Oxford: Oxford University Press

Słomczyński, K. and Shabad, G. (1997) *East European Politics and Societies*, 11 (1): 155–89.

Smart, A. (1998) Economic transformation in China: property regimes and social relations. In: J. Pickles, J. and A. Smith (eds), *Theorising Transition: The Political Economy of Post-Communist Transformations*, pp. 428–49. London: Routledge.

Smith, A. (1995) Regulation theory, strategies of enterprise integration and the political economy of regional economic restructuring in Central and Eastern Europe: the case of Slovakia. *Regional Studies*, 29: 761–72.

Smith, A. (1997) Breaking the old and constructing the new? Geographies of uneven development in Central and Eastern Europe. In: R. Lee and J. Wills (eds), *Geographies of Economies*, pp. 331–44. London: Arnold.

Smith, A. (2000) Employment restructuring and household survival in 'post-communist transition': rethinking economic practices in Eastern Europe. *Environment and Planning A*, 32: 1759–80.

Smith, A. (2002) Culture/economy and spaces of economic practice: positioning households in post-communism. *Transactions of the Institute of British Geographers*, NS 27: 232–50.

Smith, A. and Pickles, J. (1998) Introduction: theorising transition and the political economy of transformation. In: J. Pickles and A. Smith (eds), *Theorising Transition*, pp. 1–24. London: Routledge.

Smith, G. (1996) The Soviet state and nationalities policy. In: G. Smith (ed.), *The Nationalities Question in the Post-Soviet States*. London: Longman.

Smith, G. (1999) *The Post-Soviet States: Mapping the Politics of Transition*. London: Arnold.

Smith, G., Law, V., Wilson, A., Bohr, A. and Alworth, E. (1998) *Nation-building in the Post-Soviet Borderlands: The Politics of National Identities*. Cambridge: Cambridge University Press.

Smith, H. (1993) *Nicaragua: Self-determination and Survival*. London: Pluto Press.

Smith, T. (1987) *Thinking Like a Communist: State and Legitimacy in the Soviet Union, China, and Cuba*. London: WW Norton & Company.

Snooks, G. (1999) *Global Transition. A General Theory of Economic Development*. Basingstoke: Macmillan.

Sokol, M. (2001) Central and Eastern Europe a decade after the fall of state-socialism: regional dimensions of transition process. *Regional Studies*, 35: 645–55.

Solchanyk, R. (2001) *Ukraine and Russia*. Oxford: Rowman and Littlefield.

Sperling, V. (1999) *Organizing Women in Contemporary Russia. Engendering Transition*. Cambridge: Cambridge University Press.

Stark, D. (1996) Recombinant property in East European capitalism: organisational innovation in Hungary. In: D. Stark and G. Grabher (eds), *Legacies, Linkages and Localities: The Social Embeddedness of the Economic Transformation in Central and Eastern Europe*. Aldershot: Elgar.

Stern, N. (1997) The transition in Eastern Europe and the Former Soviet Union: some strategic lessons from the experience of 15 countries over six years. In: S. Zecchini (ed.), *Lessons from the Economic Transition: Central and Eastern Europe in the 1990s*, pp. 35–57. Dordrecht: Kluwer Academic.

Stiglitz, Joseph (1999) Whither reform? Ten years of the transition. Keynote Address to the World Bank Annual Bank Conference on Development Economics, Washington, DC, April 28–30.

Storey, David (2002) Territory and national identity: examples from the former Yugoslavia. *Geography*, 87 (2): 108–15.

Suchlicky, J. (2000) Castro's Cuba: continuity instead of change. In: S. Kaufmann Purcell and D. Rothkopf (eds), *Cuba: The Contours of Change*, pp. 57–80. London: Lynne Rienner Publishers.

Suddaby, C. (1997) Cuba's tourism industry. *ASCE (Association for the Study of the Cuban Economy) Annual Proceedings*, 7: 123–29. Available at: http://lanic.utexas.edu/project/asce/publications/proceedings/

Sutter, D. (1995) Settling old scores: potholes along the transition from authoritarian rule. *Journal of Conflict Resolution*, 39 (1): 110–28.

Swain, A. and Hardy, J. (1998) Globalisation, institutions, foreign investment and the reintegration of East and Central Europe and the former Soviet Union with the world economy. *Regional Studies*, 32: 587–90.

Swaminathan, S. (1999) Time, power and democratic transition. *Journal of Conflict Resolution*, 43 (2): 162–76.

Szacki, J. (1995) *Liberalism after Communism*. Budapest: Central European University Press.

Szamuely, L. and Csaba, L. (1998) Economics and systemic changes in Hungary 1945–96. In: H.-J. Wagener (ed.), *Economic Thought in Communist and Post-Communist Europe*, pp. 158–212. London: Routledge.

Szarvas, L. (1993) Transition periods in Hungary: the chances for democracy? *Journal of Theoretical Politics*, 5 (2): 267–76.

Szczerbiak, A. (2001) Polish public opinion: explaining declining support for EU membership. *Journal of Common Market Studies*, 39: 105–22.

Tan, M., Li, X., Xie, H. and Lu, C. (2005) Urban land expansion and arable land loss in China: a case study of Beijing: Tianjin-Hebei Region. *Land Use Policy*, 22: 187–96.

Tanase, S. (1999) Changing societies and elite transformation. *East European Politics and Societies*, 13: 358–63.

Thayer, C. (1992): Political reform in Vietnam: Doi Moi and the emergence of civil society. In: R. F. Miller (ed.), *The Development of Civil Society in Communist Systems*, pp. 110–29. Sydney: Allen & Unwin.

Thomson, A. (2000) *An Introduction to African Politics*. London: Routledge

Tian, X. (1999) Market orientation and regional economic disparities in China. *Post-Communist Economies*, 11 (2): 161–72.

Touraine, A. (1990) The idea of revolution. *Theory, Culture and Society*, 7: 121–41.

True, Jaqui (1999) Expanding markets and marketing gender: the integration of the post-socialist Czech Republic. *Review of International Political Economy*, 6 (3): 360–89.

True, Jaqui (2003) *Gender, Globalisation, and Postsocialism*. New York: Columbia University Press.

Trumbull, C. (2000) Economic reforms and social contradictions in Cuba. *ASCE (Association for the Study of the Cuban Economy) Annual Proceedings*, 10: 305–20. Available at: http://lanic.utexas.edu/project/asce/publications/proceedings/

Trumbull, C. (2001) Prostitution and sex tourism in Cuba. *ASCE (Association for the Study of the Cuban Economy) Annual Proceedings*, 11: 356–70. Available at: http://lanic.utexas.edu/project/asce/publications/proceedings/

Turnock, D. (2001) Location trends for foreign direct investment in East Central Europe. *Environment and Planning C*, 19: 849–80.

United Nations Development Programme (2004) *Human Development Report 2004*. New York, UNDP.

United Nations Development Programme (2003) *The Power of Decentralization*. Ukraine Human Development Report, Kiev. Available at: www.un.kiev.ua

Urban, M. (1994) The politics of identity in Russia's postcommunist transition: the nation against itself. *Slavic Review*, 53 (3): 733–65.

Urban, M. (1998) Remythologising the Russian state. *Europe-Asia Studies*, 50 (6): 969–92.

Urban, Th. (2003) Ein paar Schritte in die andere Welt [A few steps to another world]. *Die Welt*, 29 December.

van Brabant, J. M. (1990) *Remaking Eastern Europe – On the Political Economy of Transition*. Dordrecht: Kluwer Academic.

van Hoven, Bettina (2004) Women's lives in transition: 'everything gets better but nothing is good'. In: Michael Bradshaw and Alison Stenning (eds), *East Central Europe and the Former Soviet Union: The Post-Socialist States*, pp. 161–86. Harlow: Prentice Hall.

Verdery, K. (1996) *What was Socialism and What Comes Next?* Princeton, NJ: Princeton University Press.

Verdery, K. (1999) *The Political Life of Dead Bodies: Reburial and Post-socialist Change*. New York: Columbia University Press.

Vetik, R. (1998) Relationships between ethnic and security issues in the Baltic States. In: W. Ferry and R. Kanet (eds), *Post-Communist States in the World Community*, pp. 191–9. Basingstoke: Macmillan.

Vintrová, R. (2004) The CEE countries on the way into the EU: adjustment problems: institutional adjustment, real and nominal convergence. *Europe-Asia Studies*, 56: 521–41.

Wagener, H.-J. (ed.) (1998) *Economic Thought in Communist and Post-Communist Europe*. London: Routledge.

Wainright, H. (1995) Civil movements and the politics of knowledge. In: M. Mendell and K. Nielsen (eds), *Europe: Central and East*, pp. 30–55. London: Black Horse Books.

Waldron-Moore, P. (1999) Eastern Europe at the crossroads of democratic transition: evaluating support for democratic institutions, satisfaction with democratic government, and consolidation of democratic regimes. *Comparative Political Studies*, 32 (1): 32–62.

Walker, T. (2003) *Nicaragua: Living in the Shadow of the Eagle* (4th edn). Boulder, CO: Westview Press.

Wang, H., Shantong, L. and Linge, G. (1997) Regional planning: developing an indigenous framework. In: G. Linge (ed.), *China's New Spatial Economy: Heading Towards 2020*, pp. 22–45. Oxford: Oxford University Press.

Wang, Q., Guangming, S. and Zheng, Y. (2002) Changes in income inequality and welfare under economic transition: evidence from urban China. *Applied Economics Letter*, 9: 989–91.

Wanner, Catherine (1998) *Burden of Dreams: History and Identity in Post-Soviet Ukraine*. University Park, PA: Pennsylvania State University Press

Watson, A., Wu, H. and Findlay, C. (1997) Regional disparities in rural development. In: G. Linge (ed.), *China's New Spatial Economy: Heading Towards 2020*, pp. 167–90. Oxford: Oxford University Press.

Weaver, J. and Kronemer, A. (1981) Tanzanian and African socialism. *World Development*, 9 (9/10): 839–49

Webber, M. (1992) Angola: continuity and change. *Journal of Communist Studies*, 8 (2): 126–44.

Weber, H. (1981) *Nicaragua. The Sandinista Revolution*. London: Verso.

Weeks, J. (2001) A tale of two transitions: Cuba and Vietnam. In: C. Brundenius and J. Weeks (eds), *Globalization and Third World Socialism*, pp. 18–40. Basingstoke: Palgrave.

Wei, D. (1999) Regional inequality in China. *Progress in Human Geography*, 23: 331–2.

Werlau, M. (1996) Foreign investment in Cuba: the limits of commercial engagement. *ASCE (Association for the Study of the Cuban Economy) Annual Proceedings*, 6: 456–95. Available at: http://lanic.utexas.edu/project/asce/publications/proceedings/

West, B. (2002) *The Danger is Everywhere! The Insecurity of Transition in Post-Socialist Hungary*. Prospect Heights, IL: Waveland Press.

White, S., Gill, G. and Slider, D. (1993) *The Politics of Transition*. Cambridge: Cambridge University Press.

White, S., McAllister, I. and Light, M. (2002) Enlargement and the new outsiders. *Journal of Common Market Studies*, 40: 135–53.

Wiles, P. (1995) Capitalist triumphalism in the Eastern European transition. In H.-J. Chang and P. Nolan (eds), *Transformation of the Communist Economies*, pp. 46–77. Basingstoke: Palgrave.

Willerton, J. P., Jr (1992) Executive power and political leadership. In: S. White., A. Pravda and Z. Gitelman (eds), *Developments in Soviet and Post-Soviet Politics*, pp. 44–64. Durham NC: Duke University Press.

World Bank (1996) *From Plan to Market*. World Development Report.

Woube, M. and Sjöberg, Ö. (1999) Socialism and urbanization in Ethiopia, 1975–1990: a tale of two *Kebeles*. *International Journal of Urban and Regional Research*, 23 (1): 26–44.

Wu, C. T. (1987) Chinese socialism and uneven development. In: D. Forbes and N. Thrift (eds), *The Socialist Third World*, pp. 53–97. Oxford: Blackwell.

Wu, F. (2004) Urban poverty and marginalization under market transition: the case of Chinese cities. *International Journal for Urban and Regional Research*, 28 (2): 401–23.

Wu, F. and He, S. (2005) Changes in traditional urban areas and impacts of urban redevelopment: a case study of three neighbourhoods in Nanjing, China. *Tijdschrift voor Economische en Sociale Geografie*, 96 (1): 75–95.

Wu, X. (2005) *Registration Status, Labour Migration and Socio-Economic Attainment in China's Segmented Labour Markets*. Research report, Population Studies Centre, University of Michigan.

Wu Y.-S. and Sun T.-W. (1998) Four faces of Vietnamese Communism: small countries' institutional choice under hegemony. *Communist and Postcommunist Studies*, 31 (4): 381–99.

Xu, J. and Yeh, A. (2003) City profile: Guangzhou. *Cities*, 20, (5): 361–74.

Yan, X., Jia, L., Jiangping, L. and Weng, J. (2002) The development of the Chinese metropolis in the period of transition. In: J. Logan (ed.), *The New Chinese City. Globalization and Market Reform*, pp. 37–56. Oxford: Blackwell.

Yang, D. (2002) What has caused regional inequality in China? *China Economic Reform*, 13: 331–4.

Yao, S., Zhang, Z. and Hanmer, L. (2004) Growing inequality and poverty in China. *China Economic Review*, 15: 145–63.

Yao, S., Zhang, Z. and Fang, G. (2005) Rural–urban and regional inequality in output, income and consumption in China under economic reforms. *Journal of Economic Studies*, 32 (1): 2–24.

Yen, S. and Liu, J. (1998) Economic reforms and spatial income inequality in China. *Regional Studies*, 32 (8): 735–46.

Yoder, J. A. (2001) West–East integration: lessons from East Germany's accelerated transition. *East European Politics and Societies*, 15 (1): 114–38.

Young, C. (1980) *Ideology and Development*. New Haven, CT: Yale University Press.

Yurchak, A. (1999) Gagarin and the rave kids: transforming power, identity, and aesthetics in post-Soviet nightlife. In: A. M. Barker (ed.), *Consuming Russia: Popular Culture, Sex, and Society since Gorbachev*, pp. 76–109. Durham, NC: Duke University Press.

Zagoria, D. (1963) *The Sino-Soviet Conflict, 1956–1961*. Princeton, NJ: Princeton University Press.

Zecchini, S. (1997a) Transition approaches in retrospect. In: S. Zecchini (ed.), *Lessons from the Economic Transition: Central and Eastern Europe in the 1990s*, pp. 1–34. Dordrecht: Kluwer Academic.

Zecchini, S. (ed) (1997b) *Lessons from the Economic Transition: Central and Eastern Europe in the 1990s*. Dordrecht: Kluwer Academic.

Zhang, B. (1994) Corporatism, totalitarianism, and transitions to democracy. *Comparative Political Studies*, 27 (1): 108–36.

Zhang, T. (2002) Urban development and a socialist pro-growth coalition in Shanghai. *Urban Affairs Review*, 37 (4): 475–99.

Zhang, W.-W. (2000) *Transforming China. Economic Reform and its Political Implications*. Basingstoke: Macmillan.

Zhinxiang, S., Guan, X. and Linge, G. (1997) The head and tail of the dragon: Shanghai and its economic hinterland. In: G. Linge (ed.), *China's New Spatial Economy: Heading Towards 2020*, pp. 98–122. Oxford: Oxford University Press.

Zhu, J. (1999) *The Transition of China's Urban Development*. London: Praeger.

Zhu, J. (2004) Local developmental state and order in China's urban development during transition. *International Journal of Urban and Regional Research*, 28 (2): 424–47.

Zhurzhenko, Tatiana (2001) Free market ideology and the new women's identity in post-socialist Ukraine. *European Journal of Women's Studies*, 8 (1): 29–49.

Ziegler, D. J. (2002) Post-communist Eastern Europe and the cartography of independence. *Political Geography*, 21: 671–86.

Zielonka, J. (2001) How new enlarged borders will reshape the European Union. *Journal of Common Market Studies*, 39: 507–36.

Zimbalist, A. (2000) Whither the Cuban economy? In: S. Kaufmann Purcell and D. Rothkopf (eds), *Cuba: The Contours of Change*, pp. 13–30. London: Lynne Rienner Publishers.

Zimbalist, A. and Eckstein, S. (1987) Patterns of Cuban development: the first twenty-five years. *World Development*, 15 (1): 5–22.

INDEX

References to tables and figures are in **bold**.